THEORY OF ELASTICITY AND STRESS CONCENTRATION

THEORY OF ELASTICITY AND STRESS CONCENTRATION

Yukitaka Murakami
Kyushu University, Fukuoka, Japan

This edition first published 2017
© 2017 John Wiley & Sons, Ltd

Registered Office
John Wiley & Sons, Ltd, The Atrium, Southern Gate, Chichester, West Sussex, PO19 8SQ, United Kingdom

For details of our global editorial offices, for customer services and for information about how to apply for permission to reuse the copyright material in this book please see our website at www.wiley.com.

Library of Congress Cataloging-in-Publication Data

Names: Murakami, Y. (Yukitaka), 1943- author.
Title: Theory of elasticity and stress concentration / Yukitaka Murakami.
Description: Chichester, West Sussex, United Kingdom : John Wiley & Sons, Inc., 2017. | Includes index.
Identifiers: LCCN 2016029049 (print) | LCCN 2016038983 (ebook) | ISBN 9781119274094 (cloth) |
    ISBN 9781119274131 (pdf) | ISBN 9781119274100 (epub)
Subjects: LCSH: Elasticity. | Stress concentration.
Classification: LCC TA418 .M87 2017 (print) | LCC TA418 (ebook) | DDC 620.1/1232–dc23
LC record available at https://lccn.loc.gov/2016029049

A catalogue record for this book is available from the British Library.

Cover image: a_Taiga/Gettyimages

Set in 10/12pt Times by SPi Global, Pondicherry, India

1   2017

Contents

Preface

The theory of elasticity is not applied mathematics. Solving differential equations and integral equations is not the objective of the theory of elasticity. Students and young researchers, who can use the modern commercial finite element method (FEM) software, are not attracted by the classical approach of applied mathematics. This situation is not good. Students, young researchers and young engineers skip directly from the elementary theory of the strength of materials to FEM without understanding the basic principles of the theory of elasticity. The author has seen many mistakes and judgement errors made by students, young researchers and young engineers in their applications of FEM to practical problems. These mistakes and judgement errors mostly come from a lack of basic knowledge of the theory of elasticity. Firstly, this book provides the basic but very important essence of the theory of elasticity. Second, many useful and interesting applications of the basic *way of thinking* are presented and explained. Readers do not need special mathematical knowledge to study this book. They will be able to understand the new approach of the theory of elasticity which is different from the classical mathematical theory of elasticity and will enjoy solving many interesting problems without using FEM.

The basic knowledge and engineering judgement acquired in Part I will encourage the readers to enter smoothly into Part II in which various important new *ways of thinking* and simple solution methods for stress concentration problems are presented. Approximate estimation methods for stress concentration will be very useful from the viewpoint of correct boundary conditions as well as the magnitude and relative importance of numerical variables. Thus, readers will be able to quickly find approximate solutions with practically sufficient accuracy and to avoid fatal mistakes produced by FEM calculations, performed without basic knowledge of the theory of elasticity and stress concentration.

The author believes with confidence that readers of this book will be able to develop themselves to a higher level of research and structural design.

Preface for Part I: Theory of Elasticity

Part I of this book presents a new *way of thinking* for the theory of elasticity. Several good quality textbooks on this topic have already been published, but they tend to be too mathematically based. Students can become confused by the very different approaches taken towards the elementary theory of strength of materials (ETSM) and the theory of elasticity and, therefore, believe that these two cannot be easily used cooperatively.

To study this book, readers do not need special mathematical knowledge such as differential equations, integral equations and tensor analysis. The concepts of stress field and strain are the most important themes in the study of the theory of elasticity. However, these concepts are not explored in sufficient depth within ETSM in order to teach engineers how to apply simple solutions using the theory of elasticity to solve practical problems. As various examples included in this book demonstrate, this book will help readers to understand not only the difference between ETSM and the theory of elasticity but also the essential relationship between them.

In addition to the concepts of field, the concepts of infinity and infinitesimal are also important. It is natural that everyone experiences difficulties in imagining infinity or infinitesimal. As a result, we must use caution when using unbounded or very small values, as the results are sometimes unexpected. We should be aware that infinity and infinitesimal are *relative* quantities.

Once the concepts of field and those of infinity and infinitesimal are mastered, the reader will become a true engineer having true engineering judgement, even if they cannot solve the problems using lengthy and troublesome differential or integral equations. However, the existing solutions must be used fully and care must be taken at times, very large values being treated as infinitesimal and very small values as infinite values depending on the specific problem. It will be seen in many cases treated in this book that small and large are only our impressions and that approximation is not only reasonable but very important.

Part I Nomenclature

Stresses and strains in an orthogonal coordinate system (x, y, z)	Normal stress $(\sigma_x, \sigma_y, \sigma_z)$ Normal strain $(\varepsilon_x, \varepsilon_y, \varepsilon_z)$ Shear stress $(\tau_{xy}, \tau_{yz}, \tau_{zx})$ Shear strain $(\gamma_{xy}, \gamma_{yz}, \gamma_{zx})$
Stresses and strains in a cylindrical coordinate system (r, θ, z)	Normal stress $(\sigma_r, \sigma_\theta, \sigma_z)$ Normal strain $(\varepsilon_r, \varepsilon_\theta, \varepsilon_z)$ Shear stress $(\tau_{r\theta}, \tau_{\theta z}, \tau_{zr})$ Shear strain $(\gamma_{r\theta}, \gamma_{\theta z}, \gamma_{zr})$
Rotation	ω
Normal stress and shear stress in a ξ-η-ζ coordinate system	Normal stress $(\sigma_\xi, \sigma_\eta, \sigma_\zeta)$ Shear stress $(\tau_{\xi\eta}, \tau_{\eta\zeta}, \tau_{\zeta\xi})$
Remote stress	σ_0, τ_0 or $\sigma_{x\infty}, \sigma_{y\infty}, \tau_{xy\infty}$
Principal stresses	$\sigma_1, \sigma_2, \sigma_3$
Principal strains	$\varepsilon_1, \varepsilon_2, \varepsilon_3$
Direction cosines	l_i, m_i, n_i (i = 1, 2, 3)
Pressure	p or q
Concentrated force	P, Q
Body force	X, Y, Z or F_r, F_θ
Bending moment per unit length	M_x, M_y
Twisting moment per unit length	M_{xy} or M_{yx}
Twisting moment (torsional moment) or temperature	T
Torsional angle per unit length or crack propagation angle	θ_0
Surface tension	S
Airy's stress function or stress function in torsion	ϕ
Stress concentration factor	K_t
Stress intensity factor of Mode I	K_I
Stress intensity factor of Mode II	K_{II}
Stress intensity factor of Mode III	K_{III}

Radius of circle or major radius of ellipse or crack length	a
Minor radius of ellipse	b
Notch root radius or radius of curvature in membrane	ρ
Notch depth	t
Young's modulus	E
Poisson's ratio	ν
Shear modulus	G
Displacement in x, y, z coordinate system	u, v, w
	(Note: v looks the same as Poisson's ratio but is different.)
Displacement of membrane	z
Width of plate	W

Preface for Part II: Stress Concentration

Part II of this book is a compilation of the ideas on stress concentration which the author has developed over many years of teaching and research. This is not a handbook of stress concentration factors. This book guides a fundamental *way of thinking* for stress concentration. Fundamentals, typical misconceptions and new *ways of thinking* about stress concentration are presented. One of the motivations for writing this book is the concern about a decreasing basic knowledge of recent engineers about the nature of stress concentration.

It was reported in the United States and Europe [1–3] that the economic loss of fracture accidents reaches about 4% of GDP. Fracture accidents occur repeatedly regardless of the progress of science and technology. It seems that the number and severity of serious accidents is increasing. The author was involved in teaching strength of materials and theory of elasticity for many years in universities and industry and a recent impression based on the author's experience is that many engineers do not understand the fundamentals of the theory of elasticity.

How many engineers can give the correct answers to basic problems such as those in Figures 1 and 2?

The theory of elasticity lectures are likely to be abstract and mathematical. This trend is evident in the topics and emphasis of many text books. Such textbooks may be useful for some researchers but are almost useless for most practicing engineers. The author has been aware of this problem for many years and has changed the pedagogy of teaching the theory of elasticity by introducing various useful *ways of thinking* (see Part I). Engineers specializing in strength design and quality control are especially requested to acquire the fundamentals of theory of elasticity and afterwards to develop a sense about *stress concentration*. The subject is not difficult. Rather, as readers become familiar with the problems contained in this book, they will understand that the problems of stress concentration are full of interesting paradoxes.

Few accidents occur because of a numerical mistake or lack of precision in a stress analysis. A common attitude that analysis by FEM software will guarantee the correct answer and safety is the root cause of many failures. Most mistakes in the process of FEM analysis are made at the

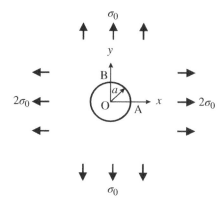

Figure 1 Stress concentration at a circular hole in a wide plate. How large is the maximum stress? (See Figure 1.2 in Example problem 1.1 in Part II, Chapter 1.)

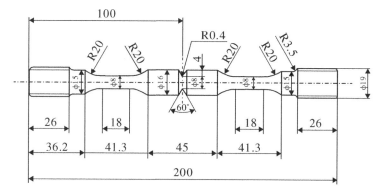

Figure 2 A cylindrical specimen for comparison of the fracture strengths at a smooth part and a notched part under tension (material is 0.13% annealed carbon steel, dimension unit is mm). Where does this specimen fracture from by tensile test? (See Figure 14.7 in the Example problem 14.1 in Part II, Chapter 14.)

beginning stage of determining boundary conditions regarding forces and displacements. Even worse, many users of FEM software are often not aware of such mistakes even after looking at strange results because they do not have a fundamental understanding of theory of elasticity and stress concentration.

The origin of fracture related accidents are mostly at the stress concentrations in a structure. As machine components and structures have various shapes for functional reasons, stress concentration cannot be avoided. Therefore, strength designers are required to evaluate stress concentration correctly and to design the shape of structures so that the stress concentration does not exceed the safety limits.

In this book, various elastic stress concentration problems are the main topic. The strains in an elastic state can be determined by Hooke's law in terms of stresses. In elastic–plastic conditions, the relationship between stresses and strains deviates from Hooke's law. Once plastic yielding occurs at a notch root, the stress concentration factor decreases compared to the elastic value and approaches one. However, the strain concentration factor increases and approaches the elastic value squared. Therefore, in elastic–plastic conditions, fatigue behavior is described in terms of strain concentration. However, if the stress and strain relationship at the notch root does not deviate much from Hooke's law or work hardening of material occurs after yielding, the description based on elastic stress concentration is valid. In general, in the case of high cycle fatigue, it is reasonable and effective for the solution of practical problems to consider only the elastic stress concentration. Thus, it is crucially important for strength design engineers to understand the nature of elastic stress concentration.

References

[1] Battelle Columbus Laboratories (1983) *Economic Effects of Fracture in the United States. Part 1: A Synopsis of the September 30, 1982, Report to NBS.* National Bureau of Standards and National Information Service, Washington, DC.

[2] Battelle Columbus Laboratories (1983) *Economic Effects of Fracture in the United States. Part 2: A Report to NBS.* National Bureau of Standards and National Information Service, Washington, DC.

[3] Commission of the European Communities (1991) Economic Effects of Fracture in Europe, Final Report, Study Contract No. 320105. Commission of the European Communities, Brussels.

Part II Nomenclature

Stresses and strains in orthogonal coordinate system (x, y, z)	Normal stress $(\sigma_x, \sigma_y, \sigma_z)$
	Normal strain $(\varepsilon_x, \varepsilon_y, \varepsilon_z)$
	Shear stress $(\tau_{xy}, \tau_{yz}, \tau_{zx})$
	Shear strain $(\gamma_{xy}, \gamma_{yz}, \gamma_{zx})$
Stresses and strains in cylindrical coordinate system (r, θ, z)	Normal stress $(\sigma_r, \sigma_\theta, \sigma_z)$
	Normal strain $(\varepsilon_r, \varepsilon_\theta, \varepsilon_z)$
	Shear stress $(\tau_{r\theta}, \tau_{\theta z}, \tau_{zr})$
	Shear strain $(\gamma_{r\theta}, \gamma_{\theta z}, \gamma_{zr})$
Normal stress and shear stress in ξ–η coordinate system	Normal stress $(\sigma_\xi, \sigma_\eta)$, shear stress $\tau_{\xi\eta}$
Remote stress	σ_0, τ_0 or $\sigma_{x\infty}, \sigma_{y\infty}, \tau_{xy\infty}$
Principal stresses	σ_1, σ_2
Pressure	p or q
Concentrated force	P, Q
Stress concentration factor	K_t
Stress concentration factor in elastic plastic state	K_σ
Strain concentration factor in elastic plastic state	K_ε
Stress intensity factor of Mode I	K_I
Stress intensity factor of Mode II	K_II
Stress intensity factor of Mode III	K_III
Radius of circle or major radius of ellipse	a
Minor radius of ellipse (or Burger's vector of dislocation)	b
Notch root radius	ρ
Notch depth	t
Young's modulus	E
Poisson's ratio	ν
Shear modulus	G
Displacement in x, y, z coordinate system	u, v, w
Shape parameter of ellipse	$R = \sqrt{(a+b)/(a-b)}$
Plastic zone size	R

Acknowledgments

This book is a joint version of two books originally published separately in Japanese by Yokendo, Tokyo: one is *Theory of Elasticity* and the other is *New Way of Thinking for Stress Concentration*. The Japanese version of *Theory of Elasticity* has been used in many universities since its publication in 1984 and it is regarded as one of the best elasticity textbooks in Japan. For the publication of the Japanese versions, the author was indebted especially to the late Prof. Tatsuo Endo and the late Prof. Makoto Isida for their invaluable comments and moreover to Mr. Kiyoshi Oikawa, the president of Yokendo Publishing Co. Ltd. for publishing the original Japanese versions of these two books and kindly approving the publication of this English version.

In advance of publication of the English version, the author gave lectures on the theory of elasticity and stress concentration at several universities in Europe and the United States. Especially at Aalto University in Finland, the author taught this subject for one complete semester and realized the importance of the new way of teaching the theory of elasticity. Throughout the author's long experience of teaching the theory of elasticity and its relationship with fatigue design, the author got useful comments and encouragements from Prof. Gary Marquis, Dean of the School of Engineering at Aalto University, Prof. Darrell Socie, University of Illinois, Prof. Stefano Beretta, Politecnico di Milano, Prof. Masahiro Endo, Fukuoka University and Prof. Hisao Matsunaga, Kyushu University. The author would like to express his sincere thanks to them.

Prof. Masahiro Endo's help is greatly appreciated for reading throughout the draft of the book and also for giving the author useful comments on the problems presented at the end of chapters.

Prof. Stefano Beretta organized the author's lecture on *Theory of Elasticity and Stress Concentration* for Italian PhD students in the summer school of the Italian Fracture Group in 2008. In the summer school, the author was deeply impressed with the students' attitudes and their strong curiosity toward the new *way of thinking* of *Theory of Elasticity*. The author would like to express his sincere thanks also to them.

The author would like to thank Ms. Hoshiko Utamaru of Kobe Materials Testing Laboratory, Co. Ltd. for her help with illustrating figures and typing equations. The author would like to thank Mr. Kazuo Takagi, the executive director of Kobe Materials Testing Laboratory, who kindly supported the author's writing work through the help of Ms. Hoshiko Utamaru's elaborative work. Without their help, the author could not have completed this book.

The author would like to thank Ms. Eeva Mikkola, PhD student of Aalto University, who also helped the author with illustrating figures and typing equations up to Chapter 6 of Part I during my lecture at Aalto University.

The author would like to thank Ms. Mari Åman, PhD student of Aalto University, who read through a draft of the book, checking typographical mistakes and giving invaluable comments from the viewpoint of students during her research stay in Kyushu University.

During the preparation of the original Japanese version of *Stress Concentration*, the author received invaluable support and help from students and colleagues. The author would like to thank Dr. H. Miyata who contributed to the preparation of the manuscript of the Japanese version when he was in the PhD course at Kyushu University. The author would like to thank Prof. S. Hashimura of Shibaura Institute of Technology, Dr. K. Toyama (a former PhD Student of Kyushu University) and Mr. A. Shiromoto (a former Graduate student of Kyushu University) and Ms. C. Narazaki (a former Graduate student of Kyushu University) for their cooperation in numerical calculations of several example problems. Without their cooperation, the author could not have completed the manuscript.

Part I

Theory of Elasticity

1

Stress

1.1 Stress at the Surface of a Body

1.1.1 Normal Stress

When a body is in a liquid of pressure p, the surface of the body is subject to the same pressure p everywhere, irrespective of the material. Naturally the pressure acts perpendicular to the curved surface, unless the surface is subject to a frictional force. However, the action of frictional force is impossible because a liquid cannot sustain shear stress.

We describe this condition by saying that the *normal stress* σ_n at the surface of the body is $-p$, that is $\sigma_n = -p$. Thus, the normal stress is the force per unit area, when a force acts perpendicular to the surface (Figure 1.1).

1.1.2 Shear Stress

When a block, of weight W, is on a flat plate, there is a minimum force, F, which is necessary to move the block (Figure 1.2). This force is expressed by the equation

$$F = \mu W \qquad (1.1)$$

where μ is the coefficient of static friction. Hence, both the bottom surface of the block and the top surface of the plate are subject to the same frictional force, F. In this situation both surfaces are subject to a shear stress τ. Denoting the average magnitude of the shear stress by τ_{ave}, we have

Theory of Elasticity and Stress Concentration, First Edition. Yukitaka Murakami.
© 2017 John Wiley & Sons, Ltd. Published 2017 by John Wiley & Sons, Ltd.

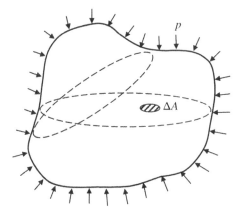

Figure 1.1

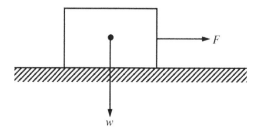

Figure 1.2

$$\tau_{\text{ave}} = \frac{F}{A} \tag{1.2}$$

where A is the area of the bottom surface of the block. In this way we can use the term *shear stress* to express the tangential force per unit area.

1.2   Stress in the Interior of a Body

If we consider the small area, ΔA, in the body shown in Figure 1.1, we can see that it is sub-jected to a force acting on the area ΔA. However, we cannot talk about the normal stress σ_n at that point yet, because we do not know either the magnitude or the direction of the internal force. However, supposing that there exists a normal component of the internal force, ΔF_n, we can use this to define the normal stress σ_n at that same point in the same way as we did at the surface by the limiting expression

$$\sigma_n = \lim_{\Delta A \to 0} \frac{\Delta F_n}{\Delta A} \tag{1.3}$$

Similarly, supposing there exists a tangential component ΔF_t of the internal force, we can define the shear stress, τ, at ΔA by a similar equation,

$$\tau = \lim_{\Delta A \to 0} \frac{\Delta F_t}{\Delta A} \tag{1.4}$$

In the previous case shown in Figure 1.1 we see that, irrespective of the position and direction of the area, the only force acting is the normal one. That is to say, $\sigma_n = -p$ and $\tau = 0$ everywhere in the body. However, in general problems the conditions are different and the stress varies from point to point and the stress field is not uniform.

When we first meet a problem, we usually have no information about the stress state inside the body, as only the stresses at the *surface* are known. Hence, we can only use this information to solve the problem. *The stresses at the surface of the body are the keys to solving the problem.* These, already known stresses (or deformations), are called *boundary conditions*.

So now we can start to use the theory of elasticity using boundary conditions, but how can we use them to obtain the stresses inside the body?

1.3 Two Dimensional Stress, Three Dimensional Stress and Stress Transformation

1.3.1 Normal Stress

When a plate of uniform thickness having an arbitrary shape (Figure 1.3) is subjected to a constant pressure, p, along its periphery Γ, the normal stress σ_n and the shear stress τ at the periphery Γ are $\sigma_n = -p$ and $\tau = 0$, respectively. However, we still cannot determine the values of σ_n and τ at an arbitrary point A. How to find the stresses at point A will be explained later.

Now let us consider a rectangular plate of uniform thickness, as illustrated in Figure 1.4. Its boundary conditions are $\sigma_n = \sigma_{x0}$ along the side BC and AD, $\sigma_n = \sigma_{y0}$ along the side AB and CD

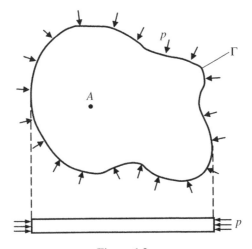

Figure 1.3

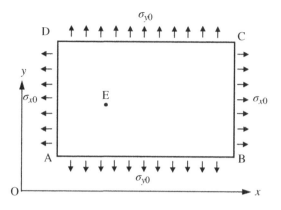

Figure 1.4

and $\tau = 0$ along all sides. The usual method of describing these conditions is to say, $\sigma_x = \sigma_{x0}$, $\tau_{xy} = 0$ along BC and AD, and $\sigma_y = \sigma_{y0}$, $\tau_{yx} = 0$ along AB and CD. The subscripts, x and y in σ_x and σ_y mean that σ_x and σ_y are normal stresses in the directions of the x and y axes respectively. The order of the double subscripts, like xy in τ_{xy}, have a universal meaning. The first indicates the face on which the stress acts and the second indicates the direction of the shear stress. Thus, τ_{xy} acts on a face normal to the x axis and acts in the y direction.

If $\sigma_{y0} = 0$, we can easily see that $\sigma_x = \sigma_{x0}$ at an arbitrary point, E, inside the plate. Likewise, if $\sigma_{x0} = 0$, then $\sigma_y = \sigma_{y0}$ everywhere in the plate. Therefore, when either σ_{x0} or σ_{y0} is not zero, we can easily see that $\sigma_x = \sigma_{x0}$ and $\sigma_y = \sigma_{y0}$ in the plate. This interpretation comes from considering the equilibrium of forces within the plate.

1.3.2 Shear Stress

Considering the case where shear stresses are the only boundary conditions, as shown in Figure 1.5, how do we find the stresses inside the plate? From the equilibrium of forces we can see that if τ_{yx0} acts along AB in the negative x direction, τ_{yx0} along CD must act in the positive x direction, otherwise the plate would not be in equilibrium. By the same reasoning we can see that the shear stresses along BC and AD will act in opposite directions.

In addition, the plate must be in equilibrium from the viewpoint of *rotation* as well. There must be no effective moments to cause rotation. This is known as the *condition of rotation*. Considering the condition of rotation we can use any z axis which pierces the plate. If we choose the z axis which pierces the plate at the point, A, then the equilibrium condition for rotation is written as follows.

$$BC \cdot \tau_{xy0} \cdot AB - CD \cdot \tau_{yx0} \cdot AD = 0 \tag{1.5}$$

and as $CD = AB$ and $AD = BC$ then this simplifies to

$$\tau_{xy0} = \tau_{yx0} \tag{1.6}$$

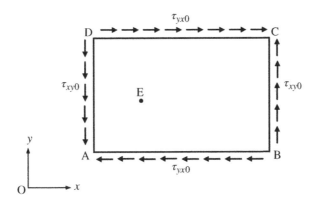

Figure 1.5

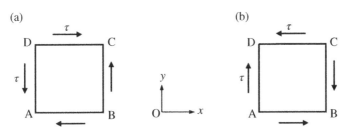

Figure 1.6

This is simple, but a very important relationship. Equation (1.6) means that *shear stresses exist only as shown in Figure 1.6a, b*. All the shear stresses are equal in magnitude and a shear stress *cannot* exist only on a single side (e.g. AB) or only on a couple of parallel sides (e.g. a couple of AB and CD). This rule also holds inside the plate. Although students and engineers sometimes underestimate this rule, an exact understanding of it will help to solve advanced problems later.

1.3.3   Stress in an Arbitrary Direction

1.3.3.1   Two Dimensional Stress Transformation

If the boundary conditions of a rectangular plate are like those shown in Figure 1.7, we can immediately see that the stresses at an arbitrary point inside the plate are $\sigma_x = \sigma_{x0}$, $\sigma_y = \sigma_{y0}$ and $\tau_{xy} = \tau_{xy0}$. However, these stresses are those defined in the x-y coordinate system. In many practical problems the stresses in different coordinate systems are needed.

Now let us determine the stresses σ_ξ, σ_η and $\tau_{\xi\eta}$ defined in the coordinate system (ξ, η) which is rotated by the angle θ from the x axis in the counterclockwise direction.

If we imagine a right angle triangle ABC as in Figure 1.8, inside the rectangular plate of Figure 1.7, the stresses σ_x, σ_y and τ_{xy} along the sides AB and AC are $\sigma_x = \sigma_{x0}$ (along AC) $\sigma_y = \sigma_{y0}$ (along AB) and $\tau_{xy} = \tau_{xy0}$ (along AB and AC) respectively. For the sake of simplicity we take the length of the side BC to be unity, that is $|BC| = 1$.

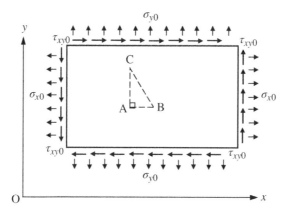

Figure 1.7

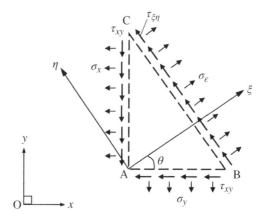

Figure 1.8

Using the *direction cosines* of the ξ and η axes with respect to the x and y axes, we can express the equilibrium conditions of the triangle ΔABC in the ξ and η directions.

Table 1.1 defines the direction cosines as follows:

$$l_1 = \cos\theta, \quad m_1 = \sin\theta$$

$$l_2 = -\sin\theta, \quad m_2 = \cos\theta$$

$$(\sigma_\xi \cdot 1) \cdot 1 = (\sigma_x \cdot l_1) \cdot l_1 + (\sigma_y \cdot m_1) \cdot m_1 + (\tau_{xy} \cdot l_1) \cdot m_1 + (\tau_{xy} \cdot m_1) \cdot l_1 \tag{1.7}$$

$$(\tau_{\xi\eta} \cdot 1) \cdot 1 = (\sigma_x \cdot l_1) \cdot l_2 + (\sigma_y \cdot m_1) \cdot m_2 + (\tau_{xy} \cdot l_1) \cdot m_2 + (\tau_{xy} \cdot m_1) \cdot l_2 \tag{1.8}$$

In the above equations, the quantities in the parentheses are forces acting on the side of the triangle, (stress · area), and the direction cosines that follow the parentheses are the operators for

Table 1.1 Direction cosine

	x	y
ξ	l_1	m_1
η	l_2	m_2

obtaining the components of the forces. A common mistake students make when using the *stress transformation* is that they multiply stresses and direction cosines *only once*. This is because they use the wrong condition of *stress* instead of *force*[1] when they are trying to solve for equilibrium conditions.

Arranging Equations 1.7 and 1.8 and adding that for σ_η gives:

$$\sigma_\xi = \sigma_x l_1^2 + \sigma_y m_1^2 + 2\tau_{xy} l_1 m_1 \tag{1.9}$$

$$\sigma_\eta = \sigma_x l_2^2 + \sigma_y m_2^2 + 2\tau_{xy} l_2 m_2 \tag{1.10}$$

$$\tau_{\xi\eta} = \sigma_x l_1 l_2 + \sigma_y m_1 m_2 + \tau_{xy}(l_1 m_2 + l_2 m_1) \tag{1.11}$$

Rewriting these equations using the angle θ in Figure 1.8 gives:

$$\left.\begin{array}{l} \sigma_\xi = \sigma_x \cos^2\theta + \sigma_y \sin^2\theta + 2\tau_{xy}\cos\theta\cdot\sin\theta \\ \sigma_\eta = \sigma_x \sin^2\theta + \sigma_y \cos^2\theta - 2\tau_{xy}\cos\theta\cdot\sin\theta \\ \tau_{\xi\eta} = (\sigma_y - \sigma_x)\cos\theta\cdot\sin\theta + \tau_{xy}(\cos^2\theta - \sin^2\theta) \end{array}\right\} \tag{1.12}$$

When students look at Equation 1.12, they often forget that it was derived using the equilibrium condition of *forces*. This is a very important equation and later we will use it often to solve various problems.

On deriving Equation 1.12, we usually draw a diagram similar to that in Figure 1.8. However, it should be noted that in Figure 1.8 we know the stresses on two sides (AB, AC) of the triangle and on only one side, (BC), is the stress unknown.

If we draw the figure like Figure 1.9 instead of Figure 1.8 then we cannot derive Equation 1.12 because we know only the stresses on the side (AB) and we do not know the stresses σ_ξ, σ_η and $\tau_{\xi\eta}$ acting on BC and AC. This is another mistake that confuses many students.

Equation 1.12 was derived assuming that the plate of Figure 1.7 was subjected to a uniform stress. However, as we did not talk about the size of the plate we can also use Equation 1.12 in the cases of non-uniform stress, simply by imagining a sufficiently small rectangle inside the arbitrarily shaped plate, over which the stresses do not vary. As we shall see, Equation 1.12 is used often in such cases.

---

[1] The author believes that the teaching of Mohr's circle is another major cause of misunderstanding. This is because the use of Mohr's circle is usually taught after the derivation of Equation 1.12 and cosine and sine appear only once in its analysis.

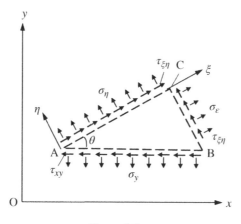

Figure 1.9

Example problem 1.1

When an arbitrarily shaped plate of uniform thickness is subjected to a constant pressure, p along its periphery, Γ, verify that the normal stress, σ, and the shear stress, τ, are $\sigma = -p$ and $\tau = 0$ throughout the plate (Figure 1.10).

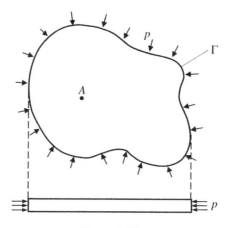

Figure 1.10

Solution

We cannot solve this problem by dividing the plane into small parts. We will solve the problem by visualizing the plate as an arbitrary shaped part within a larger, known plate instead.

Look at Figure 1.11. We know that the plate is subject to a constant pressure p along its periphery. We also know that the stresses inside the plate will be $\sigma_x = -p$, $\sigma_y = -p$ and $\tau_{xy} = 0$ when the x–y coordinate system is used. Now, drawing in the periphery Γ, as shown with the dotted line in Figure 1.11, we take the ξ axis to be normal to periphery at the arbitrary point B on the periphery Γ. Now we take an arbitrary point on the axis, the point O$'$, and define a ξ-η

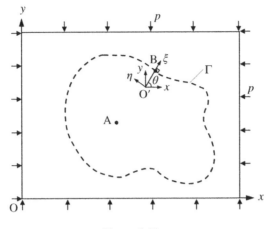

Figure 1.11

coordinate system with the origin at the point O'. When the angle between the x axis and the ξ
axis is θ, the stresses at the point B can be obtained from Equation 1.12 and are as follows:

$$\sigma_\xi = \sigma_x\cos^2\theta + \sigma_y\sin^2\theta + 2\tau_{xy}\cos\theta\cdot\sin\theta = -p\cos^2\theta - p\sin^2\theta + 0 = -p$$
$$\tau_{\xi\eta} = (\sigma_y - \sigma_x)\cos\theta\cdot\sin\theta + \tau_{xy}(\cos^2\theta - \sin^2\theta)$$
$$= (-p+p)\cos\theta\cdot\sin\theta + 0(\cos^2\theta - \sin^2\theta) = 0$$

Point B is not a special point on the periphery of the plate and so we can see that the normal
stress and shear stress will be the same everywhere along the periphery. This means that the
conditions on the boundary of the plate in Figure 1.10, correspond to those of the periphery
of the plate indicated by Γ, described within the rectangular plate in Figure 1.11. Consequently,
to determine the stress state at point A in Figure 1.10, we have only to think of the stress state of
the identical point A in Figure 1.11.

As was noted before, point B is not a special point in the rectangular plate and so the stresses
at point A can be expressed in the same way as those at point B. Thus we can see that the stresses will be $\sigma = -p$ and $\tau = 0$ at any point and direction within Γ.

This example demonstrates how we can determine the stress state inside a plate, knowing
only the boundary conditions and the stress transformation equation. We should notice the fact
that we did not mention the material of the plate besides its uniform thickness. Never forget the
conclusion gained from this problem. It is very important not only in elasticity but also in various problems of plasticity.

1.3.3.2   Three Dimensional Stress Transformation

This section describes the method to find the stresses in an arbitrary direction, when a solid,
brick-like body is subjected to a uniform stress (Figure 1.12).

Suppose that the stresses $\sigma_x, \sigma_y, \sigma_z, \tau_{xy}, \tau_{yz}, \tau_{zx}$ (note: $\tau_{yx} = \tau_{xy}, \tau_{yz} = \tau_{zy}, \tau_{xz} = \tau_{zx}$) are known
and the stresses in the $\xi-\eta-\zeta$ coordinate system are required. We denote the direction cosines
between these two coordinate systems as shown in Table 1.2.

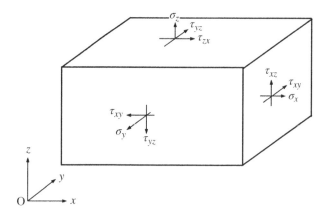

Figure 1.12

Table 1.2   Direction cosines

	x	y	z
ξ	l_1	m_1	n_1
η	l_2	m_2	n_2
ζ	l_3	m_3	n_3

Even before we find the solution it is possible to guess the form of the final solution. It is similar to the method adopted for the two dimensional (2D) case (Section 1.3.3.1) but is lengthier. So, before taking the lengthy approach, let us look at Equations 1.9 to 1.11 carefully, specifically by paying attention to the regularity of the direction cosines. In addition, we should be aware of the fact that *the 2D case is just a special case of a 3D case*, that is in 2D cases the z and ζ axes are perpendicular to the 2D plane (x–y and ξ–η planes). Thus, from these considerations and Equations 1.9 to 1.11, we can predict the following equations for the 3D case:

$$
\left.
\begin{aligned}
\sigma_\xi &= \sigma_x l_1{}^2 + \sigma_y m_1{}^2 + \sigma_z n_1{}^2 + 2\left(\tau_{xy} l_1 m_1 + \tau_{yz} m_1 n_1 + \tau_{zx} n_1 l_1\right) \\
\tau_{\xi\eta} &= \sigma_x l_1 l_2 + \sigma_y m_1 m_2 + \sigma_z n_1 n_2 + \tau_{xy}\left(l_1 m_2 + l_2 m_1\right) \\
&\quad + \tau_{yz}\left(m_1 n_2 + m_2 n_1\right) + \tau_{zx}\left(n_1 l_2 + n_2 l_1\right) \\
\tau_{\xi\zeta} &= \sigma_x l_1 l_3 + \sigma_y m_1 m_3 + \sigma_z n_1 n_3 + \tau_{xy}\left(l_1 m_3 + l_3 m_1\right) \\
&\quad + \tau_{yz}\left(m_1 n_3 + m_3 n_1\right) + \tau_{zx}\left(n_1 l_3 + n_3 l_1\right) \\
&\cdots
\end{aligned}
\right\}
\qquad (1.13)
$$

Actually, Equation 1.13 is an exact expression. The orthodox, and lengthy, derivation of Equation 1.13 was obtained by considering a triangular pyramid (tetrahedron) instead of the triangle that was used for the 2D case (Figure 1.8). We take the ξ axis to be perpendicular to the plane ABC and the area of $\triangle$ABC to be 1. Denoting the areas of $\triangle$OBC, $\triangle$OAC and

ΔOAB by A_x, A_y and A_z, respectively, we can see that $A_x = l_1$, $A_y = m_1$ and $A_z = n_1$. Using these relationships and by following a similar method based on the equilibrium condition of force, we can obtain Equation 1.13.

How the stresses of Equation 1.13 act on the plane ABC is illustrated in Figure 1.13. In Figure 1.14, we denote σ_ξ by σ and express the *resultant shear stress*, τ, on the plane ABC (ξ–η plane), as follows:

$$\tau^2 = \tau_{\xi\eta}^2 + \tau_{\xi\zeta}^2 \tag{1.14}$$

We also express the *resultant stress*, p, on the ξ–η plane by the following equation:

$$p^2 = \sigma^2 + \tau^2 \tag{1.15}$$

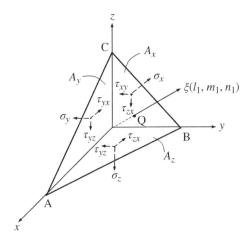

Figure 1.13

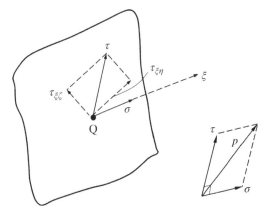

Figure 1.14

When $\tau=0$, we have $p=\sigma$ and the resultant force acts perpendicular to the $\xi-\eta$ plane. In this case, we call ξ a principal axis. As will be shown later, there are three principal axes.

1.3.4   Principal Stresses

1.3.4.1   Principal Stresses in 2D Stress State

By looking at Figures 1.7 and 1.8, we can see that in general there will be two kinds of stresses acting on the plane BC: normal stress σ_ξ and shear stress $\tau_{\xi\eta}$. However, if we vary the angle θ continuously from 0 to 2π, we will find that $\tau_{\xi\eta}$ vanishes at certain values of θ.

In Figure 1.15, p is the resultant stress, where $p^2 = \sigma_\xi^2 + \tau_{\xi\eta}^2$. If we take the length $|BC|=1$ then we have $|AB|=m_1$ and $|AC|=l_1$. Now, by considering the equilibrium conditions of the element ABC and denoting the x and y components of the force acting on the side BC by p_x and p_y, respectively, we have:

$$\left.\begin{array}{l} p_x = \sigma_x l_1 + \tau_{xy} m_1 \\ p_y = \tau_{xy} l_1 + \sigma_y m_1 \end{array}\right\} \tag{1.16}$$

As was mentioned before, at certain values of θ, $\tau_{\xi\eta}$ vanishes, so we can write

$$\left.\begin{array}{l} p_x = \sigma_\xi l_1 \\ p_y = \sigma_\xi m_1 \end{array}\right\} \tag{1.17}$$

because $p=\sigma_\xi$ at those special angles.

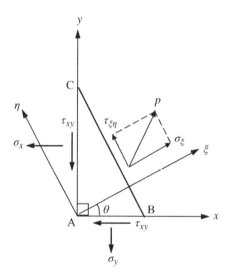

Figure 1.15

From Equations 1.16 and 1.17 and denoting σ_ξ by σ, we have

$$\left.\begin{array}{c} \sigma_x l_1 + \tau_{xy} m_1 = \sigma l_1 \\ \tau_{xy} l_1 + \sigma_y m_1 = \sigma m_1 \end{array}\right\} \tag{1.18}$$

Rewriting Equation 1.18 gives

$$\left.\begin{array}{c} (\sigma_x - \sigma) l_1 + \tau_{xy} m_1 = 0 \\ \tau_{xy} l_1 + (\sigma_y - \sigma) m_1 = 0 \end{array}\right\} \tag{1.19}$$

And by considering the properties of the direction cosines so that we know l_1 and m_1 cannot both be zero at the same angle, (i.e. $l_1^2 + m_1^2 \neq 0$) then the following equation must hold.

$$\left| \begin{array}{cc} (\sigma_x - \sigma) & \tau_{xy} \\ \tau_{xy} & (\sigma_y - \sigma) \end{array} \right| = 0 \tag{1.20}$$

Solving Equation 1.20 gives us two roots of σ. Denoting these roots by σ_1 and σ_2, we obtain,

$$\sigma_1, \sigma_2 = \frac{(\sigma_x + \sigma_y) \pm \sqrt{(\sigma_x - \sigma_y)^2 + 4\tau_{xy}^2}}{2} \tag{1.21}$$

σ_1 and σ_2 are the *principal stresses* (we define $\sigma_1 > \sigma_2$). In a 2D stress state, such as this, there are usually two principal stresses.

From the above discussion, *the principal stresses are the normal stresses acting on the planes with no shear stress.* We can see that the process to obtain Equation 1.21 from Equation 1.16 is identical with that to obtain the eigenvalues in linear algebra. This means that σ_1 and σ_2 correspond to the eigenvalues and, therefore, are the maximum and minimum normal stresses, respectively. The directions they act in are called the *principal axes*. The direction cosines for the principal axes can be obtained from Equation 1.19, that is

$$\frac{m_1}{l_1} = \tan\theta = \frac{\sigma - \sigma_x}{\tau_{xy}} \quad \text{or} \quad \theta = \tan^{-1}\left(\frac{\sigma - \sigma_x}{\tau_{xy}}\right) \tag{1.22}$$

In linear algebra, two eigenvectors, (i.e. the two principal axes), orthogonally intersect one another. This means $(l_1, m_1) \cdot (l_1', m_1') = 0$. The same result can be derived by differentiating σ_ξ with respect to θ in Equation 1.12 and using the condition $d\sigma_\xi/d\theta = 0$. However, the author recommends readers to memorize Equation 1.20 rather than Equation 1.21 or the conventional derivation by $d\sigma_\xi/d\theta = 0$, because Equation 1.20 is not only easy to memorize but it also helps us to see the physical meaning in a compact way. This method for the interpretation of the physical meaning of the principal stresses helps us to move easily into 3D problems, in which the conventional method of reduction becomes very lengthy and boring.

Example problem 1.2
Verify that the principal stresses σ_1 and σ_2 are the maximum and minimum values which normal stress can take.

Solution
Denoting the direction cosines between the principal axis 1 and the axes (ξ, η) by (l_1, l_2) and the direction cosines between the principal axis 2 and (ξ, η) by (m_1, m_2), from Equation 1.9:

$$\sigma_\xi = \sigma_1 l_1{}^2 + \sigma_2 m_1{}^2 \le \sigma_1 l_1{}^2 + \sigma_1 m_1{}^2 = \sigma_1$$
$$\sigma_\xi = \sigma_1 l_1{}^2 + \sigma_2 m_1{}^2 \ge \sigma_2 l_1{}^2 + \sigma_2 m_1{}^2 = \sigma_2$$

Hence, $\sigma_2 \le \sigma_\xi \le \sigma_1$.

Example problem 1.3
Find the principal stresses on the surface of a uniform round bar which is subjected to torsion, by determining a shear stress τ on the surface.

Solution

$$\begin{vmatrix} (\sigma_x - \sigma) & \tau_{xy} \\ \tau_{xy} & (\sigma_y - \sigma) \end{vmatrix} = \begin{vmatrix} -\sigma & \tau \\ \tau & -\sigma \end{vmatrix} = \sigma^2 - \tau^2 = 0$$

$$\sigma_1 = \tau, \quad \sigma_2 = -\tau$$

$$\theta_1 = \tan^{-1}\frac{\sigma_1 - \sigma_x}{\tau_{xy}} = \tan^{-1}\frac{\sigma_1}{\tau} = \tan^{-1}1 = \frac{\pi}{4}$$

$$\theta_2 = \tan^{-1}\frac{\sigma_2 - \sigma_x}{\tau_{xy}} = \tan^{-1}\frac{\sigma_2}{\tau} = \tan^{-1}(-1) = -\frac{\pi}{4}$$

This result means that, as shown in Figure 1.16, the stress state of $\sigma_x = \sigma_y = 0$, $\tau_{xy} = \tau$ is identical to the stress state of $\sigma_1 = \tau$ (tension) and $\sigma_2 = -\tau$ (compression) if looked from $\pm 45°$. The phenomena of shear fracture of ductile materials along the plane perpendicular to the axis of

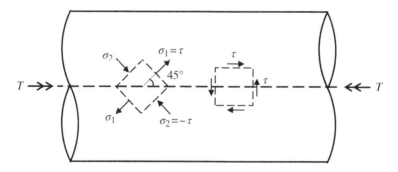

Figure 1.16

twisting moment and spiral shape fracture of brittle materials along the plane of $\sim 45°$ are related to this stress state.

1.3.4.2 Principal Stresses in 3D Stress State

In a manner similar to that used with 2D problems, we can obtain an equation which determines the three principal stresses σ_1, σ_2 and σ_3, $(\sigma_1 > \sigma_2 > \sigma_3)$ in 3D problems:

$$\begin{vmatrix} (\sigma_x - \sigma) & \tau_{xy} & \tau_{zx} \\ \tau_{xy} & (\sigma_y - \sigma) & \tau_{yz} \\ \tau_{zx} & \tau_{yz} & (\sigma_z - \sigma) \end{vmatrix} = 0 \tag{1.23}$$

Resolving Equation 1.23:

$$\sigma^3 - \left(\sigma_x + \sigma_y + \sigma_z\right)\sigma^2 + \left(\sigma_x\sigma_y + \sigma_y\sigma_z + \sigma_z\sigma_x - \tau_{xy}^2 - \tau_{yz}^2 - \tau_{zx}^2\right)\sigma$$
$$- \left(\sigma_x\sigma_y\sigma_z + 2\tau_{xy}\tau_{yz}\tau_{zx} - \sigma_x\tau_{yz}^2 - \sigma_y\tau_{zx}^2 - \sigma_z\tau_{xy}^2\right) = 0 \tag{1.24}$$

If we denote the roots by σ_1, σ_2 and σ_3, Equation 1.24 can be rewritten in the following form:

$$(\sigma - \sigma_1) \cdot (\sigma - \sigma_2) \cdot (\sigma - \sigma_3) = 0 \tag{1.25}$$

And resolving again we have

$$\sigma^3 - J_1\sigma^2 - J_2\sigma - J_3 = 0 \tag{1.26}$$

where

$$J_1 = \sigma_1 + \sigma_2 + \sigma_3$$
$$J_2 = -\left(\sigma_1\sigma_2 + \sigma_2\sigma_3 + \sigma_3\sigma_1\right)$$
$$= \frac{1}{6}\left[(\sigma_1 - \sigma_2)^2 + (\sigma_2 - \sigma_3)^2 + (\sigma_3 - \sigma_1)^2 - 2(\sigma_1 + \sigma_2 + \sigma_3)^2\right]$$
$$J_3 = \sigma_1\sigma_2\sigma_3$$

Since the principal stresses σ_1, σ_2 and σ_3 are independent of the coordinates chosen, J_1, J_2 and J_3 are constant in a certain stress state and are called the first, second and third *stress invariants*. If we note the equivalence of Equations 1.24 and 1.26, we will understand that

$$J_1 = \sigma_1 + \sigma_2 + \sigma_3 = \sigma_x + \sigma_y + \sigma_z = \sigma_\xi + \sigma_\eta + \sigma_\zeta \tag{1.27}$$

$$J_2 = -\left(\sigma_1\sigma_2 + \sigma_2\sigma_3 + \sigma_3\sigma_1\right) = -\left(\sigma_x\sigma_y + \sigma_y\sigma_z + \sigma_z\sigma_x - \tau_{xy}^2 - \tau_{yz}^2 - \tau_{zx}^2\right)$$
$$= -\left(\sigma_\xi\sigma_\eta + \sigma_\eta\sigma_\zeta + \sigma_\zeta\sigma_\xi - \tau_{\xi\eta}^2 - \tau_{\eta\zeta}^2 - \tau_{\zeta\xi}^2\right) \tag{1.28}$$

$$J_3 = \sigma_1\sigma_2\sigma_3 = \sigma_x\sigma_y\sigma_z + 2\tau_{xy}\tau_{yz}\tau_{zx} - \sigma_x\tau_{yz}{}^2 - \sigma_y\tau_{zx}{}^2 - \sigma_z\tau_{xy}{}^2$$
$$= \sigma_\xi\sigma_\eta\sigma_\zeta + 2\tau_{\xi\eta}\tau_{\eta\zeta}\tau_{\zeta\xi} - \sigma_\xi\tau_{\eta\zeta}{}^2 - \sigma_\eta\tau_{\zeta\xi}{}^2 - \sigma_\zeta\tau_{\xi\eta}{}^2 \tag{1.29}$$

We have obtained these important equations (Equations 1.27 to 1.29) without solving Equations 1.23 and 1.24. In general we cannot solve Equation 1.24 in the traditional manner, using a compass and triangles (which are used to find the principal stresses in 2D problems). This means that we cannot draw a so-called Mohr's circle from the stresses (σ_x, σ_y, σ_z, τ_{xy} τ_{yz}, τ_{zx}). However, we should not be disappointed, because Mohr's circles are not as important as they are widely assumed to be. J_1 is the quantity related to the component of hydrostatic compression of a stress state and, as explained later, it is the quantity related to the volume change in terms of strains through Hooke's law. J_2 is the quantity related to the cause of plastic deformation. The meaning of J_3 is not clear as J_1 and J_2. Equation 1.23 is simple and impressive, and as engineers know there is beauty in simplicity. This equation will also be very useful in later problems.

Example problem 1.4
Verify that the three principal axes intersect each other orthogonally in 3D problems.

Hint
Pay attention to the definition of the principal stress and the nature of shear stress.

1.3.5   Principal Shear Stresses

In the case of 2D stress states, the principal shear stress is called the maximum shear stress. Materials can fail by tensile normal stress or by shear stress depending on their ductility or brittleness. Failure of materials by shear stress occurs mostly in a plane with the maximum shear stress.

First we shall consider the maximum shear stress in a 2D stress state. From Equation 1.11, when σ_x, σ_y and τ_{xy} are known, the shear stress on the plane $\xi =$ constant, or $\eta =$ constant, can be expressed as

$$\tau_{\xi\eta} = \sigma_x l_1 l_2 + \sigma_y m_1 m_2 + \tau_{xy}(l_1 m_2 + l_2 m_1)$$

Since σ_x, σ_y and τ_{xy} are known, we can determine the principal stresses σ_1 and σ_2 and the direction cosines. Now, supposing that l_1', l_2' and so on are direction cosines defined between the ξ and η axes and the principal axes (direction of the principal stresses), we have

$$\tau_{\xi\eta} = \sigma_1 l_1' l_2' + \sigma_2 m_1' m_2', \quad \sigma_1 > \sigma_2 \tag{1.30}$$

and by considering the nature of the direction cosines (see Appendix A.1 of Part I),

$$l_1' l_2' + m_1' m_2' = 0,$$

we obtain,

$$\tau_{\xi\eta} = (\sigma_1 - \sigma_2) l_1' l_2' \tag{1.31}$$

Now we need to know the direction in which $l_1'l_2'$ becomes maximum. Since $(\sigma_1 - \sigma_2)$ is a quantity that is constant irrespective of the direction, we must make $|l_1'l_2'|$ the maximum.

Since $l'_{1^2} + l'_{2^2} = 1$, we have the absolute maximum value of $\tau_{\xi\eta}$ when $|l_1'| = |l_2'|$ and so we have the angle at which the maximum shear stress acts:

$$\theta = \pi/4 \quad \text{or} \quad (\pi/2 + \pi/4) \tag{1.32}$$

$$\tau_{\max} = \mp \frac{1}{2}(\sigma_1 - \sigma_2) \tag{1.33}$$

In Equation 1.33, the plus and minus signs in front of the parentheses are not important, they only indicate the direction of the shear stress. This result means that the maximum shear stress is in the plane at an angle of $\pi/4$ from both the principal axes. Looking at the stress state from the direction of one of the principal axes and using a similar method, we can derive the *principal shear stresses* in three dimensions that correspond to $\tau_{\max}$ in 2D problems as follows:

$$\tau_1 = \frac{1}{2}|\sigma_2 - \sigma_3|, \quad \tau_2 = \frac{1}{2}|\sigma_3 - \sigma_1|, \quad \tau_3 = \frac{1}{2}|\sigma_1 - \sigma_2| \tag{1.34}$$

One of the shear stresses is algebraically maximum, another minimum and the last one is between these. τ_1 acts on the plane which divides equally the angle made by two planes on which σ_2 and σ_3 act. The same rule applies to τ_2 and τ_3 respectively.

Problems of Chapter 1

1. A tensile test was carried out on a belt which was made by bonding the straight ends, as shown in Figure 1.17a. A tensile fracture occurred from the bonded end at 1/400 of the original strength of the belt. In order to increase the strength of the bonded end, the belt was sliced in an inclined direction and was bonded again with the same bond. Determine the approximate angle of slicing in Figure 1.17b to prevent fracture at the bonded interface. Assume that the bonded interface fractures only by tensile normal stress and the fracture is not influenced by shear stress.

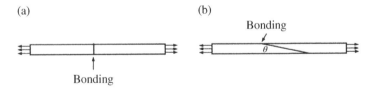

Figure 1.17

2. Determine the stress inside a cylinder which is subjected to the same internal pressure p_i and external pressure p_o, that is $p_i = p_o = -p_0$.
3. When the normal stress σ_ξ and shear stress $\tau_{\xi\eta}$ in the direction of $\theta = \theta_0$ against the axis of the bar at the outer radius of a cylindrical bar under a twisting moment are $\sigma_\xi = \sigma_0$ and

$\tau_{\xi\eta} = \tau_0$, respectively, determine the normal stress σ_η acting on the plane $\theta = \theta_0 + \pi/2$ in terms of σ_0, where θ is measured in the counterclockwise direction against the axis of the cylindrical bar.

4. When a solid body with no hole inside is under a certain stress state and is afterward subjected to a constant pressure p at the outer surface, verify that the shear stress in the interior of the body does not change regardless of the shape of the body due to the applied pressure p. (The conclusion of this problem means that the increase in hydrostatic pressure does not contribute to the deformation and fracture of materials originated by shear stress. Actually, many objects which we can deform easily on the ground can endure the deformation even under high pressure at the bottom of deep sea without changing shape.)

5. Denoting three principal stresses by σ_1, σ_2 and σ_3 ($\sigma_1 \geq \sigma_2 \geq \sigma_3$), verify that the possible value at which normal stress σ in an arbitrary direction is limited within $\sigma_3 \leq \sigma \leq \sigma_1$.

6. There is a tubular rubber balloon which has length l, diameter d ($d \ll l$) and thickness of rubber t ($t \ll d$). This balloon is subjected to an internal pressure p. Figure 1.18 shows the balloon which is subjected to a twisting moment T in addition to the internal pressure p. The twisting moment T is increased gradually. When T approaches a critical value T_c, it suddenly becomes impossible to increase T continuously. What phenomenon occurs to the balloon? Describe the phenomenon as quantitatively as possible and determine the value of T_c.

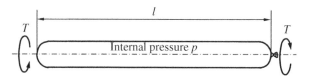

Figure 1.18   A tubular balloon subjected to an internal pressure p and a twisting moment T

2

Strain

2.1 Strains in Two Dimensional Problems

Strains are quantities which indicate a measure of the deformation of bodies. When a plate is loaded as shown in Figure 2.1, the position and shape of an arbitrarily imagined small element will change.

In two dimensional problems, as in Figure 2.1, the change in position of a point (x, y) in the body is usually described by the displacements (u, v) in the x and y directions, respectively. We can easily see that a large magnitude of u and v do not necessarily correspond to a large deformation of the plate, because it is possible to move the whole system in the x or y direction without deforming the plate. This is illustrated in columns (a) and (b) in Table 2.1.

In practical problems, the variation in the length of the sides of element ABCD in Figure 2.1 and the variation in the angles made between neighboring sides are important. The former is *normal strain* and the latter is *shear strain*.

We denote the coordinates of the four points (A, B, C, D) which compose a small rectangular element as shown in Figure 2.2. If the x coordinate of point A changes from x to $(x+u)$ after deformation of the plate, the x coordinate of point B may be written as $(x+u+dx+\partial u/\partial x \cdot dx)$ (note that ∂ is partial differentiation). Therefore, the length between A and B changes from dx to $(dx+\partial u/\partial x \cdot dx)$. So, as mentioned in the elementary theory of strength of materials, *normal strain* ε_x in the x direction is defined as

$$\varepsilon_x = \text{Increase in length/initial length}$$

$$= \frac{(dx + (\partial u/\partial x)dx) - dx}{dx} = \frac{\partial u}{\partial x} \tag{2.1}$$

Theory of Elasticity and Stress Concentration, First Edition. Yukitaka Murakami.
© 2017 John Wiley & Sons, Ltd. Published 2017 by John Wiley & Sons, Ltd.

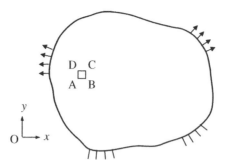

Figure 2.1

Table 2.1

Deformation	(a)	(b)	(c)	(d)	(e)	(f)
Strain	$\varepsilon_x = 0$	$\varepsilon_x = 0$	$\varepsilon_x = \dfrac{\partial u}{\partial x}$	$\varepsilon_x = 0$	$\varepsilon_x = 0$	$\varepsilon_x = \dfrac{\partial u}{\partial x}$
	$\varepsilon_y = 0$	$\varepsilon_y = 0$	$\varepsilon_y = 0$	$\varepsilon_y = \dfrac{\partial v}{\partial y}$	$\varepsilon_y = 0$	$\varepsilon_y = \dfrac{\partial v}{\partial y}$
	$\gamma_{xy} = 0$	$\gamma_{xy} = 0$	$\gamma_{xy} = 0$	$\gamma_{xy} = 0$	$\gamma_{xy} = \dfrac{\partial u}{\partial y} + \dfrac{\partial v}{\partial x}$	$\gamma_{xy} = \dfrac{\partial u}{\partial y} + \dfrac{\partial v}{\partial x}$
Rotation	$\omega_z = 0$	$\omega_z = \dfrac{1}{2}\left(\dfrac{\partial v}{\partial x} - \dfrac{\partial u}{\partial y}\right)$	$\omega_z = 0$	$\omega_z = 0$	$\omega_z = \dfrac{1}{2}\left(\dfrac{\partial v}{\partial x} - \dfrac{\partial u}{\partial y}\right)$	$\omega_z = \dfrac{1}{2}\left(\dfrac{\partial v}{\partial x} - \dfrac{\partial u}{\partial y}\right)$

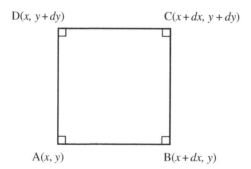

Figure 2.2

Likewise, in the y direction we have

$$\varepsilon_y = \frac{\partial v}{\partial y} \tag{2.2}$$

So, physically, the normal strain is defined as the increase in length per unit length. However, knowledge of the normal strains, ε_x and ε_y, which describe the elongation of length is not sufficient by itself to describe the deformation of the rectangle ABCD.

In general, the right angle corners at A, B, C and D do not remain right angles after deformation. For example, the rectangle may undergo deformation as shown in Figure 2.3a. In this case, points C and D move to points C′ and D′, respectively. This deformation is described by the angle γ. The angle γ is called *shear strain* and is usually written as γ_{xy}. However, in general, a rectangle will deform as shown in Figure 2.3b and not like Figure 2.3a. In such cases it may be seen that γ_{xy} is written as

$$\gamma_{xy} = \angle DAD' + \angle BAB'$$

$$= \frac{\partial u}{\partial y} + \frac{\partial v}{\partial x} \tag{2.3}$$

Now, in addition to shear strain, we can see that the body in Figure 2.3 is rotated, that is in Figure 2.3a the line AD rotates by γ but the line AB does not rotate. The definition of *rotation* for the element ABCD is reasonable if we manipulate the deformation to obtain the state of non-rotation of the element, that is, if we have equal angle displacements of B to B′ and D to D′ by removing the counter-clockwise rotation ω_z from the state of deformation. Therefore,

$$\frac{1}{2}\left(\frac{\partial u}{\partial y} + \frac{\partial v}{\partial x}\right) + \omega_z = \frac{\partial v}{\partial x} \quad \text{or} \quad \frac{1}{2}\left(\frac{\partial u}{\partial y} + \frac{\partial v}{\partial x}\right) - \omega_z = \frac{\partial u}{\partial y} \tag{2.4}$$

So we have the following equation as the definition of rotation of the element:

$$\omega_z = \frac{1}{2}\left(\frac{\partial v}{\partial x} - \frac{\partial u}{\partial y}\right) \tag{2.5}$$

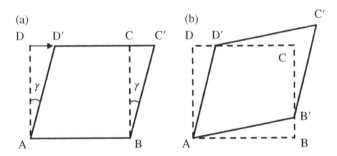

Figure 2.3

In this way we arrive at the basic patterns of deformations, illustrated in Table 2.1.

Although the deformations in Table 2.1 and Figure 2.3 are exaggerated, it must be noted that strains ε and γ of metallic materials in an elastic region are in the order of 10^{-3}. Thus, we must treat carefully the analysis of the small quantities related to strains.

2.2   Strains in Three Dimensional Problems

In the consideration of three dimensional problems, we have to regard an rectangular element, ABCD in Figure 2.1 as a single rectangular plane of a brick-like body, *within* a three dimensional body. We have additional two planes, the y–z and z–x planes. Thus, we can define the following six strains and three rotations using the displacements u, v and w in the x, y and z directions, respectively. It should be noted that six strains can be denoted by only three displacements.

$$\left.\begin{array}{l} \varepsilon_x = \dfrac{\partial u}{\partial x}, \quad \varepsilon_y = \dfrac{\partial v}{\partial y}, \quad \varepsilon_z = \dfrac{\partial w}{\partial z} \\[2ex] \gamma_{xy} = \dfrac{\partial u}{\partial y} + \dfrac{\partial v}{\partial x}, \quad \gamma_{yz} = \dfrac{\partial v}{\partial z} + \dfrac{\partial w}{\partial y}, \quad \gamma_{zx} = \dfrac{\partial w}{\partial x} + \dfrac{\partial u}{\partial z} \end{array}\right\} \tag{2.6}$$

$$\omega_z = \frac{1}{2}\left(\frac{\partial v}{\partial x} - \frac{\partial u}{\partial y}\right), \quad \omega_x = \frac{1}{2}\left(\frac{\partial w}{\partial y} - \frac{\partial v}{\partial z}\right), \quad \omega_y = \frac{1}{2}\left(\frac{\partial u}{\partial z} - \frac{\partial w}{\partial x}\right) \tag{2.7}$$

2.3   Strain in an Arbitrary Direction

2.3.1   Two Dimensional Case

In practical problems it is often necessary to determine the strains in a direction, different to the directions ε_x, ε_y, γ_{xy}, in which the strains are already known. Using the same method of denoting the direction cosines as we used for stresses and by studying Figures 2.4 and 2.5, we can

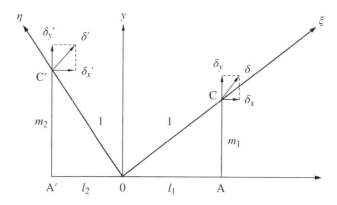

Figure 2.4

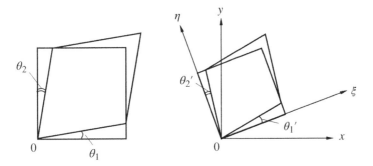

Figure 2.5

calculate the relative displacements of the point C from the point O. For the sake of simplicity we take the length OC and OC′ to be unity. So the relative displacement δ_x and δ_y of the point C to the point O are written:

$$\left.\begin{array}{l} \delta_x = \varepsilon_x l_1 + \theta_2 m_1 \\ \delta_y = \theta_1 l_1 + \varepsilon_y m_1 \end{array}\right\} \tag{2.8}$$

In the above equation l_1 and m_1 indicate the lengths of triangle sides. Since ε_ξ denotes the extension of the length OC in the direction of the ξ axis, by taking the components of δ_x and δ_y in the ξ direction we have

$$\varepsilon_\xi = \delta_x l_1 + \delta_y m_1 \tag{2.9}$$

Combining Equations 2.8 and 2.9 gives us

$$\varepsilon_\xi = \varepsilon_x l_1{}^2 + \varepsilon_y m_1{}^2 + (\theta_1 + \theta_2) l_1 m_1 = \varepsilon_x l_1{}^2 + \varepsilon_y m_1{}^2 + \gamma_{xy} l_1 m_1 \tag{2.10}$$

And since the relative displacements δ'_x and δ'_y of point C′ to O are

$$\left.\begin{array}{l} \delta'_x = \varepsilon_x l_2 + \theta_2 m_2 \\ \delta'_y = \theta_1 l_2 + \varepsilon_y m_2 \end{array}\right\}. \tag{2.11}$$

We have the following equation for ε_η:

$$\varepsilon_\eta = \varepsilon_x l_2{}^2 + \varepsilon_y m_2{}^2 + \gamma_{xy} l_2 m_2 \tag{2.12}$$

Now, from Equations 2.8 and 2.11 and by further considering Figure 2.5b, we can see that $\gamma_{\xi\eta}$ is

$$\begin{aligned} \gamma_{\xi\eta} &= \theta'_1 + \theta'_2 = (\delta_x l_2 + \delta_y m_2) + (\delta'_x l_1 + \delta'_y m_1) \\ &= 2\varepsilon_x l_1 l_2 + 2\varepsilon_y m_1 m_2 + \gamma_{xy}(l_1 m_2 + l_2 m_1) \end{aligned} \tag{2.13}$$

Finally, if we rewrite the equations for ε_ξ, ε_η and $\gamma_{\xi\eta}$ using $\cos\theta$ and $\sin\theta$ instead of l_1 and m_1 and so on, we have

$$\left.\begin{array}{l} \varepsilon_\xi = \varepsilon_x\cos^2\theta + \varepsilon_y\sin^2\theta + \gamma_{xy}\cos\theta\cdot\sin\theta \\[2mm] \varepsilon_\eta = \varepsilon_x\sin^2\theta + \varepsilon_y\cos^2\theta - \gamma_{xy}\cos\theta\cdot\sin\theta \\[2mm] \gamma_{\xi\eta} = 2\left(\varepsilon_y - \varepsilon_x\right)\cos\theta\cdot\sin\theta + \gamma_{xy}\left(\cos^2\theta - \sin^2\theta\right) \end{array}\right\} \qquad (2.14)$$

It is interesting to compare Equation 2.14 with Equation 1.12 for stress, and examine the similarities. If σ_ξ, σ_η and $\tau_{\xi\eta}$ in Equation 1.12 are replaced by ε_ξ, ε_η, and $\gamma_{\xi\eta}/2$, the equation becomes the same as Equation 2.14. It obeys the rule of tensor.

2.3.2 Three Dimensional Case

Using the same theory and method as we used for stress, we can predict the transformation x, y, z to ξ, η, ζ into three dimensional strain components by using direction cosines. Alternatively, we can reach the same conclusion by studying the similarities between the stresses and the strains where normal stresses and shear stresses correspond to normal strains and the half values of shear strains, respectively. Either way will lead to

$$\left.\begin{array}{l} \varepsilon_\xi = \varepsilon_x{l_1}^2 + \varepsilon_y{m_1}^2 + \varepsilon_z{n_1}^2 + \gamma_{xy}l_1m_1 + \gamma_{yz}m_1n_1 + \gamma_{zx}n_1l_1 \\[2mm] \varepsilon_\eta = \varepsilon_x{l_2}^2 + \varepsilon_y{m_2}^2 + \varepsilon_z{n_2}^2 + \gamma_{xy}l_2m_2 + \gamma_{yz}m_2n_2 + \gamma_{zx}n_2l_2 \\[2mm] \gamma_{\xi\eta} = 2\left(\varepsilon_xl_1l_2 + \varepsilon_ym_1m_2 + \varepsilon_zn_1n_2\right) + \gamma_{xy}\left(l_1m_2 + l_2m_1\right) \\[2mm] \qquad\quad + \gamma_{yz}\left(m_1n_2 + m_2n_1\right) + \gamma_{zx}\left(n_1l_2 + n_2l_1\right) \\[2mm] \cdots \end{array}\right\} \qquad (2.15)$$

2.4 Principal Strains

The normal strains in the coordinate system in which the shear strains are zero are called the *principal strains*. Although the principal strains can be determined in a similar way as principal stresses (see Section 1.3.4), we must take extra care in deriving them, because rotation of the coordinate system can lead to much confusion.

For example, the displacements in Equation 2.8, include rotation ω_z and can be decomposed into elements of pure deformation and pure rotation. So:

$$\left.\begin{array}{l} \delta_x = \varepsilon_xl_1 + \theta_2m_1 = \varepsilon_xl_1 + \dfrac{1}{2}(\theta_1 + \theta_2)m_1 - \dfrac{1}{2}(\theta_1 - \theta_2)m_1 \\[3mm] \delta_y = \theta_1l_1 + \varepsilon_ym_1 = \dfrac{1}{2}(\theta_1 + \theta_2)l_1 + \varepsilon_ym_1 + \dfrac{1}{2}(\theta_1 - \theta_2)l_1 \end{array}\right\} \qquad (2.16)$$

Equation 2.16 may be expressed in matrix form:

$$
\begin{Bmatrix} \delta_x \\ \delta_y \end{Bmatrix} = \begin{bmatrix} \varepsilon_x & \frac{1}{2}(\theta_1+\theta_2) \\ \frac{1}{2}(\theta_1+\theta_2) & \varepsilon_y \end{bmatrix} \begin{Bmatrix} l_1 \\ m_1 \end{Bmatrix} + \begin{bmatrix} 0 & -\frac{1}{2}(\theta_1-\theta_2) \\ \frac{1}{2}(\theta_1-\theta_2) & 0 \end{bmatrix} \begin{Bmatrix} l_1 \\ m_1 \end{Bmatrix}
$$

$$
= \begin{bmatrix} \varepsilon_x & \frac{1}{2}\gamma_{xy} \\ \frac{1}{2}\gamma_{xy} & \varepsilon_y \end{bmatrix} \begin{Bmatrix} l_1 \\ m_1 \end{Bmatrix} + \begin{bmatrix} 0 & -\omega_z \\ \omega_z & 0 \end{bmatrix} \begin{Bmatrix} l_1 \\ m_1 \end{Bmatrix}
\tag{2.17}
$$

The first term in Equation 2.17 describes pure deformation and the second term in Equation 2.17 describes pure rotation. What this equation means is that, due to the effect of rotation, there is an infinite number of possible displacement systems that have the same pure deformations, but different rotations. In order to determine the actual deformation of the body, we must obtain the principal strains from pure deformation when rotation is not present.

The manner in which we determine the principal strains is similar to that used to determine the principal stresses. Thus, the necessary condition for ξ and η to be the *principle axes*, is that the points C and C' in Figure 2.4 stay on the ξ and η axes respectively after deformation. If we write this condition for a two dimensional case, we have

$$
\begin{vmatrix} (\varepsilon_x-\varepsilon) & \frac{1}{2}\gamma_{xy} \\ \frac{1}{2}\gamma_{xy} & (\varepsilon_y-\varepsilon) \end{vmatrix} = 0
\tag{2.18}
$$

and by solving this quadratic equation, we can obtain the principal strains, ε_1 and $\varepsilon_2 (\varepsilon_1 > \varepsilon_2)$.

$$
\varepsilon_1, \varepsilon_2 = \left((\varepsilon_x+\varepsilon_y) \pm \sqrt{(\varepsilon_x-\varepsilon_y)^2+\gamma_{xy}^2} \right)/2
\tag{2.19}
$$

The mathematical procedure used to find the principal strains is similar to that used to find the principal axes of an ellipse which was a circle prior to deformation. The same method is used in three dimensional problems to find the principal axes (and hence strains) of an ellipsoid which was a sphere before deformation. Therefore, using a similar derivation, we can get the following equation from which we can calculate the three principal strains:

$$
\begin{vmatrix} (\varepsilon_x-\varepsilon) & \frac{1}{2}\gamma_{xy} & \frac{1}{2}\gamma_{zx} \\ \frac{1}{2}\gamma_{xy} & (\varepsilon_y-\varepsilon) & \frac{1}{2}\gamma_{yz} \\ \frac{1}{2}\gamma_{zx} & \frac{1}{2}\gamma_{yz} & (\varepsilon_z-\varepsilon) \end{vmatrix}
\tag{2.20}
$$

We can derive the three principal strains ε_1, ε_2 and ε_3 $(\varepsilon_1 > \varepsilon_2 > \varepsilon_3)$, and using factorization, we have the following form:

$$(\varepsilon - \varepsilon_1) \cdot (\varepsilon - \varepsilon_2) \cdot (\varepsilon - \varepsilon_3) = 0 \tag{2.21}$$

And if we note that expanding Equations 2.20 and 2.21 give the same results, we obtain

$$\left.\begin{aligned}
I_1 &= \varepsilon_x + \varepsilon_y + \varepsilon_z = \varepsilon_1 + \varepsilon_2 + \varepsilon_3 \\
I_2 &= -\left(\varepsilon_x \varepsilon_y + \varepsilon_y \varepsilon_z + \varepsilon_z \varepsilon_x - \frac{1}{4}\gamma_{xy}^{\ 2} - \frac{1}{4}\gamma_{yz}^{\ 2} - \frac{1}{4}\gamma_{zx}^{\ 2}\right) \\
&= -\left(\varepsilon_1 \varepsilon_2 + \varepsilon_2 \varepsilon_3 + \varepsilon_3 \varepsilon_1\right) \\
I_3 &= \varepsilon_x \varepsilon_y \varepsilon_z + \frac{1}{4}\gamma_{xy}\gamma_{yz}\gamma_{zx} - \frac{1}{4}\varepsilon_x \gamma_{yz}^{\ 2} - \frac{1}{4}\varepsilon_y \gamma_{zx}^{\ 2} - \frac{1}{4}\varepsilon_z \gamma_{xy}^{\ 2} \\
&= \varepsilon_1 \varepsilon_2 \varepsilon_3
\end{aligned}\right\} \tag{2.22}$$

So, principal strains are scalar quantities which are independent of choice of coordinate system; accordingly I_1, I_2 and I_3 are constants. I_1, I_2 and I_3 are called the first, second and third *strain invariants*, respectively.

Due to the fact that the angle $\pi/2$ between the two principal strain axes does not change during deformation, I_1 gives us a measure of *the change in the volume of a cube*. In the same way we can easily see that $(\varepsilon_x + \varepsilon_y)$ gives us a measure of *the change in the area* in a two dimensional problem. It is interesting to note that, although the result for strain has the same form as that obtained for stress, their derivations are totally independent of those for stress.

2.5   Conditions of Compatibility

From Equation 2.6 we can see that the six strain components $(\varepsilon_x, \varepsilon_y, \varepsilon_z, \gamma_{xy}, \gamma_{yz}, \gamma_{zx})$ are defined from the three deformation components (u, v, w). However, we still do not know if it is possible to obtain the deformation components from the strain components. If we assume the six strain components to be independent functions of x, y, z, it would appear that we could calculate the deformation components from part of the strain components by Equation 2.6 because we do not need all six strain components for the calculation of the displacements. However, by this calculation we may obtain different displacements at the identical point, depending on the choice of strains for the calculation. In such a case, the different displacement at the identical point produces the dislocation. If a problem contains a dislocation from the beginning, these solutions may be allowed. However, in most cases the existence of a dislocation, that is the discrepancy of displacements obtained from a finite number of strains, is not acceptable for the exact solution. To avoid such incompleteness in solutions, we must assume that the strains are interdependent. Their relationships are called the *compatibility conditions* and are derived as follows.

$\partial/\partial x$, $\partial/\partial y$ and $\partial/\partial z$ are represented by D_x, D_y, D_z and so the strains are written as

$$\left.\begin{aligned}
\varepsilon_x &= D_x u, \quad \varepsilon_y = D_y v, \quad \varepsilon_z = D_z w \\
\gamma_{xy} &= D_y u + D_x v, \quad \gamma_{yz} = D_z v + D_y w, \quad \gamma_{zx} = D_x w + D_z u
\end{aligned}\right\} \tag{2.23}$$

where the strains ε_x, ε_y, ε_z, γ_{xy}, γ_{yz}, γ_{zx} are considered to be the vectors in the three dimensional space (u, v, w). If we choose four arbitrary strains from six strains, they must be linearly dependent on each other.

Example
Choosing ε_x, ε_y, ε_z and γ_{xy} as the four arbitrary strains, we can write their linearly dependent relationship as follows:

$$c_1\varepsilon_x + c_2\varepsilon_y + c_3\varepsilon_z = \gamma_{xy} \tag{a}$$

$$c_1 D_x u + c_2 D_y v + c_3 D_z w = D_y u + D_x v \tag{b}$$

then

$$c_1 = \frac{D_y}{D_x}, \quad c_2 = \frac{D_x}{D_y}, \quad c_3 = 0 \tag{c}$$

substituting (c) in (a)

$$\frac{D_y}{D_x}\varepsilon_x + \frac{D_x}{D_y}\varepsilon_y = \gamma_{xy} \tag{d}$$

From this

$$D_y{}^2\varepsilon_x + D_x{}^2\varepsilon_y = D_x D_y \gamma_{xy} \tag{e}$$

Namely

$$\frac{\partial^2 \varepsilon_x}{\partial y^2} + \frac{\partial^2 \varepsilon_y}{\partial x^2} = \frac{\partial^2 \gamma_{xy}}{\partial x \cdot \partial y} \tag{f}$$

Equation (f) is one of the conditions of compatibility. There are 15 different combinations for arbitrarily chosen four strains from the total six strains. If the above method is used for each of the 15 combinations of four strains, many results are repeated. As a result, we arrive at a final set of six equations. These are the *compatibility conditions*.

$$\left. \begin{array}{l} \dfrac{\partial^2 \varepsilon_x}{\partial y^2} + \dfrac{\partial^2 \varepsilon_y}{\partial x^2} = \dfrac{\partial^2 \gamma_{xy}}{\partial x \cdot \partial y}, \quad 2\dfrac{\partial^2 \varepsilon_x}{\partial y \cdot \partial z} = \dfrac{\partial}{\partial x}\left(-\dfrac{\partial \gamma_{yz}}{\partial x} + \dfrac{\partial \gamma_{zx}}{\partial y} + \dfrac{\partial \gamma_{xy}}{\partial z}\right) \\[4mm] \dfrac{\partial^2 \varepsilon_y}{\partial z^2} + \dfrac{\partial^2 \varepsilon_z}{\partial y^2} = \dfrac{\partial^2 \gamma_{yz}}{\partial y \cdot \partial z}, \quad 2\dfrac{\partial^2 \varepsilon_y}{\partial z \cdot \partial x} = \dfrac{\partial}{\partial y}\left(\dfrac{\partial \gamma_{yz}}{\partial x} - \dfrac{\partial \gamma_{zx}}{\partial y} + \dfrac{\partial \gamma_{xy}}{\partial z}\right) \\[4mm] \dfrac{\partial^2 \varepsilon_z}{\partial x^2} + \dfrac{\partial^2 \varepsilon_x}{\partial z^2} = \dfrac{\partial^2 \gamma_{zx}}{\partial z \cdot \partial x}, \quad 2\dfrac{\partial^2 \varepsilon_z}{\partial x \cdot \partial y} = \dfrac{\partial}{\partial z}\left(\dfrac{\partial \gamma_{yz}}{\partial x} + \dfrac{\partial \gamma_{zx}}{\partial y} - \dfrac{\partial \gamma_{xy}}{\partial z}\right) \end{array} \right\} \tag{2.24}$$

Equation 2.24 is the necessary and sufficient condition that must be satisfied when the strains are defined in the manner of Equations 2.6 or 2.23[1].

Problems of Chapter 2

1. In general, the measurement of shear strain at a certain point is more difficult than the measurement of normal strain at a point. Identify the minimum information required about normal strains at the same point, to determine the shear strain without measuring shear strain.
2. Draw an x–y coordinate with origin O on a plate. Define point A on the x axis and point B on the y axis as $OA = a$ and $OB = b$. Calculate the rate of the change in the area ΔOAB when the plate is subjected to uniform strains ε_x, ε_y and γ_{xy}.
3. Verify that, although $\gamma_{xy} = 0$ in the case of Table 2.1c, the shear strain $\gamma_{\xi\eta}$ (defined in a different coordinate system ξ–η) is $\gamma_{\xi\eta} \neq 0$. Verify, however, that the rotation ω_z measured in the ξ–η coordinate system is also 0; that is $\omega_z = 0$, as in the x–y coordinate system.
4. In order to make approximate equations for expressing the stress and strain distribution in a limited area of a plate under a certain loading, normal strains in three directions ε_x, ε_y and $\varepsilon_{45°}$ at six selected points on the plate were measured by using strain gages. Regarding ε_x and ε_y, the values were successfully measured at the six points and the following approximate equations were made.

$$\varepsilon_x = a_0 + a_1 x + a_2 y + a_3 x^2 + a_4 y^2 + a_5 xy \tag{a}$$

$$\varepsilon_y = b_0 + b_1 x + b_2 y + b_3 x^2 + b_4 y^2 + b_5 xy \tag{b}$$

However, regarding $\varepsilon_{45°}$, the strain at one point could not be measured due to the failure of electric wire and only the data at five points were obtained. In such a case, judge whether there is a logical method to express γ_{xy} approximately by an approximate equation as follows:

$$\gamma_{xy} = c_0 + c_1 x + c_2 y + c_3 x^2 + c_4 y^2 + c_5 xy \tag{c}$$

5. Referring to Figure 2.6, show that the strain components in the polar coordinate (r, θ) are expressed by the following equation, where u is the displacement in the r direction and v is the displacement in the θ direction, respectively.

$$\left.\begin{array}{l} \varepsilon_r = \dfrac{\partial u}{\partial r} \\[12pt] \varepsilon_\theta = \dfrac{u}{r} + \dfrac{1}{r} \cdot \dfrac{\partial v}{\partial \theta} \\[12pt] \gamma_{r\theta} = \dfrac{1}{r} \cdot \dfrac{\partial u}{\partial \theta} + \dfrac{\partial v}{\partial r} - \dfrac{v}{r} \end{array}\right\} \tag{2.25}$$

---

[1] A more complicated, classical derivation of the compatibility equation is shown in Reference [1].

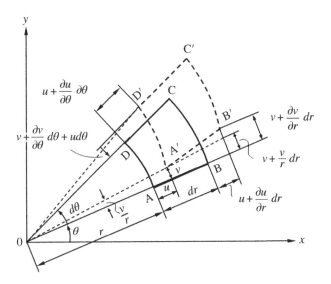

Figure 2.6

In the above equation, it must be noted that the importance of the first term in the equation of ε_θ is very often overlooked, though ε_θ is directly related with u even in the case that displacement v in the θ direction is 0.

For two dimensional axisymmetrical problems, $v=0$ and Equation 2.25 is written as a special case of Figure 2.6 as follows:

$$\left.\begin{aligned}\varepsilon_r &= \frac{\partial u}{\partial r} \\[2mm] \varepsilon_\theta &= \frac{u}{r}\end{aligned}\right\} \tag{2.26}$$

Equation 2.26 is the basic equation for axisymmetrical problems and actually will be applied for solving many 2D axisymmetrical problems. It must be noted that a very typical and common mistake for the determination of u is to pay attention to the first equation of Equation 2.26. However, the first equation is expressed in the form of a differentiation and it is not relevant to obtain u from ε_r. If we use the second equation, u can be directly obtained from ε_θ.

Reference

[1] I. S. Sokolnikoff (1956) *Mathematical Theory of Elasticity*, 2nd edn, McGraw-Hill, New York.

3

Relationship between Stresses and Strains: Generalized Hooke's Law

Here we will consider the relationship between stress and strain in the elastic region of *isotropic* and homogeneous materials. If a material is isotropic, it means that the elastic properties do not change with a change in direction. However, this is for an ideal case. In reality we cannot have completely isotropic materials. Many materials, when considered in microscopic detail, are anisotropic. However, if the microstructure is randomly formed then such materials can be treated as isotropic when dealt with on a macroscopic scale.

Homogeneous means that the elastic properties of the material do not vary from place to place. Once again this is a simplification as we cannot have completely homogeneous materials. In particular we must be careful not to take materials that have a different structural organization on a microscopic level in each crystal unit to be homogeneous.

Furthermore, if we consider materials on an even smaller scale, they become discontinuous. If we take the crystal unit to be the smallest element in the material, then when we consider the macroscopic scale, we can treat the material as a practically homogenous one.

In this book, we deal with the elasticity of isotropic and homogeneous materials only. In addition to this, the cases we will consider will be in the region of the linear relationship between stress and strain, as we will only use small values of strain.

If we assume that stress and strain are a function of one another, we have the relationship:

$$\sigma_x = f\left(\varepsilon_x, \varepsilon_y, \varepsilon_z, \gamma_{xy}, \cdots\right) \tag{3.1}$$

If $\varepsilon_x = \varepsilon_y = \varepsilon_z = \gamma_{xy} = \ldots = 0$, $\sigma_x = 0$, then by using the Taylor expansion to expand σ_x and neglecting the small terms (second degree and higher order), we will obtain the following equation:

$$\sigma_x = c_{11}\varepsilon_x + c_{12}\varepsilon_y + c_{13}\varepsilon_z + c_{14}\gamma_{xy} + c_{15}\gamma_{yz} + c_{16}\gamma_{zx} \tag{3.2}$$

Theory of Elasticity and Stress Concentration, First Edition. Yukitaka Murakami.
© 2017 John Wiley & Sons, Ltd. Published 2017 by John Wiley & Sons, Ltd.

where c_{11} to c_{16} are material constants. The other components can be expressed in the same way as follows.

$$\sigma_x = c_{11}\varepsilon_x + c_{12}\varepsilon_y + c_{13}\varepsilon_z + c_{14}\gamma_{xy} + c_{15}\gamma_{yz} + c_{16}\gamma_{zx}$$
$$\sigma_y = c_{21}\varepsilon_x + c_{22}\varepsilon_y + c_{23}\varepsilon_z + c_{24}\gamma_{xy} + c_{25}\gamma_{yz} + c_{26}\gamma_{zx}$$
$$\sigma_z = c_{31}\varepsilon_x + c_{32}\varepsilon_y + c_{33}\varepsilon_z + c_{34}\gamma_{xy} + c_{35}\gamma_{yz} + c_{36}\gamma_{zx}$$
$$\tau_{xy} = c_{41}\varepsilon_x + c_{42}\varepsilon_y + c_{43}\varepsilon_z + c_{44}\gamma_{xy} + c_{45}\gamma_{yz} + c_{46}\gamma_{zx} \qquad (3.3)$$
$$\tau_{yz} = c_{51}\varepsilon_x + c_{52}\varepsilon_y + c_{53}\varepsilon_z + c_{54}\gamma_{xy} + c_{55}\gamma_{yz} + c_{56}\gamma_{zx}$$
$$\tau_{zx} = c_{61}\varepsilon_x + c_{62}\varepsilon_y + c_{63}\varepsilon_z + c_{64}\gamma_{xy} + c_{65}\gamma_{yz} + c_{66}\gamma_{zx}$$

Rewriting these equations and expressing strain as a function of stress gives

$$\varepsilon_x = a_{11}\sigma_x + a_{12}\sigma_y + a_{13}\sigma_z + a_{14}\tau_{xy} + a_{15}\tau_{yz} + a_{16}\tau_{zx}$$
$$\varepsilon_y = a_{21}\sigma_x + a_{22}\sigma_y + a_{23}\sigma_z + a_{24}\tau_{xy} + a_{25}\tau_{yz} + a_{26}\tau_{zx}$$
$$\varepsilon_z = a_{31}\sigma_x + a_{32}\sigma_y + a_{33}\sigma_z + a_{34}\tau_{xy} + a_{35}\tau_{yz} + a_{36}\tau_{zx}$$
$$\gamma_{xy} = a_{41}\sigma_x + a_{42}\sigma_y + a_{43}\sigma_z + a_{44}\tau_{xy} + a_{45}\tau_{yz} + a_{46}\tau_{zx} \qquad (3.4)$$
$$\gamma_{yz} = a_{51}\sigma_x + a_{52}\sigma_y + a_{53}\sigma_z + a_{54}\tau_{xy} + a_{55}\tau_{yz} + a_{56}\tau_{zx}$$
$$\gamma_{zx} = a_{61}\sigma_x + a_{62}\sigma_y + a_{63}\sigma_z + a_{64}\tau_{xy} + a_{65}\tau_{yz} + a_{66}\tau_{zx}$$

where $a_{ij}(i, j = 1–6)$ are the values related to elastic moduli.

Considering the above equation, it would appear to be possible to alter ε_x by changing τ_{xy}. In reality, however, this does not occur.

Additionally if we consider the case of $a_{14} > 0$, as shown in Figure 3.1a, when τ_{xy} is positive, $\varepsilon_x > 0$, that is it extends in the x direction. Inversely if $\tau_{xy} < 0$ (Figure 3.1b), the plate contracts in the x-direction.

Now, if we look at Figure 3.1a from the back it is the same as Figure 3.1b, that is $\varepsilon_x < 0$. This means that if $a_{14} > 0$ we get two contradictory conclusions. The same happens if we take $a_{14} < 0$. Therefore the satisfactory solution is only for $a_{14} = 0$. Using the same logic we can write

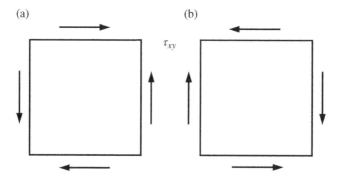

Figure 3.1

$$\left.\begin{aligned}
a_{14} &= a_{15} = a_{16} = 0 \\
a_{24} &= a_{25} = a_{26} = 0 \\
a_{34} &= a_{35} = a_{36} = 0 \\
a_{41} &= a_{42} = a_{43} = a_{45} = a_{46} = 0 \\
a_{51} &= a_{52} = a_{53} = a_{54} = a_{56} = 0 \\
a_{61} &= a_{62} = a_{63} = a_{64} = a_{65} = 0
\end{aligned}\right\}
\tag{3.5}$$

And considering an isotropic and homogeneous body, we can see

$$\left.\begin{aligned}
a_{11} &= a_{22} = a_{33} = a_1 \\
a_{12} &= a_{13} = a_{21} = a_{23} = a_{31} = a_{32} = a_2 \\
a_{44} &= a_{55} = a_{66} = a_3
\end{aligned}\right\}
\tag{3.6}$$

Thus,

$$\left.\begin{aligned}
\varepsilon_x &= a_1\sigma_x + a_2\left(\sigma_y + \sigma_z\right) \\
\varepsilon_y &= a_1\sigma_y + a_2\left(\sigma_z + \sigma_x\right) \\
\varepsilon_z &= a_1\sigma_z + a_2\left(\sigma_x + \sigma_y\right)
\end{aligned}\right\}
\tag{3.7}$$

$$\gamma_{xy} = a_3\tau_{xy}, \quad \gamma_{yz} = a_3\tau_{yz}, \quad \gamma_{zx} = a_3\tau_{zx} \tag{3.8}$$

From Equation 3.8 we can understand that when the shear stress is zero, the shear strain is also zero. We can also see that the principal axes of strain correspond to those of stress. So if the strain is in the direction of the principal strain of the axes,

$$\left.\begin{aligned}
\varepsilon_1 &= a_1\sigma_1 + a_2\left(\sigma_2 + \sigma_3\right) \\
\varepsilon_2 &= a_1\sigma_2 + a_2\left(\sigma_3 + \sigma_1\right) \\
\varepsilon_3 &= a_1\sigma_3 + a_2\left(\sigma_1 + \sigma_2\right)
\end{aligned}\right\}
\tag{3.9}$$

Using the direction cosines already defined between the principal axes and the coordinate axes, we can express shear strain in terms of normal strain and shear stress in terms of normal stress, that is from Equations 2.15 and 1.13 as follows.

$$\gamma_{xy} = 2\left(\varepsilon_1 l_1 l_2 + \varepsilon_2 m_1 m_2 + \varepsilon_3 n_1 n_2\right) \tag{3.10}$$

$$\tau_{xy} = \sigma_1 l_1 l_2 + \sigma_2 m_1 m_2 + \sigma_3 n_1 n_2 \tag{3.11}$$

Now, looking at the relationship between the first part of Equation 3.8 and Equations 3.9, 3.10 and 3.11, we find that a_3 is not independent of a_1 and a_2. So, from Equations 3.8, 3.10 and 3.11,

$$2\left(\varepsilon_1 l_1 l_2 + \varepsilon_2 m_1 m_2 + \varepsilon_3 n_1 n_2\right) = a_3\left(\sigma_1 l_1 l_2 + \sigma_2 m_1 m_2 + \sigma_3 n_1 n_2\right)$$

Substituting Equation 3.9 into the left-hand side (LHS) of the above equation,

$$\text{LHS} = 2[a_1(\sigma_1 l_1 l_2 + \sigma_2 m_1 m_2 + \sigma_3 n_1 n_2) + a_2(\sigma_1 + \sigma_2 + \sigma_3)]$$
$$(l_1 l_2 + m_1 m_2 + n_1 n_2) - a_2(\sigma_1 l_1 l_2 + \sigma_2 m_1 m_2 + \sigma_3 n_1 n_2)$$
$$= 2(a_1 - a_2) \cdot (\sigma_1 l_1 l_2 + \sigma_2 m_1 m_2 + \sigma_3 n_1 n_2)$$

Hence,

$$a_3 = 2(a_1 - a_2) \tag{3.12}$$

We can see that the number of independent elastic constants are two, where by using,

$$a_1 = 1/E, \quad a_2 = -\nu a_1 \tag{3.13}$$

Thus, we get

$$a_3 = 2(1 + \nu)/E \tag{3.14}$$

Finally, we get the following equations as the *generalized Hooke's law equations* for isotropic and homogeneous materials.

$$\left.\begin{array}{l} \varepsilon_x = \dfrac{1}{E}\left[\sigma_x - \nu(\sigma_y + \sigma_z)\right] \\[2mm] \varepsilon_y = \dfrac{1}{E}\left[\sigma_y - \nu(\sigma_z + \sigma_x)\right] \\[2mm] \varepsilon_z = \dfrac{1}{E}\left[\sigma_z - \nu(\sigma_x + \sigma_y)\right] \\[2mm] \gamma_{xy} = \tau_{xy}/G, \quad \gamma_{yz} = \tau_{yz}/G, \quad \gamma_{zx} = \tau_{zx}/G \end{array}\right\} \tag{3.15}$$

where, E is *Young's modulus*, ν is *Poisson's ratio* and G is *shear modulus*. The relationship between them is as follows.

$$G = E/2(1 + \nu) \tag{3.16}$$

In addition, Hooke's law for the cylindrical coordinate system (r, θ, z) is written as follows.

$$\left.\begin{array}{l} \varepsilon_r = \dfrac{1}{E}\left[\sigma_r - \nu(\sigma_\theta + \sigma_z)\right] \\[2mm] \varepsilon_\theta = \dfrac{1}{E}\left[\sigma_\theta - \nu(\sigma_z + \sigma_r)\right] \\[2mm] \varepsilon_z = \dfrac{1}{E}\left[\sigma_z - \nu(\sigma_r + \sigma_\theta)\right] \\[2mm] \gamma_{r\theta} = \tau_{r\theta}/G, \quad \gamma_{\theta z} = \tau_{\theta z}/G, \quad \gamma_{zr} = \tau_{zr}/G \end{array}\right\} \tag{3.17}$$

Example problem 3.1

Looking at Figure 3.2, we can see that a holed rectangular plate subjected to the stresses shown here will obviously have no shear stress on its certain planes. Using $O\Delta$ (as in $\tau_{o\Delta}$), denote the coordinates where shear stress is zero and explain why the shear stress is zero.

Solution

Since shear stress does not act on free surface, shear stresses at the free surfaces of the plate, which are the upper and lower ends of the plate, the both sides of the plate and the circumferential edge of the hole, are obviously zero.

Moreover, since shear stress is related to shear strain by Hooke's law, the shear stress on the line, along which shear strain is obviously zero, is zero, even at nonfree surface. Such lines can be found as follows. Since shear strain is a change in the right angle which is 90 degrees in a coordinate before deformation, if the right angle is kept unchanged after deformation, the shear strain is zero and as the result shear stress is zero, according to Hooke's law. To practice this judgement concretely and easily, we have only to put a small *cross mark* (+) at the point in question. We can judge, if shear strain exists or not by distortion of the cross mark.

Following the procedure explained above, the solution for Figure 3.2 can be derived as follows:

$$\left.\begin{array}{l} \tau_{xy}=0 \quad \text{along} \quad y=\pm L \\ \tau_{xy}=0 \quad \text{along} \quad x=\pm W \\ \tau_{r\theta}=0 \quad \text{at the circumferential edge of the hole} \end{array}\right\} \text{Reason : Free surface}$$

$$\left.\begin{array}{l} \tau_{xy}=0 \quad \text{along} \quad x \text{ axis} \\ \tau_{xy}=0 \quad \text{along} \quad y \text{ axis} \end{array}\right\} \begin{array}{l} \text{Reason : Cross mark on these axes are} \\ \text{not distorted due to the symmetry,} \\ \text{that is the shear strain is zero.} \end{array}$$

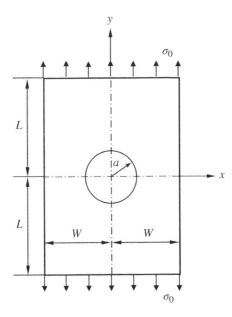

Figure 3.2

This problem may look easy, but in the author's experience, most students are not good at solving this kind of problem, because this problem requires the ability to judge the location of zero-shear stress based on an intuitive ability without computational numerical analysis. If we know all the stress components after computing the stress field by software, such as FEM, and we determine the principal axes as explained in Chapter 1 of Part I, we can indicate the location and direction of the zero-shear stress. However, it is important to judge this kind of problem intuitively in practical situations without time- and effort-consuming numerical computation. Such an ability will be useful for solving problems of a higher level and also to solve three dimensional problems.

Hooke's law is very important not only for the form of Equation 3.15 but also for the role of bridging the nature of stresses in Chapter 1 and strains in Chapter 2 of Part I. Hooke's law will be used in many problems treated in this book in later chapters.

Problems of Chapter 3

1. Derive Hooke's law of Equation 3.17 in the cylindrical coordinate (r, θ, z).
2. A cylindrical bar of a diameter d was subjected to twisting moment T. The strain ε in the direction of angle θ against the cylinder axis on the surface of the cylinder was measured by strain gage. The value of ε was $\varepsilon = \varepsilon_0$. Answer the following questions for $\theta = 30°$, $d = 50$ mm, $\varepsilon_0 = 500 \times 10^{-6}$, Young's modulus $E = 206$ GPa and Poisson's ratio $\nu = 0.3$.

    1. Determine the normal stress σ in the direction of θ.
    2. Determine T.

3. Figure 3.3 shows a rectangular plate containing a rigid circular plate and a circular hole. When tensile stress $\sigma_y = \sigma_0$ is applied at the ends of the plate $y = \pm L$, indicate the lines

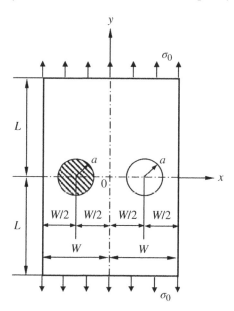

Figure 3.3

where shear stress is judged obviously 0. Regarding the zero-shear stress, indicate the coordinate in the form of $\tau_{o\Delta}$ and explain the reason for zero shear stress.

4. The surface of a cylindrical bar of material A with diameter d is plated with material B with a very thin layer. When the material A is applied a tensile stress σ in the axial direction, determine the axial stress σ_x, radial stress σ_r and circumferential stress σ_θ in material B. Use the materials' elastic moduli (Young's modulus, Poisson's ratio) for materials A and B as (E_A, ν_A) and (E_B, ν_B), respectively.

5. Figure 3.4 shows a glass container which is composed with two spherical glass bulbs connected with a glass pipe. The dimensions of the parts are as follows:

Radius of the glass bulbs: R. Diameter of the glass pipe: d. Length of the glass pipe: l. Thickness of glass: h. Assume $h \ll d \ll l$.

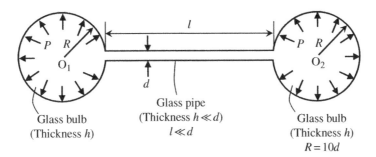

Figure 3.4

When pressure p is applied in the glass container, answer the following questions. The Young's modulus and Poisson's ratio of the glass are E and ν, respectively.

1. Calculate the axial stress σ_a and circumferential stress σ_t of the glass pipe.
2. Calculate the variation δ of the length l of the glass pipe.

4

Equilibrium Equations

A small element inside a larger loaded body will be subjected to forces from the surrounding material. This element may also be subjected to body forces (X, Y, Z). These forces maintain the equilibrium of the element.

Isolating the small element shown in Figure 4.1, we can calculate the forces in the x direction from the product of stresses and areas on which they act, as shown in Figure 4.2.

Taking body forces into account, we can express equilibrium of force in the x direction as follows:

$$\left(\sigma_x + \frac{\partial \sigma_x}{\partial x}dx\right)dydz - \sigma_x dydz + \left(\tau_{xy} + \frac{\partial \tau_{xy}}{\partial y}dy\right)dzdx - \tau_{xy}dzdx$$

$$+ \left(\tau_{zx} + \frac{\partial \tau_{zx}}{\partial z}dz\right)dxdy - \tau_{zx}dxdy + Xdxdydz = 0 \tag{4.1}$$

Rearranging gives:

$$\left(\frac{\partial \sigma_x}{\partial x} + \frac{\partial \tau_{xy}}{\partial y} + \frac{\partial \tau_{zx}}{\partial z} + X\right)dxdydz = 0 \tag{4.2}$$

And so,

$$\frac{\partial \sigma_x}{\partial x} + \frac{\partial \tau_{xy}}{\partial y} + \frac{\partial \tau_{zx}}{\partial z} + X = 0 \tag{4.3}$$

Theory of Elasticity and Stress Concentration, First Edition. Yukitaka Murakami.
© 2017 John Wiley & Sons, Ltd. Published 2017 by John Wiley & Sons, Ltd.

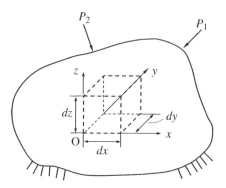

Figure 4.1

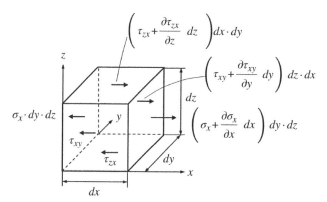

Figure 4.2

The equilibrium equations of force for the y and z directions are derived using the same method. These equations take the form:

$$\left.\begin{array}{l} \dfrac{\partial \sigma_x}{\partial x} + \dfrac{\partial \tau_{xy}}{\partial y} + \dfrac{\partial \tau_{zx}}{\partial z} + X = 0 \\[3mm] \dfrac{\partial \tau_{xy}}{\partial x} + \dfrac{\partial \sigma_y}{\partial y} + \dfrac{\partial \tau_{yz}}{\partial z} + Y = 0 \\[3mm] \dfrac{\partial \tau_{zx}}{\partial x} + \dfrac{\partial \tau_{yz}}{\partial y} + \dfrac{\partial \sigma_z}{\partial z} + Z = 0 \end{array}\right\} \qquad (4.4)$$

Examples of the body forces (X, Y, Z) are gravitational force, centrifugal force, electromagnetic force and so forth, which act per unit volume.

In three dimensional problems, there are three *equilibrium equations* and six unknowns (σ_x, σ_y, σ_z, τ_{xy}, τ_{yz}, τ_{zx}). The equilibrium equations for two dimensional problems are obtained by omitting the z terms from Equation 4.4.

$$\left.\begin{array}{l} \dfrac{\partial \sigma_x}{\partial x} + \dfrac{\partial \tau_{xy}}{\partial y} + X = 0 \\[3mm] \dfrac{\partial \tau_{xy}}{\partial x} + \dfrac{\partial \sigma_y}{\partial y} + Y = 0 \end{array}\right\}$$ (4.5)

In two dimensional problems there are two equations and three unknowns. We must be aware of the difference in the number of equations and unknowns between two and three dimensional problems.

In symmetrical problems, it is usually better to use the cylindrical coordinate system for expressing the equilibrium conditions. The cylindrical coordinate representation of Equation 4.5 is as follows.

$$\dfrac{\partial \sigma_r}{\partial r} + \dfrac{1}{r}\dfrac{\partial \tau_{r\theta}}{\partial \theta} + \dfrac{\sigma_r - \sigma_\theta}{r} + F_r = 0$$

$$\dfrac{1}{r}\dfrac{\partial \sigma_\theta}{\partial \theta} + \dfrac{\partial \tau_{r\theta}}{\partial r} + \dfrac{2\tau_{r\theta}}{r} + F_\theta = 0$$ (4.6)

where F_r and F_θ are the body forces in the r and θ directions respectively.

Example problem 4.1
When a cantilever beam is subjected to a load P at its free end, as shown Figure 4.3, determine the shear stress distribution τ_{xy} on a cross section away from the loaded point.

Solution
The equilibrium equations (Equation 4.5) for this problem are written as follows:

$$\dfrac{\partial \sigma_x}{\partial x} + \dfrac{\partial \tau_{xy}}{\partial y} = 0$$ (4.7)

$$\dfrac{\partial \tau_{xy}}{\partial x} = 0$$ (4.8)

In Figure 4.3, we define the bending moment positive when the moment acts to bend the cantilever convex upwards.

$$M = P(l - x)$$ (4.9)

The stress σ_x in the x direction by the bending moment is calculated as:

$$\sigma_x = -\dfrac{M}{Z} \cdot \dfrac{2y}{h} = -\dfrac{2P}{hZ}(l-x)y; \; Z = \dfrac{bh^2}{6}$$ (4.10)

Substituting σ_x of Equation 4.10 into Equation 4.7,

$$\dfrac{\partial \tau_{xy}}{\partial y} = -\dfrac{2P}{hZ}y$$ (4.11)

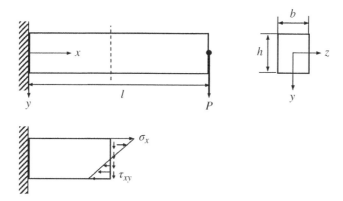

Figure 4.3

Therefore,

$$\tau_{xy} = -\frac{P}{hZ}y^2 + f(x)$$ (4.12)

From Equations 4.8 and 4.12, it follows that $f(x)$ = constant. Since $\tau_{xy} = 0$ at $y = \pm h/2$, the shear stress distribution is obtained as follows.

$$\tau_{xy} = \frac{P}{hZ}\left(\frac{h^2}{4} - y^2\right)$$ (4.13)

From this result, the maximum shear stress τ_{xymax} occurs at $y = 0$ as

$$\tau_{xymax} = \frac{3P}{2bh}$$ (4.14)

It should be noted that τ_{xymax} is 1.5 times larger than the average value $\tau_{xyave} = P/(bh)$.

Problems of Chapter 4

1. Derive Equation 4.6.
2. Since the equilibrium equations (Equation 4.4) are written with stresses, and as the stresses are related with strains by Hooke's law, we can obtain the *equilibrium equations expressed by strains*. Derive the equilibrium equations with strains. Furthermore, derive the *equilibrium equations expressed by displacements*.
3. Determine the distribution of shear stress for the cantilever of Example problem 4.1 when the shape of a section of the cantilever is an isosceles triangle (height h and bottom side $2a$).

5

Saint Venant's Principle and Boundary Conditions

5.1 Saint Venant's Principle

Figure 5.1 shows the stress distribution in a rectangular plate to which a concentrated tensile force is applied at the center of the end faces. The stress distribution was calculated by the finite element method software made by the author, using the mesh shown in Figure 5.1b to model the shaded area of Figure 5.1a. Due to the symmetry of the plate, modeling of this part combined with the boundary conditions completely describes the plate. The stress distributions at three cross sections, A–A, B–B and C–C, are shown in Figure 5.1c by the solid lines. The mean stress is shown in each case by a broken line (constant throughout the length of the plate). The important point is that, as we move from the end of the plate to the middle, the stress distribution becomes less uneven. In this case, the section C–C, where $L = 2W$, is sufficiently far from the loading point for the stress distribution to be approximately uniform and equal to the mean stress in the plate.

Figure 5.2 shows the stress distribution for the case where point loads are applied at the corners of the plate. The stress distribution at the cross section A–A for this case differs from that of Figure 5.1, in that the peak stresses are at the edge rather than in the center. However, the stress distribution at section C–C is similar, with approximately uniform distribution to that of C–C in Figure 5.1. The total tensile load in this example is twice that of Figure 5.1, but if the magnitude of the force P is halved then the stress levels at C–C are the same for both examples.

From these two examples we can see that, if the resultants of the applied static loads are equal (although the distribution of the load and the stress distribution in the plate, close to the point at which the load is applied, may be utterly different), then the stress distribution in a location far removed from the load application will be approximately the same. This phenomenon is known as *Saint Venant's principle*.

The examples in Figures 5.1 and 5.2 are comparatively simple, but it can easily be imagined that other cases, where the loading system may include bending moments or twisting moments, can be analyzed in the same way. In general, Saint Venant's principle states that if the loading

Theory of Elasticity and Stress Concentration, First Edition. Yukitaka Murakami.
© 2017 John Wiley & Sons, Ltd. Published 2017 by John Wiley & Sons, Ltd.

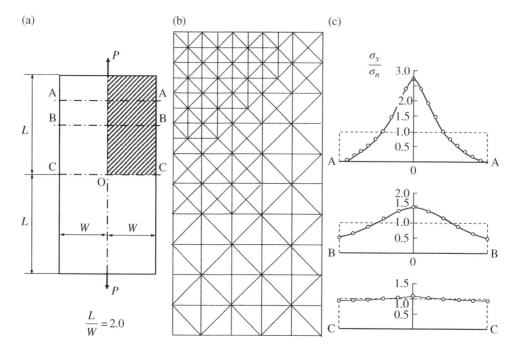

Figure 5.1 (a) $E = 196$ GPa, $v = 0.3$, $t = 1.0$ mm. (b) Number of elements: 190; number of nodes: 115. (c) $\sigma_n = P/2W$

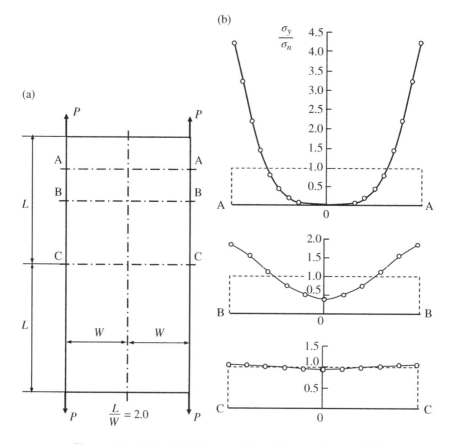

Figure 5.2 (a) $E = 196$ GPa, $v = 0.3$, $t = 1.0$ mm. (b) $\sigma_n = P/W$

system in a certain region is replaced by another statically equivalent loading system, then although the stress in the neighborhood of the load may be very different, we can say that the stresses at a distant point, removed from the load by a *representative dimension* of the structure, will be unchanged, ignoring small effects. In the examples of Figures 5.1 and 5.2, such a representative dimension is the plate width, $2W$.

The contents of Saint Venant's principle may seem rather vague, but in fact the principle has many important applications. For example, when an experiment is being designed, it may be possible to achieve the desired stress in the test part of a sample by a variety of statically equivalent loading systems, so the most convenient system can be chosen. The same method of calculation can be applied to computational analyses.

5.2 Boundary Conditions

As mentioned in Chapter 1, before solving a problem we consider the known conditions, characteristics of the problem, or *boundary conditions*. In many cases, certain characteristics of the outer surface of a body are known, and these are known as the external boundary conditions. In some instances however, obvious characteristics of the body itself exist, such that if the body is divided into smaller parts in a certain way, new faces can be observed for which certain conditions must hold. These may also be considered to be boundary conditions. For example, in the problem of Figures 5.1 and 5.2, the symmetry of the plate means that the boundary conditions are easily defined for any of the quarters of the plate obtained by division along the center lines. Boundary conditions can be conveniently divided into two types for consideration: *stress boundary conditions* and *displacement boundary conditions*.

5.2.1 Stress Boundary Conditions

This term is used to describe cases where the known boundary conditions are the forces external to the body. A typical example of a stress boundary condition is a free surface where the resultant of the external forces is zero.

Figure 5.3 shows a common two dimensional fixed plane stress problem, in which the given conditions are the normal stress and the shear stress at the boundary. The following notation is used:

$$\sigma_\xi = \sigma_{n0}, \quad \tau_{\xi\eta} = \tau_{nt0} \tag{5.1}$$

Alternatively, considering the forces exerted on a unit length of the boundary, the forces in the x and y directions are denoted by X^* and Y^*, and the following relationship can be written:

$$\sigma_x l + \tau_{xy} m = X^*, \quad \tau_{xy} l + \sigma_y m = Y^* \tag{5.2}$$

In the above equations, l and m are the direction cosines (see Table 1.2) in the x–y coordinate system relating to the surface of the body.

However, in reality, many types of problems exist for which, when we try to solve them, we are unable to satisfy the boundary conditions perfectly. In such cases, various approximate

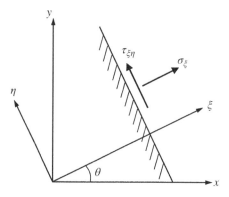

Figure 5.3

methods can be used. We can choose a representative point on the boundary and satisfy the boundary conditions at the point (the point collocation method), or we can integrate the stress distribution over a particular section of the boundary (the resultant force method) and so on. The validity of such approximate methods can be understood by considering Saint Venant's principle. Even though the boundary conditions are not perfectly satisfied, if we obtain the boundary stress by numerical calculation using a statistically equivalent loading system, then, although the stress distribution near to the boundary itself will deviate from the exact solution, we can assume that the stress distribution at a point far removed from the boundary will be approximately correct.

5.2.2   Displacement Boundary Conditions

When the boundary conditions are defined by the displacement at a boundary, we call it the *displacement boundary condition*. A typical example of this condition is the case of a completely fixed boundary (no displacement). In general, nonzero displacement is given at a boundary. When we are unable to satisfy the boundary conditions perfectly, various approximate methods similar to the cases of stress boundary condition can be used. We can choose a representative point on the boundary and satisfy the boundary conditions at the point (the point collocation method), or we can calculate the average displacement over a particular section of the boundary and so on. Considering that the displacement at a boundary is correlated with the integration of strain fields of the total structure rather than the detailed strain fields near the boundary, we can say that the obtained solutions of stresses and strains far removed from the boundary are close to the exact solution. The validity of such approximate methods can also be understood by considering Saint Venant's principle.

5.2.3   Mixed Boundary Conditions

When the boundary conditions are defined by a combination of the displacement and stress at a boundary, we call it a *mixed boundary condition*. In most cases, the stress boundary conditions and the displacement boundary conditions are defined separately at separate points. However,

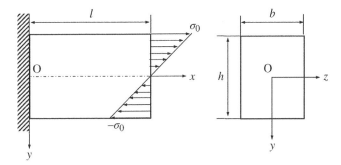

Figure 5.4

some problems contain both conditions in a coupled manner. One such example is a composite, such as the fiber reinforced plastics, which contains fibers or inclusions having elastic moduli different from the matrix material. In such problems, although the stresses (normal stresses and shear stresses) and the displacements at the boundary are unknown, we satisfy the condition that all these quantities must be continuous at the boundary. When we discuss this problem, we need to pay attention to Hooke's laws which are separately applied to two materials facing or bonded to each other at the boundary.

Example problem 5.1
Pick up all the boundary conditions for the cantilever in Figure 5.4 as a three dimensional problem.

Solution

$$1. \ \sigma_x = -\frac{2y}{h}\sigma_0, \ \ \tau_{xy} = 0, \ \ \tau_{zx} = 0 \ (\tau_{xz} = 0) \ \ \text{at} \ \ x = l$$

$$2. \ \sigma_y = 0, \ \ \tau_{xy} = 0, \ \ \tau_{yz} = 0 \ \ \ \text{at} \ \ y = \pm\frac{h}{2}$$

$$3. \ \sigma_z = 0, \ \ \tau_{zx} = 0, \ \ \tau_{yz} = 0 \ \ \ \text{at} \ \ z = \pm\frac{b}{2}$$

$$4. \ u = 0, \ \ v = 0^*, \ \ w = 0 \ \ \ \text{at} \ \ x = 0$$

$$^*dv/dx = 0 \ \text{is not a correct condition.}$$

Problems of Chapter 5

1. Figure 5.5a, b shows the stress distributions at three sections in a cylinder which has the same sectional dimension as the plate of Figure 5.1a and is subjected to concentrated point forces at both ends of the cylinder. In the case of Figure 5.5a, the point force P is applied at the center of the end of the cylinder. In case of Figure 5.5b, the total point force P is applied

at the circumference of the end of the cylinder. Indicate the difference in the stress distributions between the plate and cylinder, and explain the reason.

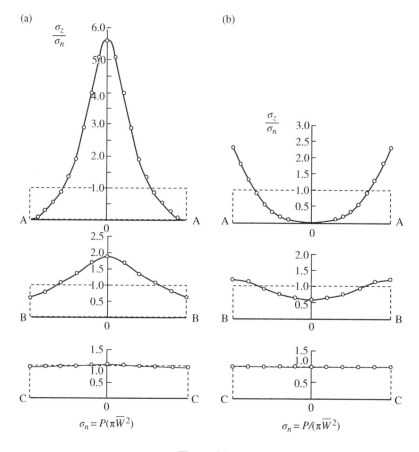

Figure 5.5

2. Comparing the stress distribution between parts (a) and (b) in Figure 5.6, indicate where in the cantilevers the stress distributions show a difference. Explain in what condition of the length of cantilever l the displacements of both cases show a big difference.

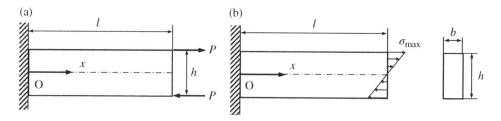

Figure 5.6   In part (b) $\sigma_{max} = 6Pl/(bh)$

3. Figure 5.7 shows a plate which is subjected to a distributed loading at the part of the boundary AB with length s of the plate with thickness b. The loading varies linearly from σ_1 to σ_2 at the boundary from A to B.

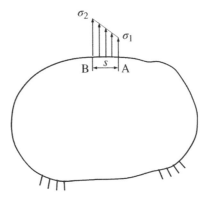

Figure 5.7

When we replace the boundary condition at AB by one single concentrated force P and one single concentrated moment M which together act at the midpoint of AB, determine the relevant values for P and M.

4. When these concentrated forces are applied to a plate, as shown in Figure 5.8 (the same plate as Figures 5.1a and 5.2b), draw the approximate stress distributions at the sections

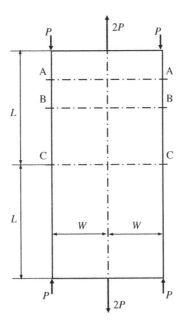

Figure 5.8

A–A, B–B and C–C, where the thickness t of the plate is unity and P is the value over unit thickness. Refer to Figures 5.1c and 5.2b for illustration.

5. Figure 5.9 shows a rectangular plate of a material A containing a different material B. The plate is subjected to a remote tensile stress σ_0 at the edge of the plate. Young's modulus and Poisson's ratio of the plate A are E_A and ν_A. Young's modulus and Poisson's ratio of the material B are E_B and ν_B.

When we consider the stresses and strains at the interface of the two materials A and B, judge if the following equations are correct or wrong. The coordinates ξ and η are taken to the normal and tangential directions respectively at the interface of the two materials.

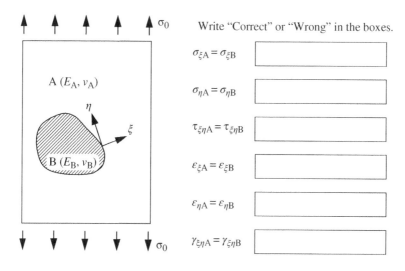

Write "Correct" or "Wrong" in the boxes.

$\sigma_{\xi A} = \sigma_{\xi B}$

$\sigma_{\eta A} = \sigma_{\eta B}$

$\tau_{\xi\eta A} = \tau_{\xi\eta B}$

$\varepsilon_{\xi A} = \varepsilon_{\xi B}$

$\varepsilon_{\eta A} = \varepsilon_{\eta B}$

$\gamma_{\xi\eta A} = \gamma_{\xi\eta B}$

Figure 5.9

6

Two Dimensional Problems

The solutions for two dimensional (2D) problems were developed first by closed form stress function methods, by numerical methods and recently by the finite element method (FEM). Although the closed form solutions obtained in the history of theory of elasticity are limited to relatively simple problems, learning and understanding such solutions is very important, not only for effective applications of the solutions to many practical problems but also for checking the validity of solutions obtained by numerical methods such as FEM.

This chapter describes the basic solutions for a cylinder with a hole, stress concentration due to a circular hole, stress concentration due to an elliptical hole, crack problems and other basic problems. The problems of stress concentration are very important for the discussion of strength and fracture of various structures (see Part II).

In the following, we separate 2D problems into two typical cases. One is called the *Plane Stress Problem* and the other is called the *Plane Strain Problem*. These two cases correspond to the two extreme stress and strain conditions in 2D problems. If we study the characteristics of these two cases, we can understand the real cases which occur in the stress and strain conditions between Plane Stress and Plane Strain.

6.1 Plane Stress and Plane Strain

6.1.1 Plane Stress

When we take an x–y coordinate within the plane of a plate and take a z axis in the direction of the plate thickness, plane stress is defined by the case that the stresses σ_x, σ_y and τ_{xy} are in question and other stresses are all zero. If the thickness of a plate is thin and both surfaces of the plate are free of stresses (a free surface), we may regard that the plate is in plane stress condition. The stress condition of a thin-walled pressurized cylinder can be considered approximately close to a plane stress condition, because the internal pressure p is relatively very small

Theory of Elasticity and Stress Concentration, First Edition. Yukitaka Murakami.
© 2017 John Wiley & Sons, Ltd. Published 2017 by John Wiley & Sons, Ltd.

compared to the axial stress σ_z and circumferential stress σ_θ, and then the stress in the direction of the thickness can be ignored. Thus, many problems which are not in a perfectly plane stress condition are analyzed by assuming them to be approximately in plane stress.

In a plane stress condition, Hooke's law is written as follows:

$$\left.\begin{array}{l}\varepsilon_x = \dfrac{1}{E}\left(\sigma_x - \nu\sigma_y\right), \quad \varepsilon_y = \dfrac{1}{E}\left(\sigma_y - \nu\sigma_x\right), \quad \varepsilon_z = -\dfrac{\nu}{E}\left(\sigma_x + \sigma_y\right) \\[2mm] \gamma_{xy} = \tau_{xy}/G, \quad \sigma_z = 0, \quad \tau_{yz} = 0, \quad \tau_{zx} = 0 \end{array}\right\} \tag{6.1}$$

6.1.2 Plane Strain

Plane strain is defined by the case that the strain in one direction (e.g. $\varepsilon_z = 0$) is completely restricted. For example, we can imagine problems such as a cylinder having a fixed displacement at both ends and a river dam having fixed ends in both sides to be approximately in a plane strain condition. Other typical examples we meet in practical problems are a case of a thick plate which have locally very different deformation due to a stress concentration such as at a notch root. If we have a large strain in the x-direction, we may have a large strain also in the z-direction due to the effect of Poisson's ratio (see Equation 6.1). However, this does not occur, because the strains at the point apart from the notch root are smaller than those at the notch root and restrict free deformation at the notch root. This case cannot be regarded as a perfect problem of plane strain, but is very often regarded as an approximately plane strain problem.

Hooke's law for a plane strain condition can be written as follows:

$$\left.\begin{array}{l}\varepsilon_x = \dfrac{1}{E}\left\{\sigma_x - \nu\left(\sigma_y + \sigma_z\right)\right\} \\[2mm] \varepsilon_y = \dfrac{1}{E}\left\{\sigma_y - \nu\left(\sigma_z + \sigma_x\right)\right\} \\[2mm] \varepsilon_z = \dfrac{1}{E}\left\{\sigma_z - \nu\left(\sigma_x + \sigma_y\right)\right\} = 0, \quad \text{namely } \sigma_z = \nu\left(\sigma_x + \sigma_y\right) \\[2mm] \gamma_{xy} = \tau_{xy}/G, \quad \gamma_{yz} = 0, \quad \gamma_{zx} = 0 \end{array}\right\} \tag{6.2}$$

Considering the third equation in Equation 6.2, the relationship between stresses and strains can be expressed as follows:

$$\varepsilon_x = \dfrac{1}{E^*}\left(\sigma_x - \nu^*\sigma_y\right), \quad \varepsilon_y = \dfrac{1}{E^*}\left(\sigma_y - \nu^*\sigma_x\right), \quad \gamma_{xy} = \tau_{xy}/G^* \tag{6.3}$$

However, we keep the following expression for the equivalent Young's modulus E^* and Poisson's ratio ν^*.

$$E^* = \dfrac{E}{1-\nu^2}, \quad \nu^* = \dfrac{\nu}{1-\nu}, \quad G^* = \dfrac{E^*}{2\left(1+\nu^*\right)} = \dfrac{E}{2\left(1+\nu\right)} = G \tag{6.4}$$

Thus, if the stresses σ_x, σ_y and τ_{xy} are the same in a plane stress problem and in a plane strain problem, the strains ε_x and ε_y under a plane strain condition are equal to those of ε_x and ε_y under a plane stress condition for a different material having the moduli E^* and ν^* given in Equation 6.4. Especially, it is to be noted that the equivalent Young's modulus E^* is $1/(1 - \nu^2)$ times higher than E. For most steels, $\nu = \sim 0.3$ and the displacement and deflection in a plane strain condition are approximately 10% smaller than in a plane stress condition.

Example problem 6.1
Figure 6.1a shows an elastic material, having Young's modulus E and Poisson's ratio ν, which is put in a rigid groove having a width $2a$, $(-a, a)$ in the x coordinate, height h in the z coordinate and being infinite in the y coordinate without any restriction of movement. A pressure p is applied on the top flat surface of the elastic material. Assume there is no clearance and no friction at the interface between the elastic material and the rigid groove. Determine the stresses σ_x, σ_y, σ_z and the apparent Young's modulus E' as defined by σ_z/ε_z.

Figure 6.1b shows a similar problem in a cylindrical coordinate. A cylindrical elastic material is put in a cylindrical rigid hole without clearance and friction and subjected to pressure p on the top flat surface of the elastic material. Determine the stresses σ_r, σ_θ, σ_z and the apparent Young's modulus E' as defined by σ_z/ε_z.

Solution
In Figure 6.1a, $\sigma_z = -p$, $\sigma_y = 0$; $\tau_{xy} = 0$ and $\tau_{zx} = 0$ at $x = \pm a$; and $\varepsilon_x = 0$ at $-a \leq x \leq a$. From these conditions, $\sigma_x = -\nu p = \nu \sigma_z$.

Substituting the above σ_x and σ_y into $\varepsilon_z = (1/E)\left[\sigma_z - \nu\left(\sigma_y + \sigma_x\right)\right]$:

$$E' = \sigma_z/\varepsilon_z = E/\left(1 - \nu^2\right).$$

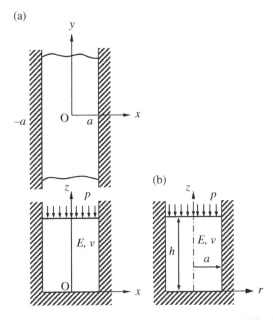

Figure 6.1 Elastic material (a) in a rigid groove and (b) in a rigid cylindrical hole

In Figure 6.1b, the known boundary conditions are $\sigma_z = -p$, $\varepsilon_\theta = 0$ (see Equation 2.25); $\tau_{r\theta} = 0$ and $\tau_{rz} = 0$ at $r = \pm a$.

Therefore, $\sigma_r = -\sigma_0$ (unknown yet) is expected at $r = a$. In this case, the stresses are $\sigma_r = \sigma_\theta = -\sigma_0$ everywhere inside the cylinder (see Example problem 6.1). Considering $\varepsilon_\theta = 0$ at $r = a$,

$$\varepsilon_\theta = \frac{1}{E}[\sigma_\theta - \nu(\sigma_r + \sigma_z)] = 0, \text{ namely, } -\sigma_0 - \nu(-\sigma_0 - p) = 0$$

Therefore,

$$\sigma_0 = \frac{\nu}{1-\nu} p$$

Then, the stresses are written as follows everywhere in the cylinder:

$$\sigma_r = \sigma_\theta = -\frac{\nu}{1-\nu} p = \frac{\nu}{1-\nu} \sigma_z$$

From Hooke's law:

$$\varepsilon_z = \frac{1}{E}[\sigma_z - \nu(\sigma_r + \sigma_\theta)] = \frac{1}{E} \cdot \frac{1-\nu-2\nu^2}{1-\nu} \sigma_z$$

Finally, we have for Figure 6.1b:

$$E' = \frac{\sigma_z}{\varepsilon_z} = \frac{1-\nu}{1-\nu-2\nu^2} E$$

6.2   Basic Conditions for Exact Solutions: Nature of Solutions

Let us seek the conditions for the exact solution when we try to solve a problem. When we consider the stresses as the solution we want to find, it is natural that the solution must satisfy the *equilibrium conditions* and the *boundary conditions*. However, can we regard the solution satisfying these two conditions to be the exact solution? We can not necessarily say the solution is exact, because there is no guarantee that the deformation can be determined uniquely by these stresses.

For example, let us look at a problem like Figure 6.2. The boundary conditions of this problem can be expressed as follows:

$$\left. \begin{array}{ll} \sigma_x = 0, \quad \tau_{xy} = 0 & \text{at} \quad x = 0 \\[2mm] \sigma_x = \sigma_0, \quad \tau_{xy} = -2\sigma_0(y/a) & \text{at} \quad x = a \\[2mm] \sigma_y = 0, \quad \tau_{xy} = 0 & \text{at} \quad y = 0 \\[2mm] \sigma_y = \sigma_0, \quad \tau_{xy} = -2\sigma_0(x/a) & \text{at} \quad y = a \end{array} \right\} \quad (6.5)$$

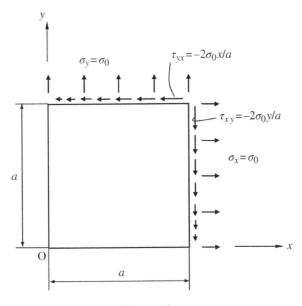

Figure 6.2

These boundary conditions can be satisfied by assuming the stresses in the following forms:

$$\sigma_x = \sigma_0(x/a)^2, \quad \sigma_y = \sigma_0(y/a)^2, \quad \tau_{xy} = -2\sigma_0(xy/a^2) \tag{6.6}$$

Following Equation 4.5, the equilibrium conditions for these stresses are given by:

$$\frac{\partial \sigma_x}{\partial x} + \frac{\partial \tau_{xy}}{\partial y} = 0, \quad \frac{\partial \tau_{xy}}{\partial x} + \frac{\partial \sigma_y}{\partial y} = 0 \tag{6.7}$$

Substituting Equation 6.6 into Equation 6.7, we can see that the equilibrium conditions are satisfied irrespective of the values of x and y, namely also at internal points of the plate.

Now, we check whether the stresses of Equation 6.6 are actually the exact ones in the plate of Figure 6.2. Referring to Hooke's law, we have:

$$\left. \begin{array}{l} \varepsilon_x = \dfrac{1}{E}\left(\sigma_x - \nu\sigma_y\right) = \dfrac{\sigma_0}{E} \cdot \dfrac{x^2 - \nu y^2}{a^2} \\[3mm] \varepsilon_y = \dfrac{1}{E}\left(\sigma_y - \nu\sigma_x\right) = \dfrac{\sigma_0}{E} \cdot \dfrac{y^2 - \nu x^2}{a^2} \\[3mm] \gamma_{xy} = \dfrac{\tau_{xy}}{G} = -\dfrac{2\sigma_0}{G} \cdot \dfrac{xy}{a^2} \end{array} \right\} \tag{6.8}$$

Considering $\varepsilon_x = \partial u/\partial x$ and $\varepsilon_y = \partial v/\partial y$, and integrating these equations, we obtain

$$u = \frac{\sigma_0}{E} \cdot \frac{1}{a^2}\left(\frac{x^3}{3} - \nu x y^2\right) + f(y) \tag{6.9}$$

$$v = \frac{\sigma_0}{E} \cdot \frac{1}{a^2}\left(\frac{y^3}{3} - \nu x^2 y\right) + g(x) \tag{6.10}$$

where $f(y)$ is a function only of y or a constant and $g(x)$ is a function only of x or a constant. Therefore,

$$\gamma_{xy} = \frac{\partial u}{\partial y} + \frac{\partial v}{\partial x} = -\frac{4\nu\sigma_0}{Ea^2}xy + \frac{\partial f(y)}{\partial y} + \frac{\partial g(x)}{\partial x} \tag{6.11}$$

However, we can easily see that γ_{xy} of (Equation 6.11) does not coincide with γ_{xy} of Equation 6.8. The meaning of this discrepancy is equivalent to that the *conditions of compatibility* (Equation 2.24) given in the following are not satisfied.

$$\frac{\partial^2 \varepsilon_x}{\partial y^2} + \frac{\partial^2 \varepsilon_y}{\partial x^2} = \frac{\partial^2 \gamma_{xy}}{\partial x \cdot \partial y} \tag{6.12}$$

As described above, the stresses which satisfy only the boundary conditions and equilibrium conditions cannot be the exact solution. To avoid such a contradiction with respect to displacements, the coincidence of γ_{xy} of Equation 6.8 with γ_{xy} of Equation 6.11 is necessary. As the general requirement to solve this difficulty, the compatibility conditions must be satisfied. Now, we understand that the exact solution is required to satisfy the following three conditions:

1. Equilibrium condition
2. Compatibility condition
3. Boundary condition

Let us seek the way to satisfy all these three conditions in two dimensional problems.

6.3 Airy's Stress Function

The equilibrium equation, Equation 6.7, having two equations, contains three unknowns (σ_x, σ_y, τ_{xy}). These unknowns are functions of the coordinates (x, y) and expressed as follows:

$$\sigma_x = f(x,y), \quad \sigma_y = g(x,y), \quad \tau_{xy} = h(x,y) \tag{6.13}$$

Our final goal is to determine these three functions. We can see from Equation 6.7 that these three functions are not independent of each other. Since Equation 6.7 has two equations coupled with three unknown quantities, we can choose only one among the unknown functions of Equation 6.13 to treat a problem. If we choose $f(x, y)$ as the representative of three, τ_{xy} can be expressed as follows:

$$\tau_{xy} = -\int \frac{\partial f(x,y)}{\partial x} dy + G(x) \tag{6.14}$$

where $G(x, y)$ is a function only of x.

However, the generally acceptable and more useful expression than Equation 6.13 is the introduction of a new function $\phi(x, y)$ and express σ_x in the form $\sigma_x = \partial^2 \phi / \partial y^2$. Expressing σ_x in this manner, we can obtain expressions for τ_{xy} and σ_y using ϕ automatically from Equation 6.7. Thus, we have:

$$\sigma_x = \frac{\partial^2 \phi}{\partial y^2}, \quad \sigma_y = \frac{\partial^2 \phi}{\partial x^2}, \quad \tau_{xy} = -\frac{\partial^2 \phi}{\partial x \cdot \partial y} \tag{6.15}$$

When we express the three unknowns σ_x, σ_y, τ_{xy} in two dimensional problems as in the form of Equation 6.15, we call the function $\phi(x, y)$ *Airy's stress function*.

Now we understand that the form of Equation 6.15 satisfies the equilibrium conditions, because we determined the form using the equilibrium conditions. The next thing we need to do is to derive a general expression for satisfying the compatibility conditions in terms of $\phi(x, y)$.

Although the compatibility conditions of Equation 6.12 are written in terms of strains, we can replace strains by stresses using Hooke's law:

$$\frac{1}{E} \left[\frac{\partial^2 (\sigma_x - \nu \sigma_y)}{\partial y^2} + \frac{\partial^2 (\sigma_y - \nu \sigma_x)}{\partial x^2} \right] = \frac{2(1+\nu)}{E} \cdot \frac{\partial^2 \tau_{xy}}{\partial x \cdot \partial y} \tag{6.16}$$

Inserting Equation 6.15 into Equation 6.16, we obtain:

$$\frac{\partial^4 \phi}{\partial x^4} + 2 \frac{\partial^4 \phi}{\partial x^2 \cdot \partial y^2} + \frac{\partial^4 \phi}{\partial y^4} = 0 \tag{6.17}$$

Equation 6.17 can be written as follows, too:

$$\left(\frac{\partial^2}{\partial x^2} + \frac{\partial^2}{\partial y^2} \right) \cdot \left(\frac{\partial^2 \phi}{\partial x^2} + \frac{\partial^2 \phi}{\partial y^2} \right) = \nabla^2 \cdot \nabla^2 \phi = \nabla^4 \phi = 0 \tag{6.18}$$

where ∇^2 is Laplacian defined by $\nabla^2 = \partial^2 / \partial x^2 + \partial^2 / \partial y^2$. Thus, the stress function ϕ must be a *bi-harmonic function*. There are an unlimited number of bi-harmonic functions. But the exact solution we want to find is the only one which is a bi-harmonic function and at the same time satisfies the boundary condition.

Here, it should be noted that Equation 6.18 does not include Young's modulus E and Poisson's ratio ν. This means that, if the boundary conditions of a problem are given only by forces or stresses, the solution does not include E and ν. Therefore, the stresses are not influenced by these moduli.[1]

This property of Equation 6.18 is the basis of the validity of the photo elasticity method. We can also understand from Equation 6.3 that, if the boundary conditions of a plane strain problem are given by forces or stresses, the solution $(\sigma_x, \sigma_y, \tau_{xy})$ of a plane strain problem is equal to those of a plane stress problem with the same boundary conditions.

Since Equation 6.18 is a linear differential equation, the sum of two stress functions ϕ_1 and ϕ_2 [i.e. $(\phi_1 + \phi_2)$] is also a stress function which expresses the stress state obtained by superposing the individual stress states of ϕ_1 and ϕ_2.

This nature of stress function is called the *principle of superposition*. The principle of superposition is very useful for obtaining the solution of a complex problem by the combination of solutions for multiple simple problems and also for obtaining approximate solutions for practical problems. Many interesting examples of applications of the principle of superposition will be explained later in this chapter.

Example problem 6.2

Explain the stress distribution given by the following stress functions. Show the equations requested for the coefficients a, b, c, ..., if necessary.

    i. $\phi = ax^2 + bxy + cy^2$
    ii. $\phi = ax^3 + bx^2y + cxy^2 + dy^3$
    iii. $\phi = ax^4 + bx^3y + cx^2y^2 + dxy^3 + ey^4$
    iv. $\phi = ax^5 + bx^4y + cx^3y^2 + dx^2y^3 + exy^4 + fy^5$

Solution

    i. $\sigma_x = 2c$, $\sigma_y = 2a$, $\tau_{xy} = -b$
    ii. $\sigma_x = 2cx + 6dy$, $\sigma_y = 6ax + 2by$, $\tau_{xy} = -2bx - 2cy$
    iii. $\sigma_x = 2cx^2 + 6dxy + 12ey^2$, $\sigma_y = 12ax^2 + 6bxy + 2cy^2$, $\tau_{xy} = -3bx^2 - 4cxy - 3dy^2$,
        where $e = -(a + c/3)$ is a necessary relationship from Equation 6.17.
    iv. $\sigma_x = 2cx^3 + 6dx^2y + 12exy^2 + 20fy^3$,

        $\sigma_y = 20ax^3 + 12bx^2y + 6cxy^2 + 2dy^3$,
        $\tau_{xy} = -4bx^3 - 6cx^2y - 6dxy^2 - 4ey^3$,

where $e = -(5a + c)$ and $f = -(b + d)/5$ are necessary relationships from Equation 6.17.

By combinations of the above stress functions, solutions for a rectangular region having various complex boundary conditions can be composed.

6.4 Hollow Cylinder

There are many practical problems of circular plate and hollow cylinder. To solve these problems, it is convenient to express the stress function ϕ in the polar coordinate (r, θ). Therefore,

---

[1] There is an exceptional case in which holes are contained in a plate and the forces applied along the boundaries of holes are not in equilibrium condition (see Chapter 10 of Part II).

it is necessary to express the differential Equation 6.18 in the polar coordinate as follows (see Appendix of this chapter).

$$\left(\frac{\partial^2}{\partial r^2} + \frac{1}{r} \cdot \frac{\partial}{\partial r} + \frac{1}{r^2} \cdot \frac{\partial^2}{\partial \theta^2} \right) \cdot \left(\frac{\partial^2 \phi}{\partial r^2} + \frac{1}{r} \cdot \frac{\partial \phi}{\partial r} + \frac{1}{r^2} \cdot \frac{\partial^2 \phi}{\partial \theta^2} \right) = 0 \tag{6.19}$$

The stresses are also expressed with polar coordinate as follows.

$$\sigma_r = \frac{1}{r} \cdot \frac{\partial \phi}{\partial r} + \frac{1}{r^2} \cdot \frac{\partial^2 \phi}{\partial \theta^2}, \quad \sigma_\theta = \frac{\partial^2 \phi}{\partial r^2}, \quad \tau_{r\theta} = -\frac{\partial}{\partial r} \left(\frac{1}{r} \cdot \frac{\partial \phi}{\partial \theta} \right) \tag{6.20}$$

Now we start solving problems from the simplest case of Equation 6.19, that is, the case of ϕ being a function of only r. Thus, Equation 6.19 becomes

$$\left(\frac{d^2}{dr^2} + \frac{1}{r} \cdot \frac{d}{dr} \right) \cdot \left(\frac{\partial^2 \phi}{\partial r^2} + \frac{1}{r} \cdot \frac{d\phi}{dr} \right) = 0 \tag{6.21}$$

Expanding the above equation, we have

$$\frac{\partial^4 \phi}{\partial r^4} + \frac{2}{r} \cdot \frac{\partial^3 \phi}{\partial r^3} - \frac{1}{r^2} \cdot \frac{\partial^2 \phi}{\partial r^2} + \frac{1}{r^3} \cdot \frac{d\phi}{dr} = 0 \tag{6.22}$$

Here, replacing the variable r by t by putting $r = e^t$ (or $t = \log r$), Equation 6.19 can be expressed as:

$$D^2 (D-2)^2 \phi = 0 \tag{6.23}$$

where $D = d/dt$.

The general solution of Equation 6.23 is expressed as

$$\phi = C_1' + C_2't + (C_3' + C_4't)e^{2t} \tag{6.24}$$

Replacing t by r, we obtain the following equation:

$$\phi = A \cdot \log(r) + Br^2 \cdot \log(r) + Cr^2 + D \tag{6.25}$$

where A, B, C and D are the constants to be determined by the boundary conditions. Therefore, the stresses are derived by Equation 6.20 as follows, excluding the terms related to θ:

$$\left. \begin{aligned} \sigma_r &= \frac{1}{r} \cdot \frac{d\phi}{dr} = 2C + \frac{A}{r^2} + B[1 + 2\log(r)] \\[2mm] \sigma_\theta &= \frac{d^2\phi}{dr^2} = 2C - \frac{A}{r^2} + B[3 + 2\log(r)] \end{aligned} \right\} \tag{6.26}$$

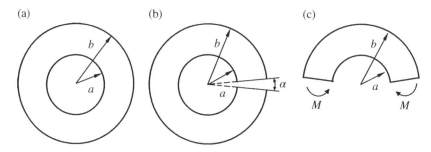

Figure 6.3

Equation 6.26 can be used for the solutions of the problems shown in Figure 6.3. The unknown constants A, B and C are determined considering the boundary conditions of the problems of Figure 6.3. Three conditions are necessary and sufficient.

For the problem of Figure 6.3a, three boundary conditions are written as follows:

Stress boundary conditions: (1) $\sigma_r = \sigma_1$ (we can decide) at $r = a$.
           (2) $\sigma_r = \sigma_2$ (we can decide) at $r = b$.
Displacement boundary conditions: u and v are the function of only r. This is the special condition for Figure 6.3a.

Namely, $v = 0$ and u is the displacement only to be prescribed in terms of radial coordinate r. Considering Equation (2.25):

$$\varepsilon_r = du/dr \tag{6.27}$$

$$\varepsilon_\theta = u/r \tag{6.28}$$

If we calculate du/dr from Equation (6.28), the value must coincide with Equation (6.27). This is the condition required to displacement. ε_r and ε_θ can be expressed with stresses of Equation (6.26) through Hooke's law and one equation which must be satisfied by the constants A, B and C is obtained as follows.

From Equations (6.27) and (6.28):

$$\varepsilon_r - \frac{d(r\varepsilon_\theta)}{dr} = \frac{1}{E}(\sigma_r - \nu\sigma_\theta) - \frac{1}{E}(\sigma_\theta - \nu\sigma_r) - \frac{r}{E}\left(\frac{d\sigma_\theta}{dr} - \nu\frac{d\sigma_r}{dr}\right) = -\frac{4B}{E} = 0 \tag{6.29}$$

Then,

$$B = 0 \tag{6.30}$$

Thus, the solution for the problem of Figure 6.3a is expressed by the following equations.

$$
\left.\begin{aligned}
\sigma_r &= 2C + \frac{A}{r^2} \\
\sigma_\theta &= 2C - \frac{A}{r^2}
\end{aligned}\right\}
\quad\rightarrow\quad
\left.\begin{aligned}
\sigma_r &= C_1 + \frac{C_2}{r^2} \\
\sigma_\theta &= C_1 - \frac{C_2}{r^2}
\end{aligned}\right\}
\tag{6.31}
$$

Equation 6.31 can be used for the solution of a hollow cylinder subjected to internal and external pressure.

In case of the problems of Figure 6.3b, c, v is a function of θ and B is not equal to 0.

Example problem 6.3

Determine the stresses in a hollow cylinder with inner radius a and outer radius b under internal pressure p_i and external pressure p_o.

Solution

C_1 and C_2 can be determined by applying the boundary conditions of $\sigma_r = -p_i$ at $r = a$, and $\sigma_\theta = -p_o$ at $r = b$ to 6.31.

$$\left.\begin{aligned}\sigma_r &= \frac{p_i a^2 - p_o b^2}{b^2 - a^2} + \frac{a^2 b^2 (p_o - p_i)}{b^2 - a^2} \cdot \frac{1}{r^2} \\[2mm] \sigma_\theta &= \frac{p_i a^2 - p_o b^2}{b^2 - a^2} - \frac{a^2 b^2 (p_o - p_i)}{b^2 - a^2} \cdot \frac{1}{r^2}\end{aligned}\right\} \tag{6.32}$$

Problem 6.4.1   When a hollow cylinder of the internal radius a and the outer radius b is subjected to external pressure p_o, determine the decrease in the inner radius and outer radius.

Problem 6.4.2   There is a very wide plate containing a small circular hole with radius a. When the plate is subjected to a uniform tensile stress $\sigma_x = \sigma_0$ and $\sigma_y = \sigma_0$ at far removed from the circular hole, determine the stresses σ_r and σ_θ at the edge of the circular hole.

6.5   Stress Concentration at a Circular Hole

When a wide plate containing a circular hole is subjected to a tensile stress far removed from the hole, the stress at the edge of the circular hole becomes higher than in other places. This phenomenon is called *stress concentration*. Since the failure of structures and machine components often initiates from sites of stress concentration, it is very important to make clear the intensity of stress concentration in terms of failure accidents in real cases. This section will investigate the stress concentration due to a circular hole with radius a, contained in an infinite plate subjected to a uniform remote tensile stress $\sigma_x = \sigma_0$, as shown Figure 6.4. Although there is no infinite plate, this is a convenient assumption to solve the problem mathematically. If the size of a plate is sufficiently large in comparison with the size of a circular hole, the solution obtained based on the infinite plate assumption can be used as an approximately exact solution.

The problem of Figure 6.4 was solved for the first time by G. Kirsch (1898). The solution will be introduced in the following.

Solving the problem of Figure 6.4 is equivalent to finding the stress function which satisfy the boundary conditions and $\nabla^4 \phi = 0$ (Equation (6.18)). If we consider Saint Venant's principle, we can ignore the influence of the circular hole on the remote stresses. Then, we can write the boundary conditions as follows.

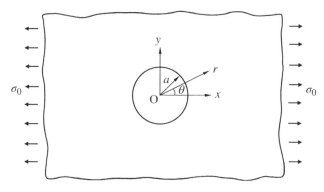

Figure 6.4

Boundary conditions:
1. Remote stresses: $\sigma_x = \sigma_0$, $\sigma_y = 0$, $\tau_{xy} = 0$ (in orthogonal coordinate).
2. Stresses at the edge of the hole $(r = a) : \sigma_r = 0$, $\tau_{r\theta} = 0$ (in polar coordinate).

It is not convenient to find the solution ϕ by dealing with the boundary conditions expressed with different coordinates such as (1) and (2). Since it is presumed that satisfying condition (2) will be the key in finding the solution, it is better to rewrite boundary condition (1) in a polar coordinate. Thus, rewriting boundary condition (1) by Equation 1.12 as follows is denoted by (1′):

Boundary condition:
1′. Remote stress $(\text{at } r = \infty)$, $\sigma_r = \sigma_0\cos^2\theta = \sigma_0/2 + \sigma_0\cos2\theta/2$, $\tau_{r\theta} = -\sigma_0\sin2\theta/2$

Considering the form of the boundary conditions (1′) and (2), the problem of Figure 6.4 can be solved by the superposition, as in Figure 6.5. Since the problem of Figure 6.5b is substantially the same as the problem of the previous section (the problem of a hollow cylinder), we know the solution.
The problem of Figure 6.5c can be solved as follows.

Basic equation: $\nabla^4\phi = 0$
Boundary conditions: (I) $\sigma_r = 0$, $\tau_{r\theta} = 0$ at $r = a$
(II) $\sigma_r = \sigma_0\cos2\theta/2$, $\tau_{r\theta} = -\sigma_0\sin2\theta/2$ at $r = \infty$

Namely, the problem is reduced to solving Equation 6.19 with the boundary conditions (I) and (II).

$$\left.\begin{array}{l} \sigma_r = \dfrac{1}{r}\cdot\dfrac{\partial\phi}{\partial r} + \dfrac{1}{r^2}\cdot\dfrac{\partial^2\phi}{\partial\theta^2} = \dfrac{1}{2}\sigma_0\cos2\theta \\[3mm] \tau_{r\theta} = -\dfrac{\partial}{\partial r}\left(\dfrac{\partial\phi}{r\partial\theta}\right) = -\dfrac{1}{2}\sigma_0\sin2\theta \end{array}\right\} \qquad (6.33)$$

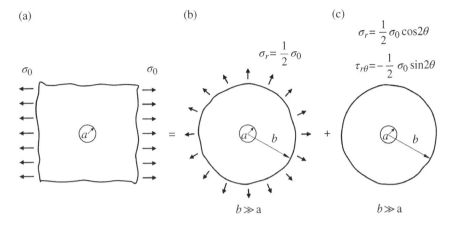

Figure 6.5

Looking at the form of Equation 6.33 carefully, we can understand that ϕ must have the following form:

$$\phi = f(r) \cdot \cos 2\theta \tag{6.34}$$

where $f(r)$ is a function of r. Introducing ϕ from Equation 6.34 into Equation 6.19 and excluding $\cos 2\theta$, which is attached to the whole equation, we obtain the following differential equation:

$$\left(\frac{d^2}{dr^2} + \frac{1}{r} \cdot \frac{d}{dr} - \frac{4}{r^2} \right) \cdot \left(\frac{d^2 f}{dr^2} + \frac{1}{r} \cdot \frac{df}{dr} - \frac{4f}{r^2} \right) = 0 \tag{6.35}$$

Expanding this equation, we have

$$\frac{d^4 f}{dr^4} + \frac{2}{r} \cdot \frac{d^3 f}{dr^3} - \frac{9}{r^2} \cdot \frac{d^2 f}{dr^2} + \frac{9}{r^3} \cdot \frac{df}{dr} = 0 \tag{6.36}$$

Solving this equation in the same way as Equation 6.22:

$$f(r) = C_1 r^2 + C_2 r^4 + \frac{C_3}{r^2} + C_4 \tag{6.37}$$

Then,

$$\phi = \left(C_1 r^2 + C_2 r^4 + \frac{C_3}{r^2} + C_4 \right) \cos 2\theta \tag{6.38}$$

$$\sigma_r = \frac{1}{r} \cdot \frac{\partial \phi}{\partial r} + \frac{1}{r^2} \cdot \frac{\partial^2 \phi}{\partial \theta^2} = -\left(2C_1 + \frac{6C_3}{r^4} + \frac{4C_4}{r^2} \right) \cos 2\theta \tag{6.39}$$

$$\sigma_\theta = \frac{\partial^2 \phi}{\partial r^2} = \left(2C_1 + 12C_2 r^2 + \frac{6C_3}{r^4}\right)\cos2\theta \tag{6.40}$$

$$\tau_{r\theta} = -\frac{\partial}{\partial r}\left(\frac{1}{r}\cdot\frac{\partial\phi}{\partial\theta}\right) = \left(2C_1 + 6C_2 r^2 - \frac{6C_3}{r^4} - \frac{2C_4}{r^2}\right)\sin2\theta \tag{6.41}$$

We can determine the constants C_1, C_2, C_3 and C_4 from the boundary conditions (I) and (II). Then,

$$C_1 = -\frac{1}{4}\sigma_0, \quad C_2 = 0, \quad C_3 = -\frac{a^4}{4}\sigma_0, \quad C_4 = \frac{a^2}{2}\sigma_0 \tag{6.42}$$

Putting these constants into Equations 6.39 to 6.41, the solution for Figure 6.5c is obtained. Superposing this solution with the solution of Figure 6.5b, we can obtain the solution of Figure 6.5a. Thus, the stress distribution of Figure 6.5a is given by the following equations:

$$\left.\begin{array}{l}
\sigma_r = \dfrac{\sigma_0}{2}\left(1 - \dfrac{a^2}{r^2}\right) + \dfrac{\sigma_0}{2}\left(1 + \dfrac{3a^4}{r^4} - \dfrac{4a^2}{r^2}\right)\cos2\theta \\[4mm]
\sigma_\theta = \dfrac{\sigma_0}{2}\left(1 + \dfrac{a^2}{r^2}\right) - \dfrac{\sigma_0}{2}\left(1 + \dfrac{3a^4}{r^4}\right)\cos2\theta \\[4mm]
\tau_{r\theta} = -\dfrac{\sigma_0}{2}\left(1 - \dfrac{3a^4}{r^4} + \dfrac{2a^2}{r^2}\right)\sin2\theta
\end{array}\right\} \tag{6.43}$$

Equation 6.43 is very important, because we can apply this solution to many practical problems. We can calculate the stresses at the edge of the circular hole using this solution as follows:

$$\sigma_r = 0 \quad \text{and} \quad \tau_{r\theta} = 0 \quad \text{at} \quad r = a \tag{6.44}$$

$$\sigma_\theta = \sigma_0 - 2\sigma_0\cos2\theta \quad \text{at} \quad r = a \tag{6.45}$$

Equation 6.44 is natural, because it is the same as the boundary conditions from which we started seeking solution. Equation 6.45 expresses the variation of σ_θ along the hole edge, and σ_θ has a maximum value at $\theta = \pm\pi/2$ and a minimum value at $\theta = 0$, π. Then,

$$\sigma_{\theta max} = 3\sigma_0 \quad \text{at} \quad \theta = \pm\frac{\pi}{2} \tag{6.46}$$

$$\sigma_{\theta min} = -\sigma_0 \quad \text{at} \quad \theta = 0, \pi \tag{6.47}$$

Now, we understand that the stress is concentrated at the hole edge and the maximum value is three times higher than the remote stress σ_0. We use the term "*Stress Concentration Factor K_t*" to express the magnitude of the stress concentration as $K_t = 3$. The result of the stress concentration and the stress distribution around the hole is illustrated as Figure 6.6.

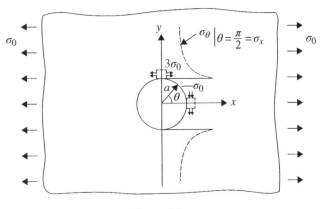

Figure 6.6

The dotted line is the distribution of the stress σ_θ, namely σ_x, along the y axis and is expressed as follows:

$$\sigma_\theta = \sigma_x = \sigma_0 \left(1 + \frac{a^2}{2r^2} + \frac{3a^4}{2r^4} \right) \quad \text{along } y \text{ axis } (r \geq a) \tag{6.48}$$

The above solutions will be more extendedly applied to many other practical solutions using the principle of the superposition. Equation 6.47 means that the value of σ_θ (or σ_y on the x axis) at the point where the edge of the circular hole intersects with the x axis is compressive. This compressive stress is likely to be often ignored compared with the maximum tensile stress $3\,\sigma_0$. However, when we apply the principle of superposition, we must not forget the role of this compressive stress.

Example problem 6.4
Show that the solution of Problem 6.4.2 can be obtained by the superposition of the solution of Figure 6.4.

Solution
The boundary condition at the circumference of the circular hole is $\sigma_r = 0$. Since $\sigma_x = \sigma_0$ and $\sigma_y = \sigma_0$ at a remote distance, the stress σ_θ at the circumference of the hole can be obtained by the superposition using Equation 6.45.

$$\sigma_\theta = (\sigma_0 - 2\sigma_0 \cos 2\theta) + \left\{ \sigma_0 - 2\sigma_0 \cos 2\left(\theta - \frac{\pi}{2} \right) \right\} = 2\sigma_0$$

Thus, the stress at the circumference of the hole is $\sigma_\theta = 2\sigma_0$ regardless of θ.

Example problem 6.5
When a thin cylindrical tube containing a small circular hole at the side, as shown in Figure 6.7, is subjected to a twisting moment T, estimate the maximum tensile stress occurs at the edge of

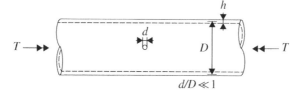

Figure 6.7

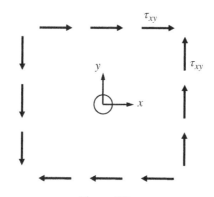

Figure 6.8

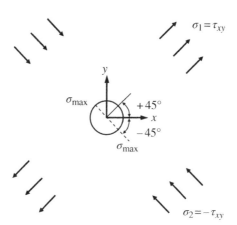

Figure 6.9

the circular hole. Assume that the diameter of the circular hole is very small compared to the diameter of the tube, D.

Solution
Denoting the shear stress acting in the thin cylindrical tube without a hole by τ_{xy}, the small circular hole is regarded to be approximately equivalent to exist in a wide plate which is subjected to τ_{xy} at a remote distance, as in Figure 6.8.

In order to replace this problem by another problem made by the superposition of the problem of Figure 6.4, we calculate the principal stresses.

From Example problem 1.3 in Chapter 1 of Part I, $\sigma_1 = \tau_{xy}$ and $\sigma_2 = -\tau_{xy}$, where σ_1 acts in a $+45°$ direction and σ_2 acts in a $-45°$ direction. Therefore, as shown in Figure 6.9, the maximum stress σ_{max} occurs at the locations of $-45°$ and $+135°$ at the circumference of the hole. Consequently,

$$\left.\begin{array}{l} \sigma_{max} = 3\sigma_1 - \sigma_2 = 4\tau_{xy} \quad at \quad \theta = -45°, \ +135° \\ \sigma_{min} = 3\sigma_2 - \sigma_1 = -4\tau_{xy} \quad at \quad \theta = +45°, \ -135° \end{array}\right\} \tag{6.49}$$

where θ is measured from the axis of the pipe and from the size of the pipe and twisting moment T, $\tau_{xy} = 2T/(pD^2h)$.

Problem 6.5.1 Confirm Saint Verant's principle in the stress distribution of Equation 6.48. Verify that the resultant force calculated by the second and third term of Equation 6.48 is equal to the force which is sustained by the part which existed when the circular hole did not exist.

Problem 6.5.2 When a rotating disk having a uniform thickness and radius b rotates at an angular velocity ω, the stress distribution inside the disk is given by the following equations:

$$\sigma_r = \frac{3+\nu}{8}\rho\omega^2\left(b^2 - r^2\right)$$

$$\sigma_\theta = \frac{3+\nu}{8}\rho\omega^2 b^2 - \frac{1+3\nu}{8}\rho\omega^2 r^2$$

where ν is Poisson's ratio and ρ is the density of the disk material.

When a small hole is drilled at the center of the disk, estimate the stress at the edge of the small hole. Assume the diameter of the hole is very small compared with the radius of the disk, b.

Problem 6.5.3 There is a wide plate which contains a circular hole, A, with radius a. We take the origin of the x–y coordinate at the center of the hole, A. There is another small hole, B, at $y = \sqrt{3}a$ on the y axis. Assume that the radius of hole B is very small compared with the radius a of hole A. Estimate the maximum stress which occurs at the edge of the small hole, B, when the wide plate is subjected to remote tensile stress $\sigma_x = \sigma_0$.

6.6 Stress Concentration at an Elliptical Hole

Figure 6.10 shows a wide plate which contains an *elliptical hole* and is subjected to a remote tensile stress $\sigma_y = \sigma_0$.

The stress distribution of σ_y along the x axis ($x \geq a$) is expressed by the following equation (see Part II):

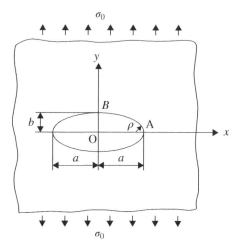

Figure 6.10

$$\sigma_y = \sigma_0 \left[\begin{array}{l} \dfrac{1}{\xi^2-1}\left(\xi^2+\dfrac{a}{a-b}\right) - \dfrac{1}{\left(\xi^2-1\right)^2}\left\{\dfrac{1}{2}\left(\dfrac{a-b}{a+b}-\dfrac{a+3b}{a-b}\right)\xi^2 - \dfrac{(a+b)b}{(a-b)^2}\right\} \\[3mm] -\dfrac{4\xi^2}{\left(\xi^2-1\right)^3}\left(\dfrac{b}{a+b}\xi^2 - \dfrac{b}{a-b}\right)\dfrac{a}{a-b} \end{array} \right] \tag{6.50}$$

where

$$\xi = \frac{x+\sqrt{x^2-c^2}}{c}, \quad c=\sqrt{a^2-b^2} \tag{6.51}$$

σ_y has the maximum value at the point A $(x=a)$. We denote it by σ_{ymax}, and then,

$$\sigma_{ymax} = \left(1+\frac{2a}{b}\right)\sigma_0 \tag{6.52}$$

Thus, the stress concentration factor K_t is

$$K_t = 1+\frac{2a}{b} = 1+2\sqrt{\frac{t}{\rho}} \tag{6.53}$$

Where, ρ is the root radius $(\rho=b^2/a)$ of the elliptical hole at point A, denoting $t=a$.

It is interesting to note that the stress σ_x at point B is expressed as follows irrespective of the aspect ratio of the ellipse, b/a:

$$\sigma_{xB} = -\sigma_0 \tag{6.54}$$

As mentioned in the section on the stress concentration of a circular hole, not only Equation 6.52 but also Equation 6.54 are important from the viewpoint of practical applications.

The solution of the stress concentration of an elliptical hole is often used for the approximate estimation of stress concentration for *holes* and *notches* whose shapes are not completely elliptical. For example, the hole shown in Figure 6.11 can be approximated by a dotted elliptical hole and the notch shown in Figure 6.12 can be approximated by a dotted semi-elliptical notch. This approximation is based on the rational consideration of the stress fields and the boundary conditions of these holes and notches. The key of the approximation is to keep the root radius of a hole or notch at the point in question equal to that of the dotted line. This way of thinking is called "*The Concept of Equivalent Ellipse*" [1,2].

Problem 6.6.1   Figure 6.13 shows an elliptical hole in an infinite plate. Figure 6.14 shows a semi-elliptical notch in a semi-infinite plate.

The geometrical shape of the half $(x > 0)$ of Figure 6.13 is completely the same as Figure 6.14. However, the stress distributions are not completely the same. Describe the difference between the stresses in the half plate $(x > 0)$ of Figure 6.13 and those of Figure 6.14, and explain the reason why the difference in conditions along the y axis does not cause a big difference in the stress concentrations between these two plates.

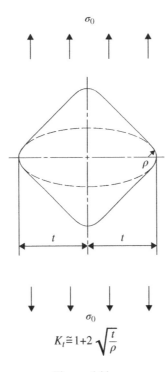

$$K_t \cong 1 + 2\sqrt{\frac{t}{\rho}}$$

Figure 6.11

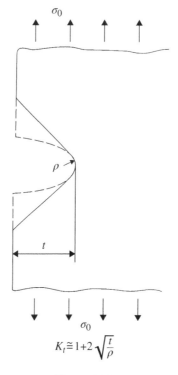

$$K_t \cong 1+2\sqrt{\frac{t}{\rho}}$$

Figure 6.12

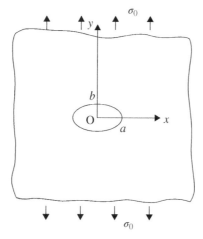

Figure 6.13

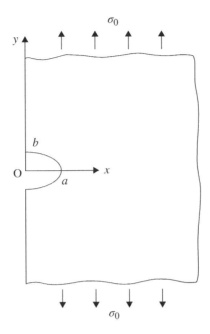

Figure 6.14

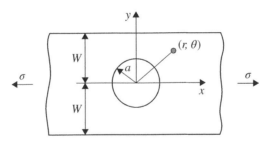

Figure 6.15

Problem 6.6.2   A wide plate containing an elliptical hole expressed by $x^2/a^2 + y^2/b^2 = 1$ $(a > b)$ is subjected to remote stress $\sigma_x = \sigma_0$, $\sigma_y = 2\sigma_0$, $\tau_{xy} = 0$. Determine the maximum stress at the edge of the elliptical hole.

6.7   Stress Concentration at a Hole in a Finite Width Plate

Figure 6.15 shows a finite width plate with width $2W$ containing a circular hole with radius a at the center of the plate which is subjected to remote tensile stress σ. This problem was solved for the first time by R. C. J. Howland (1930).

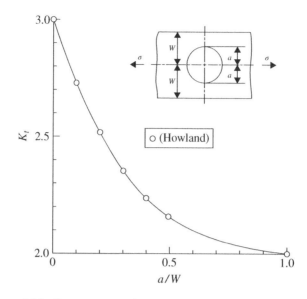

Figure 6.16   Stress concentration factor K_t for a circular hole in a strip

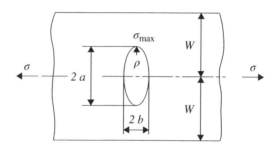

Figure 6.17

Figure 6.16 shows the result of the solution for $a/W \le 0.5$. It must be noted that the stress concentration factor K_t is defined as the relative value to the mean stress at the net section of the plate, $2(W - a)$.

Then, the maximum stress at the edge of the circular hole is defined by

$$\sigma_{max} = K_t \frac{W}{W-a}\sigma \tag{6.55}$$

Figure 6.17 shows a finite width problem with width $2W$ containing an elliptical hole at the center of the plate which is subjected to remote tensile stress σ. This problem was solved for the first time by M. Isida (first in 1955 and later in 1990s).

Figure 6.18 shows the result of the solution for $a/W \le 0.95$.

The definition of the stress concentration factor K_t is the same as the case for a circular hole, that is, K_t is defined by the mean stress at the net section. It is interesting to see that K_t

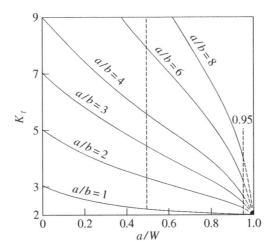

Figure 6.18   Stress concentration factor K_t for an elliptical hole in a strip (M. Isida, 1955, and later in the 1990s)

approaches to 2 ($K_t \rightarrow 2$) as a/W approaches 1.0 ($a/W \rightarrow 1.0$), irrespective of the value of a/b. For long years (and even nowadays) it has been believed that $K_t \rightarrow 1$ as $a/W \rightarrow 1.0$; and some papers have been published incorrectly insisting $K_t \rightarrow 1$.

Howland calculated K_t up to $a/W = 0.5$, beyond which he could not obtain reliable numerical values due to the manual calculating machine used. Isida calculated K_t for an elliptical hole up to $a/W = 0.95$ (Figure 6.18), beyond which he could not obtain reliable values, even with a large-scale computer in the 1990s. Even today, it is difficult to determine the exact K_t for the limiting case of $a/W \rightarrow 1.0$, even by using modern FEM software, if we do not know the exact solution in advance (see Appendix A.4 of Part II).

Although stress concentration factors for many cases are published in various handbooks, it is commonly accepted to define K_t based on the mean stress at the net section. If we misunderstand this rule and regard K_t is defined by remote stress σ, the fatigue design based on the calculation will cause a fatal accident.

Problem 6.7.1   Figure 6.19 shows a hollow cylinder which contains a circular hole of diameter d at the center between the inner wall of radius a and the outer wall of radius b. Estimate the approximate maximum stress at the edge of the circular hole for the case of $a = 140\,$mm, $b = 260\,$mm, $d = 60\,$mm, internal pressure $p_i = 19.6\,$MPa.

6.8   Stress Concentration at a Crack

Since most failure accidents are caused by existing *cracks* or by the extension of newly nucleated cracks, the analysis of stress concentration and stress distribution at cracks is very important for the prevention of fracture accidents.

A most commonly accepted mechanics model of crack is the extremely slender ellipse in which the shorter radius of the ellipse is reduced to zero. Namely, an ellipse becomes a crack

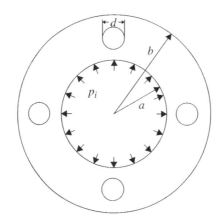

Figure 6.19

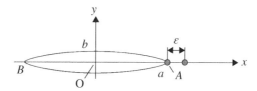

Figure 6.20

by putting $b \rightarrow 0$, as shown in Figure 6.20. This model of crack is called a *Griffith crack*. Griffith (1920, 1922) studied the fracture problem of glass for the first time from the viewpoint of the energy release rate calculated by the stress distribution around an extremely slender elliptical hole based on C. E. Inglis' solution (1913) for an elliptical hole. Since it is easily understood that the stress in the vicinity of crack tip is extremely high compared to the remote stress, we pay attention to the stress at a point small distance ε away from the crack tip A.

By putting $b \rightarrow 0$ in Equation 6.50, the stress at $(a+\varepsilon, 0)$ is expressed as follows:

$$\sigma_y = \frac{\xi^2 + 1}{\xi^2 - 1}\sigma_0 = \frac{\sigma_0 x}{\sqrt{x^2 - a^2}} = \frac{\sigma_0(a+\varepsilon)}{\sqrt{2a\varepsilon + \varepsilon^2}} \tag{6.56}$$

Considering $\varepsilon/a \ll 1$ in the very near vicinity of crack,

$$\sigma_y \cong \frac{\sigma_0\sqrt{a}}{\sqrt{2\varepsilon}} \tag{6.57}$$

Now, we can understand that the stress in the vicinity of crack increases in proportion to *the inverse of $\sqrt{\varepsilon}$*. The common expression of this nature of stress singularity at the crack tip is that "the crack tip stress has the *singularity* of $1/\sqrt{\varepsilon}$ or $1/\sqrt{r}$ or $r^{-0.5}$." According to Equation 6.57, if two cracks having different sizes have the same value of $\sigma_0\sqrt{a}$, the stress distributions in the vicinity of the tips of these two cracks become identical. However, Equation 6.57 means that, according to the linear theory of elasticity, putting $\varepsilon \to 0$ results $\sigma_y \to \infty$. Of course, this cannot occur. Increasing σ_y at the crack tip, before a fracture starts, the material at the crack tip will deviate from an elastic condition and in most cases plastic deformation will occur. Namely, when we compare two cracks, A and B, having different lengths, as shown in Figure 6.21, identical phenomena are expected to occur at the crack tips of these two cracks under the condition of $\sigma_{0A}\sqrt{a_A} = \sigma_{0B}\sqrt{a_B}$, even if we do not know the real phenomena at the crack tips. However, this expectation will be justified only when the plastic zone size at a crack tip is sufficiently small compared with the crack length, in other words only when the *condition of small scale yielding* is satisfied. From these considerations, the quantity $\sigma_0\sqrt{a}$ in Equation 6.57 is very important when we discuss the fracture condition.

Focusing on the singular stress distribution at a crack tip, a systematic approach to fracture problems has been developed by G. R. Irwin et al. since 1957, giving birth to a new field, named *Fracture Mechanics*.

In general, the displacement of a crack can be classified to three patterns as shown in Figure 6.22a–c.

These three crack displacement patterns or modes are called *Mode I, Mode II* and *Mode III*, respectively. The stress and displacement in the vicinity of a crack tip for all these modes are expressed in terms of the coordinate system in Figure 6.23 as follows.

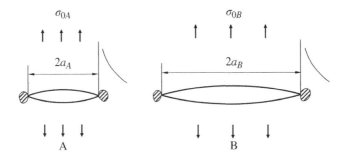

Figure 6.21 Crack tips of two cracks (A, B) under the condition $\sigma_{0A}\sqrt{a_A} = \sigma_{0B}\sqrt{a_B}$

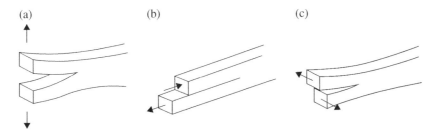

(a)   (b)   (c)

Figure 6.22 The displacement of a crack can be classified to three modes: (a) opening, (b) in plane shear, (c) out of plane shear

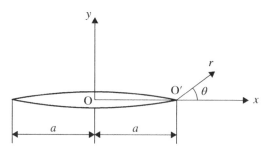

Figure 6.23

Mode I (opening mode)

$$
\left.
\begin{aligned}
\sigma_x &= \frac{K_{\mathrm{I}}}{\sqrt{2\pi r}}\cos\frac{\theta}{2}\left(1-\sin\frac{\theta}{2}\cdot\sin\frac{3\theta}{2}\right) \\[2mm]
\sigma_y &= \frac{K_{\mathrm{I}}}{\sqrt{2\pi r}}\cos\frac{\theta}{2}\left(1+\sin\frac{\theta}{2}\cdot\sin\frac{3\theta}{2}\right) \\[2mm]
\tau_{xy} &= \frac{K_{\mathrm{I}}}{\sqrt{2\pi r}}\cos\frac{\theta}{2}\cdot\sin\frac{\theta}{2}\cdot\cos\frac{3\theta}{2} \\[2mm]
u &= \frac{K_{\mathrm{I}}}{2G}\sqrt{\frac{r}{2\pi}}\cos\frac{\theta}{2}\left(k-1+2\sin^2\frac{\theta}{2}\right) \\[2mm]
v &= \frac{K_{\mathrm{I}}}{2G}\sqrt{\frac{r}{2\pi}}\sin\frac{\theta}{2}\left(k+1-2\cos^2\frac{\theta}{2}\right)
\end{aligned}
\right\}
\tag{6.58}
$$

Mode II (in plane shear mode)

$$
\left.
\begin{aligned}
\sigma_x &= -\frac{K_{\mathrm{II}}}{\sqrt{2\pi r}}\sin\frac{\theta}{2}\left(2+\cos\frac{\theta}{2}\cdot\cos\frac{3\theta}{2}\right) \\[2mm]
\sigma_y &= \frac{K_{\mathrm{II}}}{\sqrt{2\pi r}}\sin\frac{\theta}{2}\cdot\cos\frac{\theta}{2}\cdot\cos\frac{3\theta}{2} \\[2mm]
\tau_{xy} &= \frac{K_{\mathrm{II}}}{\sqrt{2\pi r}}\cos\frac{\theta}{2}\left(1-\sin\frac{\theta}{2}\cdot\sin\frac{3\theta}{2}\right) \\[2mm]
u &= \frac{K_{\mathrm{II}}}{2G}\sqrt{\frac{r}{2\pi}}\sin\frac{\theta}{2}\left(k+1+2\cos^2\frac{\theta}{2}\right) \\[2mm]
v &= -\frac{K_{\mathrm{II}}}{2G}\sqrt{\frac{r}{2\pi}}\cos\frac{\theta}{2}\left(k-1-2\sin^2\frac{\theta}{2}\right)
\end{aligned}
\right\}
\tag{6.59}
$$

Mode III (out of plane shear mode)

$$
\left.
\begin{aligned}
\tau_{xz} &= -\frac{K_{III}}{\sqrt{2\pi r}}\sin\frac{\theta}{2}, \quad \tau_{yz} = \frac{K_{III}}{\sqrt{2\pi r}}\cos\frac{\theta}{2} \\[2mm]
w &= \frac{K_{III}}{G}\sqrt{\frac{r}{2\pi}}\sin\frac{\theta}{2}
\end{aligned}
\right\}
\tag{6.60}
$$

where G is shear modulus, ν is Poisson's ratio, $k = (3-\nu)/(1+\nu)$ for plane stress, $k = 3-4\nu$ for plane strain.

K_I, K_{II} and K_{III} in Equations 6.58 to 6.60 are called the *stress intensity factors* for Mode I, Mode II and Mode III, respectively.

Thus, if the remote stress σ_y is $\sigma_y = \sigma_0$, we express the Mode I stress intensity factor by

$$
K_I = \sigma_0\sqrt{\pi a}
\tag{6.61}
$$

Similarly for Mode II, if the remote shear stress τ_{xy} is $\tau_{xy} = \tau_0$,

$$
K_{II} = \tau_0\sqrt{\pi a}
\tag{6.62}
$$

And also for Mode III, if the remote shear stress τ_{yz} is $\tau_{yz} = \tau_0$,

$$
K_{III} = \tau_0\sqrt{\pi a}
\tag{6.63}
$$

The expression of Equation 6.61 can be understood by comparing Equations 6.57 and 5.58. Although the importance of the quantity of the numerator, $\sigma_0\sqrt{a}$ (Equation 6.57), in the fracture problem was already described, in the current fracture mechanics it is common to add a constant, $\sqrt{\pi}$, to this numerator and call it the stress intensity factor.

The stress intensity factor K expressed in Equations 6.61 to 6.63 present the simplest case. However, in general, the stress intensity factor K varies depending on the shapes of structures and boundary conditions. For such cases, it is common to express K as the following form:

$$
K_I = F\sigma\sqrt{\pi a}
\tag{6.64}
$$

where F is the function to be determined considering shapes of structures and boundary conditions. The values or equations of the function F for various cases are collected in the handbooks of stress intensity factors [3–5].

When Mode I and Mode II coexist (Mixed mode), it is convenient for practical use to express the stress distribution in the vicinity of the crack tip in terms of a polar coordinate, (r, θ), using Equations 6.58 and 6.59, together with the stress transformation equation, Equation 1.12. The equation is given by Equation 6.65.

$$\left. \begin{aligned}
\sigma_r &= \frac{K_I}{\sqrt{2\pi r}}\left(\frac{5}{4}\cos\frac{\theta}{2}-\frac{1}{4}\cos\frac{3\theta}{2}\right)+\frac{K_{II}}{\sqrt{2\pi r}}\left(-\frac{5}{4}\sin\frac{\theta}{2}+\frac{3}{4}\sin\frac{3\theta}{2}\right) \\
\sigma_\theta &= \frac{K_I}{\sqrt{2\pi r}}\left(\frac{3}{4}\cos\frac{\theta}{2}+\frac{1}{4}\cos\frac{3\theta}{2}\right)+\frac{K_{II}}{\sqrt{2\pi r}}\left(-\frac{3}{4}\sin\frac{\theta}{2}-\frac{3}{4}\sin\frac{3\theta}{2}\right) \\
\tau_{r\theta} &= \frac{K_I}{\sqrt{2\pi r}}\left(\frac{1}{4}\sin\frac{\theta}{2}+\frac{1}{4}\sin\frac{3\theta}{2}\right)+\frac{K_{II}}{\sqrt{2\pi r}}\left(\frac{1}{4}\cos\frac{\theta}{2}+\frac{3}{4}\cos\frac{3\theta}{2}\right)
\end{aligned} \right\} \tag{6.65}$$

When a crack is inclined at angle β to the direction of the remote tensile stress as Figure 6.24, K_I and K_{II} are expressed as follows:

$$K_I = \sigma\sqrt{\pi a}\cdot\cos^2\beta \tag{6.66}$$

$$K_{II} = \sigma\sqrt{\pi a}\cdot\cos\beta\cdot\sin\beta \tag{6.67}$$

In problems of a mixed mode loading with K_I and K_{II}, the angle θ_0 at which the stress σ_θ has the maximum value near the point O' of Figure 6.24 is important for predicting the fracture direction and path of brittle materials [6]. The value of θ_0 can be calculated from Equation 6.65 and must satisfy the following equation:

$$K_I\sin\theta_0 + K_{II}(3\cos\theta_0-1)=0 \tag{6.68}$$

The roots of Equation 6.68 are calculated as follows.

$$\tan\frac{\theta_0}{2} = \frac{1\pm\sqrt{1+8\gamma^2}}{4\gamma}, \quad \gamma=\frac{K_{II}}{K_I} \tag{6.69}$$

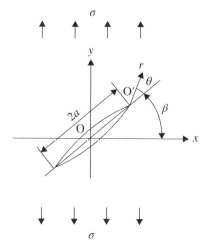

Figure 6.24

The value of these two roots which gives the larger value of σ_θ is the angle θ_0.
Figure 6.25 shows some typical examples of stress intensity factors K_I.

Problem 6.8.1 Estimate the stress intensity factor K_I for the short crack in Figure 6.26 using the stress concentration factor for a circular hole and the K_I for Figure 6.25a. Assume $a << R$.

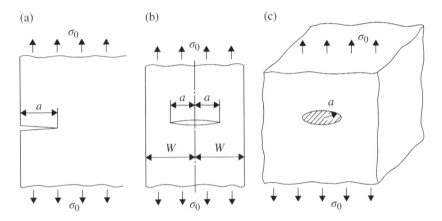

Figure 6.25 Typical examples of stress intensity factor K_I. (a) Edge crack in a semi-infinite plate $(K_I = 1.12\sigma_0\sqrt{\pi a})$. (b) Center crack in a finite width plate $[K_I = F(\lambda)\sigma_0\sqrt{\pi a}, F(\lambda) = \left(1 - 0.025\lambda^2 + 0.06\lambda^4\right)\sqrt{\sec(\pi\lambda/2)}, \lambda = a/W]$. (c) Penny-shaped crack in an infinite body $[K_I = 2/\pi(\sigma_0\sqrt{\pi a})]$

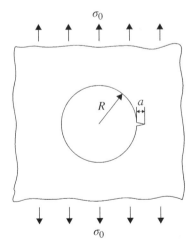

Figure 6.26

6.9   Stress Field due to a Point Force Applied at the Edge of a Semi-Infinite Plate

The solutions for the stress field produced by a *point force* (or *concentrated force*) applied at the edge of a wide plate as shown in Figures 6.27 and 6.28 can be widely used to obtain solutions for various practical problems by application of the superposition method. These problems were solved for the first time by Melan in 1932.

The stress functions for these problems can be determined in the following manner [7].

We assume the thickness of the plate is unity.

As already described, the solution can be obtained by solving the differential equation $\nabla^4\phi = 0$ (Equation 6.18) for the given boundary conditions. The boundary conditions for the problem of Figure 6.27 are written as follows.

1. $\sigma_\theta = 0$, $\tau_{r\theta} = 0$ at $\theta = \pm \pi/2$
2. The resultant forces F_x and F_y which act along a semi-circular contour with radius r surrounding the origin have the following values *irrespective of r*:

$$F_x = 0, \quad F_y = -P$$

Considering the boundary conditions and the form of Equation 6.20, we can understand that ϕ must satisfy the following equations:

$$\frac{\partial\phi}{\partial r} = \text{constant}, \quad \frac{1}{r}\cdot\frac{\partial\phi}{\partial r} = \text{constant} \quad \text{at} \quad \theta = \pm\frac{\pi}{2} \tag{6.70}$$

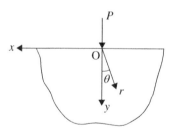

Figure 6.27

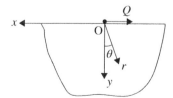

Figure 6.28

In order to satisfy these conditions, we write ϕ in the following form:

$$\phi = r \cdot f(\theta) \tag{6.71}$$

Putting ϕ of Equation (6.71) into the equation $\nabla^4 \phi$ in the polar coordinate, we have

$$\left(\frac{\partial^2}{\partial r^2} + \frac{1}{r} \cdot \frac{\partial}{\partial r} + \frac{1}{r^2} \cdot \frac{\partial^2}{\partial \theta^2} \right) \cdot \left(\frac{\partial^2 \phi}{\partial r^2} + \frac{1}{r} \cdot \frac{\partial \phi}{\partial r} + \frac{1}{r^2} \cdot \frac{\partial^2 \phi}{\partial \theta^2} \right) = \frac{1}{r^3} \left(\frac{d^4 f}{d\theta^4} + 2 \frac{d^2 f}{d\theta^2} + f \right) = 0$$

Namely,

$$\frac{d^4 f}{d\theta^4} + 2 \frac{d^2 f}{d\theta^2} + f = 0 \tag{6.72}$$

The solution can be expressed with unknown constants A, B, C and D as follows:

$$f(\theta) = A \cdot \cos\theta + B \cdot \sin\theta + C\theta \cdot \cos\theta + D\theta \cdot \sin\theta \tag{6.73}$$

Then, from Equations 6.71 and 6.73, we have

$$\phi = r(A \cdot \cos\theta + B \cdot \sin\theta + C\theta \cdot \cos\theta + D\theta \cdot \sin\theta) \tag{6.74}$$

Using this equation with Equation 6.20, we have

$$\left. \begin{array}{l} \sigma_r = \dfrac{1}{r}(-2C\sin\theta + 2D\cos\theta) \\[2mm] \sigma_\theta = 0, \quad \tau_{r\theta} = 0 \end{array} \right\} \tag{6.75}$$

Looking at Figure 6.27, we understand that σ_r must be an even function of θ, and then $C = 0$.

$$\sigma_r = \frac{2D}{r} \cos\theta \tag{6.76}$$

From boundary condition (2):

$$\int_{-\pi/2}^{\pi/2\sigma} \cos\theta \cdot r d\theta = -P \tag{6.77}$$

$$\int_{-\pi/2}^{\pi/2} \sigma_r \sin\theta \cdot r d\theta = 0 \tag{6.78}$$

From Equations 6.76 and 6.77:

$$D = -\frac{P}{\pi} \tag{6.79}$$

On the other hand, Equation 6.78 is satisfied with the form of Equation 6.76. Consequently, the stress function and the equation of the stress distribution for the problem of Figure 6.27 are summarized as follows:

$$\phi = -\frac{P}{\pi} r\theta \cdot \sin\theta \tag{6.80}$$

$$\sigma_r = -\frac{2P}{\pi} \cdot \frac{\cos\theta}{r}, \quad \sigma_\theta = 0, \quad \tau_{r\theta} = 0 \tag{6.81}$$

This stress distribution is very specific, because stresses other than σ_r are all 0. This specific stress distribution is called the *simple radial distribution*.

The problem of Figure 6.28 can also be obtained by the similar method.

The solution is:

$$\phi = \frac{Q}{\pi} r\theta \cdot \cos\theta \tag{6.82}$$

$$\sigma_r = -\frac{2Q}{\pi} \cdot \frac{\sin\theta}{r}, \quad \sigma_\theta = 0, \quad \tau_{r\theta} = 0 \tag{6.83}$$

This stress distribution is also the simple radial distribution. The superposition of Figures 6.27 and 6.28 gives also the simple radial distribution.

Problem 6.9.1   Although the solution for the problem of Figure 6.29 in which an inclined point force is applied at the edge of semi-infinite plate can be obtained by the superposing the solutions for Figures 6.27 and 6.28, verify that the same solution can be obtained by putting boundary condition (2) as $F_x = -R \cdot \sin\alpha$ and $F_y = -R \cdot \cos\alpha$ in the procedure of solving the problem of Figure 6.29.

Example problem 6.6
Determine the stress in a long wedge having the tip angle α (see Figure 6.30) which is subjected to a point force F_0 in the direction of the center line of the wedge.

Solution 1

The boundary conditions are written as follows.

1. $\sigma_\theta = 0, \quad \tau_{r\theta} = 0$ at $\theta = \pm\alpha/2$

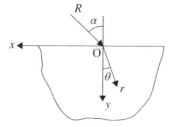

Figure 6.29

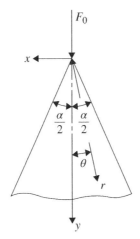

Figure 6.30

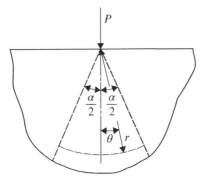

Figure 6.31

2. The resultant forces F_x and F_y which act along a partially circular contour with radius r surrounding the origin have the following values irrespective of r:

$$F_x = 0, \quad F_y = F_0$$

If we imagine a wedge with the tip angle α inside a semi-infinite plate, as illustrated with dotted lines in Figure 6.31, we can see the boundary conditions along the dotted lines are the same as the boundary conditions of Figure 6.30. Considering that the stress distribution of σ_r inside the angle $-\alpha/2 \le \theta \le \alpha/2$ is given by Equation 6.81, we determine the value of the force P so that the resultant force produced by σ_r inside the angle be equal to $-F_0$.

$$\int_{-\alpha/2}^{\alpha/2} \sigma_r \cos\theta \cdot r d\theta = -F_0 \tag{6.84}$$

Then,

$$P = \frac{\pi}{(\alpha + \sin\alpha)} F_0 \tag{6.85}$$

Thus, the stress inside the wedge of Figure 6.30 is given by

$$\sigma_r = -\frac{2F_0}{(\alpha + \sin\alpha)} \cdot \frac{\cos\theta}{r}, \quad \sigma_\theta = 0, \quad \tau_{r\theta} = 0 \tag{6.86}$$

Solution 2   Use the method of stress function as applied to Figure 6.27.

Example problem 6.7
Determine the stress distribution σ_x, σ_y and τ_{xy} in the x–y coordinate for the problem of Figure 6.27.

Solution

$$\sigma_x = \frac{2P}{\pi} \cdot \frac{\cos\theta \cdot \sin^2\theta}{r}, \quad \sigma_y = -\frac{2P}{\pi} \cdot \frac{\cos^3\theta}{r}, \quad \tau_{xy} = \frac{2P}{\pi} \cdot \frac{\cos^2\theta \cdot \sin\theta}{r} \tag{6.87}$$

or,

$$\sigma_x = -\frac{2P}{\pi} \cdot \frac{x^2 y}{r^4}, \quad \sigma_y = -\frac{2P}{\pi} \cdot \frac{y^3}{r^4}, \quad \tau_{xy} = -\frac{2P}{\pi} \cdot \frac{xy^2}{r^4} \tag{6.88}$$

Problem 6.9.2   Figure 6.32a shows a *concentrated moment* applied at the edge of a semi-infinite plate. Determine the solution as the limiting case of Figure 6.32b ($s \to 0$, $Ps = M$).

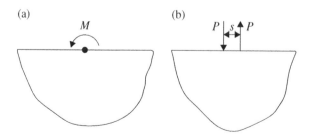

(a)   (b)

Figure 6.32

Problem 6.9.3 The solution for the case in which a *distributed load* is applied at the edge of a semi-infinite plate can be obtained by replacing P of Figure 6.27 by $q(\xi)d\xi$ and applying the integral (see Figure 6.33). Using this idea, derive the stress distribution equation due to a simple distributed load.

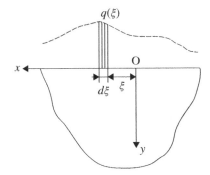

Figure 6.33

6.10   Circular Disk Subjected to Concentrated Force

The solution for the problem of a circular disk under two equal concentrated forces P per unit thickness applied at the opposite ends of the diameter (Figure 6.34a) is useful for the strength analysis of cylindrical concrete, rolls of steel making mills, roller bearings and so on.

The solution for Figure 6.34a can be solved by a surprising idea of superposing parts b–d of Figure 6.34 [8]. The problems in Figure 6.34b and c are the same as the problem of Figure 6.27. The normal stresses in the direction of r_1 and r_2 at the imaginary boundary with radius d are expressed by:

$$\sigma_{r1} = -\frac{2P}{\pi} \cdot \frac{\cos\theta_1}{r_1} \tag{6.89}$$

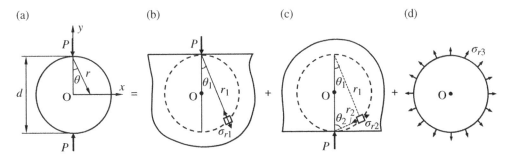

Figure 6.34

$$\sigma_{r2} = -\frac{2P}{\pi} \cdot \frac{\cos\theta_2}{r_2} \tag{6.90}$$

Considering $\theta_1 + \theta_2 = \pi/2$, $r_1 = d\cos\theta_1$ and $r_2 = d\cos\theta_2$, we obtain

$$\sigma_{r1} = -\frac{2P}{\pi d}, \quad \sigma_{r2} = -\frac{2P}{\pi d} \tag{6.91}$$

Since $\sigma_{r1} = \sigma_{r2}$ and no shear stress exists in the small element at (r_1, θ_1) and (r_2, θ_2), superposing Figure 6.34b and c produces a compressive stress of $-2P/\pi d$ in the radial direction of the circular disk acting along the outer periphery of the circular disk except for the point of the concentrated forces. To vanish the compressive stress, we superpose the tensile stress $\sigma_{r3} = 2P/\pi d$ along the boundary of the disk of Figure 6.34d.

Consequently, as the result of the superposition of Figure 6.34b–d, we obtain the stress σ_x along the y axis ($x = 0$) as follows:

$$\sigma_x = \frac{2P}{\pi d} \tag{6.92}$$

And the stress σ_y along the x axis is given as follows.

$$\sigma_y = -\frac{4P}{\pi} \cdot \frac{\cos^3\theta}{r} + \frac{2P}{\pi d} = \frac{2P}{\pi d}\left[1 - \frac{4d^4}{(d^2 + 4x^2)^2}\right] \tag{6.93}$$

Problems of Chapter 6

1. Figure 6.35 shows a combined cylinder which is composed with a solid cylinder A with radius a and a hollow cylinder B with inner radius a and outer radius b. Cylinder A is inserted inside cylinder B without clearance. There exist no initial stresses in A and B.

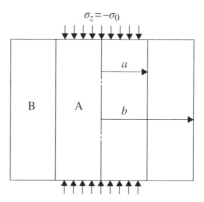

Figure 6.35

When only the end face of cylinder A is subjected to an axial compressive stress $\sigma_z = -\sigma_0$, calculate the circumferential stress σ_θ produced at the inner wall of cylinder B. Assume that there is no frictional stress between A and B. Young's modulus is E and Poisson's ratio is ν.

2. When we use the shrink fit technique to fit a cylinder having the outer radius $\sim d$ into a hollow cylinder having the inner radius d and outer radius D, determine the stresses σ_r and σ_θ at the inner wall of the hollow cylinder for the shrink fit size Δ [= the diameter of the inner cylinder $(\sim d)$ – the inner diameter of the hollow cylinder (d)].

    Young's modulus is E and Poisson's ratio is ν. Consider that $\Delta/d < < 1$.

3. Determine the approximate stress at point A of the holes shown in Figure 6.36.

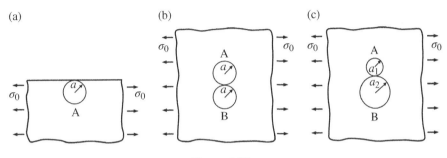

Figure 6.36

4. When (as shown in Figure 6.37) a circular disk contains a small circular hole at the center and is subjected to radial concentrated forces at the opposite ends of the diameter, estimate the stresses at point A (the intersection of the loading axis and the circular hole) and point B (the point 90° moved from A along the circumference of the circular hole). Compare the estimation with the numerical results in Table 6.1. Note that the concentrated force is defined as the quantity per unit thickness.

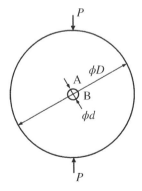

Figure 6.37

Table 6.1

$$K_t = \frac{\sigma_\theta}{2P/\pi D}$$

d/D	K_{tA}	K_{tB}
0.05	6.095	−10.03
0.1	6.385	−10.11
0.2	7.598	−10.53
0.3	9.856	−11.56
0.4	13.68	−13.70
0.5	20.31	−17.80
0.6	32.84	−25.72
0.7	60.45	−42.99
0.8	140.7	−92.07

5. Figure 6.38 shows a wide plate having a slender column on the edge of the plate. The width of the column is $2a$ and the height is h. A force P (per unit thickness) is applied at the tip of the column. When a circular hole exists at (x, y) removed far from the bottom of the column (i.e. $\sqrt{x^2 + y^2} >> 2a$), estimate the maximum stress at the edge of the circular hole.

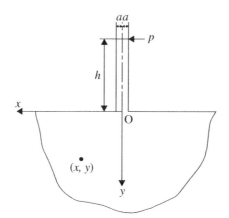

Figure 6.38

6. When $\sigma_x = \sigma_0$, $\sigma_y = \sigma_0$, $\tau_{xy} = \tau_0$ and $\beta = 30°$ in Figure 6.24, determine the stress intensity factors K_I and K_{II}.
7. When a crack exists at an incline angle β in a brittle material, as shown in Figure 6.24, the direction of crack propagation can be predicted by assuming that the crack propagates along the plane where tangential tensile normal stress σ_θ becomes maximum.
    Estimate the direction θ_0 when $\beta = \pi/4$. If necessary, use the values of Table 6.2.

Table 6.2

α	$\sin^{-1}\alpha$	$\cos^{-1}\alpha$	$\tan^{-1}\alpha$
0.1	0.100	1.47	0.100
0.2	0.201	1.37	0.197
0.3	0.305	1.27	0.291
0.4	0.412	1.16	0.381
0.5	0.524	1.05	0.464
0.6	0.644	0.927	0.540
0.7	0.775	0.795	0.610
0.8	0.927	0.643	0.675
0.9	1.12	0.451	0.733

8. When pipe A under internal pressure p is reinforced by pipe B, as shown in Figure 6.39, determine the circumferential tensile stress σ_θ at the inner wall ($r=a$) of pipe A and the circumferential tensile stress σ_θ at the inner wall ($r=b$) of pipe B, where pipe A is fitted to the inside of pipe B without any initial stress and with no clearance. The Young's modulus and the Poisson's ratio are E_A, ν_A and E_B, ν_B, respectively, for pipes A and B.

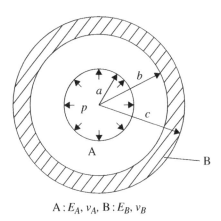

A : E_A, ν_A, B : E_B, ν_B

Figure 6.39

Appendix of Chapter 6

Transformation of coordinate systems in differentiation from x–y coordinate to r–θ coordinate:

$$x = r\cos\theta, \quad y = r\sin\theta$$

$$r^2 = x^2 + y^2, \quad \theta = \arctan\frac{y}{x}$$

$$\frac{\partial r}{\partial x} = \frac{x}{r} = \cos\theta, \quad \frac{\partial r}{\partial y} = \frac{y}{r} = \sin\theta$$

$$\frac{\partial \theta}{\partial x} = -\frac{y}{r^2} = -\frac{\sin\theta}{r}, \quad \frac{\partial \theta}{\partial y} = \frac{x}{r^2} = \frac{\cos\theta}{r}$$

$$\frac{\partial f}{\partial x} = \frac{\partial f}{\partial r} \cdot \frac{\partial r}{\partial x} + \frac{\partial f}{\partial \theta} \cdot \frac{\partial \theta}{\partial x} = \cos\theta \frac{\partial f}{\partial r} - \frac{\sin\theta}{r} \frac{\partial f}{\partial \theta}$$

$$\frac{\partial^2 f}{\partial x^2} = \left(\cos\theta \frac{\partial}{\partial r} - \frac{\sin\theta}{r} \frac{\partial}{\partial \theta}\right)\left(\cos\theta \frac{\partial f}{\partial r} - \frac{\sin\theta}{r} \frac{\partial f}{\partial \theta}\right)$$

$$= \cos^2\theta \frac{\partial^2 f}{\partial r^2} - \cos\theta \sin\theta \frac{\partial}{\partial r}\left(\frac{1}{r} \cdot \frac{\partial f}{\partial \theta}\right)$$

$$-\frac{\sin\theta}{r} \cdot \frac{\partial}{\partial \theta}\left(\cos\theta \frac{\partial f}{\partial r}\right) + \frac{\sin\theta}{r^2} \cdot \frac{\partial}{\partial \theta}\left(\sin\theta \frac{\partial f}{\partial \theta}\right)$$

$$= \cos^2\theta \frac{\partial^2 f}{\partial r^2} - 2\frac{\sin\theta\cos\theta}{r} \cdot \frac{\partial^2 f}{\partial \theta \partial r} + \frac{\sin^2\theta}{r} \cdot \frac{\partial f}{\partial r}$$

$$+ 2\frac{\sin\theta\cos\theta}{r^2} \cdot \frac{\partial f}{\partial \theta} + \frac{\sin^2\theta}{r^2} \cdot \frac{\partial^2 \phi}{\partial \theta^2}$$

$$\frac{\partial^2 f}{\partial y^2} = \sin^2\theta \frac{\partial^2 f}{\partial r^2} + 2\frac{\sin\theta\cos\theta}{r} \cdot \frac{\partial^2 f}{\partial \theta \partial r} + \frac{\cos^2\theta}{r} \cdot \frac{\partial \phi}{\partial r}$$

$$- 2\frac{\sin\theta\cos\theta}{r^2} \cdot \frac{\partial f}{\partial \theta} + \frac{\cos^2\theta}{r^2} \cdot \frac{\partial^2 f}{\partial \theta^2}$$

$$\left(\frac{\partial^2}{\partial x^2} + \frac{\partial^2}{\partial y^2}\right)f = \left(\frac{\partial}{\partial r^2} + \frac{1}{r} \cdot \frac{\partial}{\partial r} + \frac{1}{r^2} \cdot \frac{\partial^2}{\partial \theta^2}\right)f$$

$$\frac{\partial^2 f}{\partial x^2} + \frac{\partial^2 f}{\partial y^2} = \frac{\partial^2 f}{\partial r^2} + \frac{1}{r} \cdot \frac{\partial f}{\partial r} + \frac{1}{r^2} \cdot \frac{\partial^2 f}{\partial \theta^2}$$

Thus,

$$\nabla^4 \phi = \left(\frac{\partial^2}{\partial x^2} + \frac{\partial^2}{\partial y^2}\right)\left(\frac{\partial^2 \phi}{\partial x^2} + \frac{\partial^2 \phi}{\partial y^2}\right)$$

$$= \left(\frac{\partial^2}{\partial r^2} + \frac{1}{r} \cdot \frac{\partial}{\partial r} + \frac{1}{r^2} \cdot \frac{\partial^2}{\partial \theta^2}\right)\left(\frac{\partial^2 \phi}{\partial r^2} + \frac{\partial \phi}{r \partial r} + \frac{1}{r^2} \cdot \frac{\partial^2 \phi}{\partial \theta^2}\right)$$

References

[1] F. Hirano (1950) *Trans. Japan Soc. Mech. Eng.* **16**:52.
[2] F. Hirano (1951) *Trans. Japan Soc. Mech. Eng.* **17**:12.

[3] Y. Murakami (ed.) (1987) *Stress Intensity Factors Handbook*, vols 1, 2, Pergamon, Oxford.

[4] Y. Murakami (ed.) (1992) *Stress Intensity Factors Handbook*, vols 3, Soc. Materials Sci., Japan, Kyoto and Pergamon Press, Oxford.

[5] Y. Murakami (ed.) (2001) *Stress Intensity Factors Handbook*, vol 4, 5, Soc. Materials Sci., Japan, Kyoto and Elsevier, Oxford.

[6] F. Erdogan and G.C. Sih (1963) *Trans. ASME Ser. D*, **85**:519.

[7] L. D. Landau and E. M. Lifshitz (1970) *Theory of Elasticity (A Course of Theoretical Physics*, vol **7**), Pergamon, New York.

[8] S. P. Timoshenko and J. N. Goodier (1982) *Theory of Elasticity*, 3rd edn, McGraw Hill, New York, p. 122.

7

Torsion of a Bar with Uniform Section

Torsion problems of bars and columns with uniform section are encountered in many machine components such as power transmission shafts and many other structures. Torsion problems of thin walled sections are important in aircraft and ships, though the torsional rigidity of these structures is very different in *closed section* and *open section*. Although the solution method is a little bit different from the plane problems explained in Chapter 6, the three conditions of equilibrium condition, compatibility condition and boundary condition must be satisfied as the basis for an exact solution, as well as in 2D problems.

7.1 Torsion of Cylindrical Bars

Although this problem is basic and is usually treated in the so-called *strength of materials*, the solution will be explained in the following for reference with other problems having noncircular sections.

Since $\tau_{\theta z} = \tau_{z\theta}$ in Figure 7.1, $\tau_{z\theta}$ will be used in the following equations as the representative shear stress. When a cyclindrical bar having diameter d is subjected to twisting moment T, the twisting moment dT supported by an annular region having the ring width dr at radius r can be written as

$$dT = 2\pi r^2 \tau_{z\theta} dr \tag{7.1}$$

Denoting $\tau_{z\theta}$ at $r = d/2$ by τ_{max},

$$\tau_{z\theta} = 2r/d \, \tau_{max} \tag{7.2}$$

Theory of Elasticity and Stress Concentration, First Edition. Yukitaka Murakami.
© 2017 John Wiley & Sons, Ltd. Published 2017 by John Wiley & Sons, Ltd.

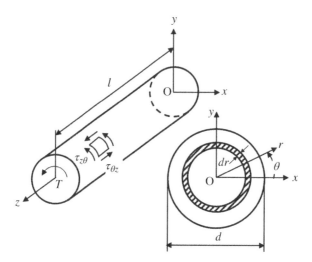

Figure 7.1

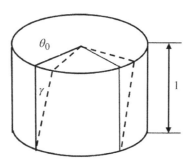

Figure 7.2

Thus,

$$T = \int dT = \int_0^{d/2} \frac{4\pi}{d} r^3 \tau_{max} dr = \frac{\pi d^3}{16} \tau_{max} \tag{7.3}$$

$$\tau_{max} = 16T/\pi d^3 = T/Z_p \quad (Z_p = d^3/16) \tag{7.4}$$

Denoting the *twisting angle per unit length* by θ_0, the twisting angle φ can be expressed with reference to Figure 7.2, which represents the twisting deformation for unit length.

$$\theta_0 = \frac{\gamma}{d/2} = \frac{\tau_{max}}{G(d/2)} = \frac{T}{GZ_p(d/2)} = \frac{T}{GI_p} \tag{7.5}$$

$$I_p = \frac{\pi d^4}{32} : \text{Polar moment of inertia} \tag{7.6}$$

$$\varphi = l\theta_0 = \frac{Tl}{GI_P}$$ (7.7)

Equations 7.4 and Equation 7.7 can be applied only to circular sections.

If section shapes are not circular, the polar moment of inertia and section modulus must not be used to calculate shear stress and twisting angle.

7.2   Torsion of Bars Having Thin Closed Section

Shear stress and twisting angle for torsion of bars having a *thin closed section* under twisting moment T as in Figure 7.3a can be obtained as follows.

Let us imagine a closed pipe like Figure 7.3a and cut it by a plane which includes straight lines across the thickness at A and B. Imagine the part cut by the plane having a unit length like Figure 7.3b. The cut plane should be subjected to shear stress in the direction indicated by arrows in Figure 7.3b. The equilibrium condition of the cut part in the longitudinal direction (z direction) can be written only with the shear stress in the longitudinal direction at A and B as follows.

$$\tau_A h_A 1 = \tau_B h_B 1$$ (7.8)

where h_A and h_B are the thicknesses at A and B, respectively. Since A and B are not particular points, Equation 7.8 means that the following equation holds everywhere along the circumference of the pipe:

$$\tau h = \text{constant}$$ (7.9)

The relationship of Equation 7.9 is called the *condition of constant shear flow*.

In order to obtain the relationship among twisting moment, shear stress and shape of section, let us consider the torsional moment dT with respect to the z axis passing a point O produced by the shear stress τ acting at small part of the section ds along the tangential direction, as shown in Figure 7.4.

(a)   (b)

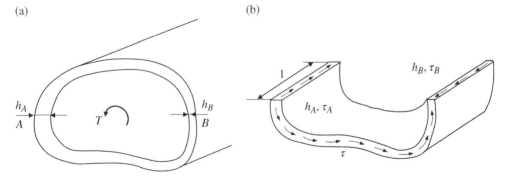

Figure 7.3

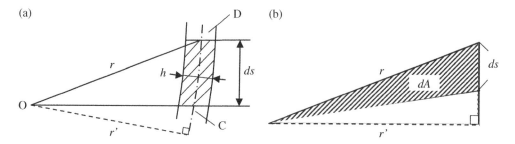

Figure 7.4

$$dT = \tau h ds \cdot r' \tag{7.10}$$

Considering Equation 7.9,

$$T = \oint \tau h r' ds = \tau h \oint r' ds \tag{7.11}$$

Since $r'ds$ is equal to twice the area of $\triangle OCD$, $\oint r'ds$ expresses twice the total internal area A covered by the center line at the thin section. Namely, we have

$$T = 2\tau h A \tag{7.12}$$

The relationship between T and the twisting angle per unit length, θ_0, can be obtained by relating work done by T equal to the strain energy[1]. Namely, paying attention to the quantity per unit length,

$$\oint \frac{\tau^2}{2G} h ds = \frac{1}{2} T \theta_0 \tag{7.13}$$

From Equations 7.12 and 7.13,

$$\theta_0 = \frac{T}{4A^2 G} \oint \frac{ds}{h} \tag{7.14}$$

If the thickness of a pipe is constant (h_0) along all periphery, the following relationship is derived by denoting the total periphery length by s_0.

$$\theta_0 = \frac{T s_0}{4A^2 G h_0} \tag{7.15}$$

---

[1] The concept of strain energy is explained in Chapter 8. Here, the basic knowledge of strength of materials is sufficient for understanding this equation.

7.3 Saint Venant's Torsion Problems

Problems in which bars with a uniform cross section are subjected to twisting moment at the both ends are called Saint Venant's torsion problems. The solution of the problems is based on the following three assumptions.

1. If the twisting angle φ is defined by the values from the original coordinate $z = 0$, the twisting angles at $z = z_1$ and $z = z_2$ follow the following equation:

$$\varphi_2/\varphi_1 = z_2/z_1 \qquad (7.16)$$

2. The shape of the deformed cross section is the same along the length of the bar irrespective of z.
3. If the twisting moment is applied in statically equivalent way, the effect can be equivalent at the section apart from the edge in the distance of the diameter or the representative dimension of the bar, regardless of ways of application of twisting moment (see Chapter 5 of Part I).

Assumption 3 assumes that not only shear stress but also axial normal stresses may occur at the cross section near the edge of the bar where twisting moment is applied. However, since the resultant forces calculated by integrating these normal stresses over the cross section vanish, the disturbance is not transferred to the cross section far from the end of the bar. Therefore, assumptions 1 and 2 are also rational.

In order to solve torsional problems, Saint Venant started the analysis by assuming the deformation of a cross section based on assumptions 1–3. The solution method is named *Saint Venant's semi-inverse method*. The method will be introduced in the following.

Let us define the coordinate system as Figure 7.5 where the z axis is in the direction of the axis of a bar and the origin O is taken at the stationary point on the cross section coordinate x–y. Denoting the displacements in the x, y and z directions by u, v and w, the assumption of Saint Venant is expressed by

$$\left. \begin{array}{l} u = -\theta_0 z y \\ v = \theta_0 z x \\ w = \theta_0 \psi(x, y) \end{array} \right\} \qquad (7.17)$$

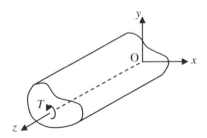

Figure 7.5

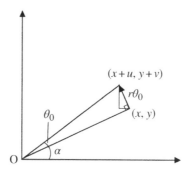

Figure 7.6

where θ_0 is the twisting angle per unit length and $\psi(x, y)$ is a function which expresses the warp of the cross section plane, being $\psi(x,y) = 0$ for the case of a circular cross section. Although Navier assumed that shear stresses should be proportional to the distance from the origin O, it resulted in failure.

Saint Venant's assumption is expected to be reasonable, though it must be examined by judging if it satisfies the three basic conditions of theory of elasticity, that is equilibrium condition, compatibility condition and boundary condition (Figure 7.6).

Strains and stresses corresponding to the displacement of Equation 7.17 are expressed as follows.

$$
\left.
\begin{aligned}
\varepsilon_x &= \varepsilon_y = \varepsilon_z = \gamma_{xy} = 0 \\
\gamma_{zx} &= \frac{\partial w}{\partial x} + \frac{\partial u}{\partial z} = \theta_0 \left(\frac{\partial \psi}{\partial x} - y \right) \\
\gamma_{yz} &= \frac{\partial w}{\partial y} + \frac{\partial v}{\partial z} = \theta_0 \left(\frac{\partial \psi}{\partial y} + x \right)
\end{aligned}
\right\}
\tag{7.18}
$$

$$
\left.
\begin{aligned}
\sigma_x &= \sigma_y = \sigma_z = \tau_{xy} = 0 \\
\tau_{zx} &= G\theta_0 \left(\frac{\partial \psi}{\partial x} - y \right) \\
\tau_{yz} &= G\theta_0 \left(\frac{\partial \psi}{\partial y} + x \right)
\end{aligned}
\right\}
\tag{7.19}
$$

Since the strains expressed in the form of Equation 7.18 satisfy the compatibility equation (Equation 2.24), the remaining task is reduced to how to determine the function $\psi(x, y)$ which satisfies the equilibrium condition (Equation 4.4) and the boundary condition.

7.4 Stress Function in Torsion

Exact solutions in elasticity must satisfy the equilibrium condition, the compatibility condition and the boundary condition. These conditions will be examined in order.

7.4.1 Equilibrium Condition

The equilibrium conditions in the x and y directions are naturally satisfied.
    The equilibrium condition in the z direction is expressed as

$$\frac{\partial \tau_{zx}}{\partial x} + \frac{\partial \tau_{yz}}{\partial y} + \frac{\partial \sigma_z}{\partial z} + Z = 0 \tag{7.20}$$

Considering $\sigma_z = 0$ and $Z = 0$, Equation 7.20 is reduced to

$$\frac{\partial \tau_{zx}}{\partial x} + \frac{\partial \tau_{yz}}{\partial y} = 0 \tag{7.21}$$

The final goal is to find two functions: $\tau_{zx} = f(x, y)$ and $\tau_{yz} = g(x, y)$. Since we have one condition of the equilibrium equation (Equation 7.21), we can express τ_{zx} and τ_{yz} by one function. Therefore, without using $f(x, y)$, we use one function $\phi(x, y)$ by which we put

$$\tau_{zx} = \frac{\partial \phi}{\partial y} \quad \text{and} \quad \tau_{yz} = -\frac{\partial \phi}{\partial x} \tag{7.22}$$

Here, we call ϕ the *stress function in torsion*.

7.4.2 Compatibility Equation

From Equations 2.24 and 7.18, we have

$$\frac{\partial}{\partial x} \left(-\frac{\partial \gamma_{yz}}{\partial x} + \frac{\partial \gamma_{zx}}{\partial y} \right) = 0 \tag{7.23}$$

$$\frac{\partial}{\partial y} \left(\frac{\partial \gamma_{yz}}{\partial x} - \frac{\partial \gamma_{zx}}{\partial y} \right) = 0 \tag{7.24}$$

$$\frac{\partial}{\partial z} \left(\frac{\partial \gamma_{yz}}{\partial x} + \frac{\partial \gamma_{zx}}{\partial y} \right) = 0 \tag{7.25}$$

Other equations of compatibility condition are automatically satisfied. Since γ_{zx} and γ_{yz} are independent from z, Equation 7.25 is naturally satisfied. Thus, from Equations 7.23 and 7.24,

$$-\frac{\partial \gamma_{yz}}{\partial x} + \frac{\partial \gamma_{zx}}{\partial y} = C' \tag{7.26}$$

namely,

$$-\frac{\partial \tau_{yz}}{\partial x} + \frac{\partial \tau_{zx}}{\partial y} = C \tag{7.27}$$

To assure constant values in Equations. 7.26 and 7.27, the strains and stresses must have the forms of Equations 7.18 and 7.19 which are derived by Saint Venant's assumption (Equation 7.19).

Substituting Equation 7.22 into Equation 7.27, we have

$$\frac{\partial^2 \phi}{\partial x^2} + \frac{\partial^2 \phi}{\partial y^2} = C \tag{7.28}$$

By substituting Equation 7.19 into Equation 7.27, we have

$$C = -2G\theta_0 \tag{7.29}$$

Thus,

$$\frac{\partial^2 \phi}{\partial x^2} + \frac{\partial^2 \phi}{\partial y^2} = -2G\theta_0 \tag{7.30}$$

Consequently, the exact solution is the one which satisfies Equation 7.30 and the boundary conditions.

7.4.3   Boundary Conditions

In order to investigate the relationship between the stress function ϕ and the boundary conditions, let us pay attention to a small triangular element which includes the boundary as in Figure 7.7. Since the outer surface of the bar AA′BB′ is a free surface, no stresses are acting. Therefore, the sum of the forces in the z direction acting on $O_1AA′O_2$ and the force acting in the z direction on $O_1O_2B′B$ should be zero. Namely, taking the length of the small element in the z direction unity, we have

$$\tau_{zx}l \cdot 1 + \tau_{yz}m \cdot 1 = 0 \tag{7.31}$$

where l and m are the direction cosines of the normal direction fixed at the boundary with respect to the x and y axes.

Defining s in the counterclockwise direction, l and m are expressed as follows:

$$l = \frac{dy}{ds}, \quad m = -\frac{dx}{ds} \tag{7.32}$$

Therefore, substituting Equation 7.32 into Equation 7.31, we have

$$\tau_{zx}\frac{dy}{ds} - \tau_{yz}\frac{dx}{ds} = 0 \tag{7.33}$$

Now, using the stress function of Equation 7.22, we have

$$\frac{\partial \phi}{\partial y} \cdot \frac{dy}{ds} + \frac{\partial \phi}{\partial x} \cdot \frac{dx}{ds} = \frac{\partial \phi}{\partial s} = 0 \text{ (along a boundary)} \tag{7.34}$$

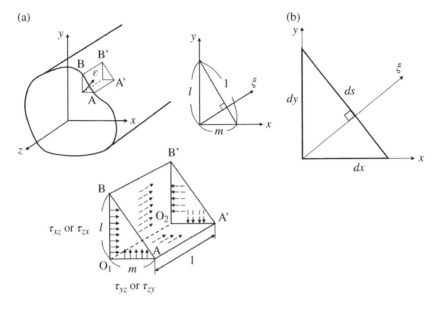

Figure 7.7

Consequently, assuming C to be an arbitrary constant value, we can write

$$\phi = C \quad \text{(along a boundary)} \tag{7.35}$$

Although C is an arbitrary constant, for simplicity we choose $C = 0$ for problems having one boundary.

In order to determine the final form of ϕ, the relationship between ϕ and twisting moment T in addition to Equations 7.30 and 7.35 must be formulated. Considering Figure 7.8 and $\phi = 0$ along boundary, the relationship can be derived as follows, where $y^+{}_B$ and $y^-{}_B$ are the y coordinates of the boundary and $x^+{}_B$ and $x^-{}_B$ are the x coordinates of the boundary.

$$T = \iint \left(-y\tau_{zx}dx \cdot dy + x\tau_{yz}dx \cdot dy \right)$$

$$= \iint \left(-y\tau_{zx} + x\tau_{yz} \right) dx \cdot dy$$

$$= -\iint \left(y\frac{\partial \phi}{\partial y} + x\frac{\partial \phi}{\partial x} \right) dx \cdot dy \tag{7.36}$$

$$= \int \left\{ [y\phi]_{y_B^-}^{y_B^+} - \int \phi dy \right\} dx - \int \left\{ [x\phi]_{x_B^-}^{x_B^+} - \int \phi dx \right\} dy$$

$$= 2\iint \phi dx \cdot dy$$

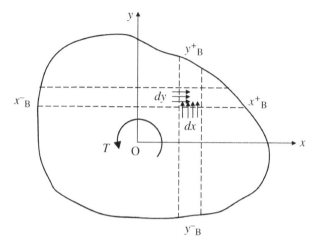

Figure 7.8

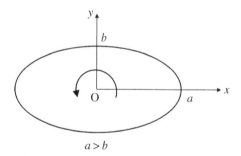

$a > b$

Figure 7.9 Torsion of a bar with an elliptical section

Namely,

$$T = 2\int_A \phi dA, \quad A : \text{Area of the section of a bar} \tag{7.37}$$

Example Torsion of a bar having an elliptical section (Figure 7.9)
The torsion of a bar having an elliptical section expressed by $x^2/a^2 + y^2/b^2 - 1 = 0$ can be solved by assuming ϕ in the following form:

$$\phi = C_0 \left(\frac{x^2}{a^2} + \frac{y^2}{b^2} - 1 \right) \tag{7.38}$$

where it should be noted that $C = 0$ in Equation 7.35 is taken into consideration in assuming Equation 7.38. C_0 is determined by considering Equations 7.30 and 7.37 as follows:

$$C_0 = \frac{a^2 b^2}{2(a^2 + b^2)} \cdot C \tag{7.39}$$

$$C = -2G\theta_0 = -\frac{2T(a^2 + b^2)}{\pi a^3 b^3} \tag{7.40}$$

It is interesting to note where the maximum shear stresses occurs on the elliptical section of Figure 7.9.

7.5 Membrane Analogy: Solution of Torsion Problems by Using the Deformation of Pressurized Membrane

L. Prandtl (1903) found that the differential equation for expressing the deformation of a pressurized soap membrane is analogous to that for the torsion problem expressed by stress function and many torsion problems can be solved by the theory and experiments of soap membrane.

If a soap membrane having the frame shape Γ is pressurized with pressure q, the soap membrane deforms like a hill as Figure 7.10a. Referring to Figure 7.10b, we can write the equilibrium of a small membrane element $dxdy$ as

$$\frac{1}{\rho_x} = -\frac{\partial^2 z}{\partial x^2}, \quad \frac{1}{\rho_y} = -\frac{\partial^2 z}{\partial y^2} \tag{7.41}$$

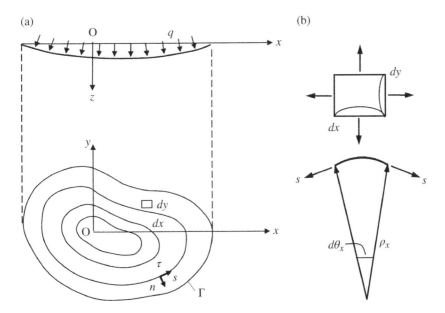

Figure 7.10   Deformation of pressurized soap membrane

where z is the displacement of the membrane.

Considering the following relationships,

$$d\theta_x = \frac{dx}{\rho_x}, \quad d\theta_y = \frac{dy}{\rho_y} \tag{7.42}$$

and denoting the surface tension by S, we have

$$\frac{dx}{\rho_x} Sdy + \frac{dy}{\rho_y} Sdx - qdx \cdot dy = 0 \tag{7.43}$$

Consequently, from Equations 7.41 and 7.43,

$$\frac{\partial^2 z}{\partial x^2} + \frac{\partial^2 z}{\partial y^2} = -\frac{q}{S} \tag{7.44}$$

Equation 7.44 has the same form as Equation 7.30 which the stress function for torsion problems must satisfy with the conditions $dz/dx \ll 1$ and $dz/dy \ll 1$.

From the above analysis, the correspondences between torsion problems and soap membrane problems are summarized in Table 7.1.

From these correspondences the following relationships are derived.

$$\frac{\phi}{2G\theta_0} = \frac{z}{q/S} \tag{7.45}$$

$$\frac{\partial\phi/\partial n}{2G\theta_0} = \frac{\partial z/\partial n}{q/S} \tag{7.46}$$

$$-\frac{\partial z}{\partial n} = \frac{\tau}{2G\theta_0} \cdot \frac{q}{S} \tag{7.47}$$

Table 7.1 Correspondences between torsion problems and soap membrane problems

Torsion problem	Membrane problem
ϕ	z
$\phi = 0$ on boundary	$z = 0$ along on the same frame
$-2G\theta_0$	$-\dfrac{q}{S}$
$T = 2\displaystyle\iint \phi dx \cdot dy$	$2\displaystyle\iint z dx \cdot dy$ (= twofold volume occupied by membrane)
$-\dfrac{\partial\phi}{\partial n}$ (stress in direction of equiheight line)	$-\dfrac{\partial z}{\partial n}$ (gradient of membrane in normal direction to equiheight line)
	$z = $ constant and $-\dfrac{\partial z}{\partial s} = 0$ along equiheight line

Now, considering the following relationship along an equiheight closed loop,

$$-\oint \frac{\partial z}{\partial n} S ds = qA, \quad A = \text{area occupied by equiheight closed loop} \tag{7.48}$$

the following equation can be derived with Equation 7.47:

$$\oint \tau ds = 2G\theta_0 A \tag{7.49}$$

The following section will show various applications of the membrane analogy.

7.6 Torsion of Bars Having a Thin Unclosed Cross Section

Bars having a cross section with a boundary expressed with one curve, such as Figure 7.11a is called a bar with an *unclosed cross (or open) section*. On the other hand, a cross section having two or more closed boundary curves, such as Figure 7.11b is called a *closed section*.

Torsion problems with an unclosed cross section can be successfully solved by the application of *membrane analogy*.

Example 1 Torsion of a plate having a narrow rectangular cross section
In order to apply the membrane analogy to a plate having a narrow rectangular cross section, such as Figure 7.12, we imagine a narrow rectangular frame over which a soap membrane is covered. Then, we imagine that the membrane is pressurized from the other side of membrane with a pressure q.

In this case, the equiheight curves will be like those expressed by Figure 7.12.

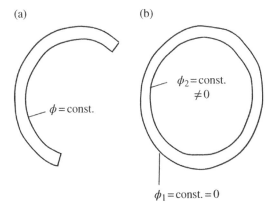

(a)   (b)

$\phi = \text{const.}$

$\phi_2 = \text{const.}$
$\neq 0$

$\phi_1 = \text{const.} = 0$

Figure 7.11 (a) Unclosed section. (b) Closed section

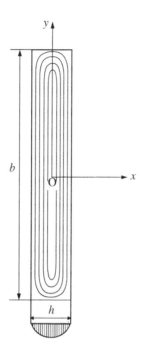

Figure 7.12   Bar with narrow rectangular section

Expressing this state by equation, we have

$$\frac{\partial^2 z}{\partial x^2} + \frac{\partial^2 z}{\partial y^2} = -\frac{q}{S}$$

(7.50)

Now, since d^2z/dy^2 can be regarded as $d^2z/dy^2 \sim 0$ over almost all the cross section, Equation 7.50 can be approximately reduced to

$$\frac{d^2 z}{dx^2} = -\frac{q}{S}$$

(7.51)

By integrating this equation,

$$\frac{dz}{dx} = -\frac{q}{S}x + C_1$$

(7.52)

Considering $dz/dx = 0$ at $x = 0$, $C_1 = 0$.
Therefore,

$$z = -\frac{q}{2S}x^2 + C_2$$

(7.53)

Since $z = 0$ at $x = \pm h/2$, $C_2 = qh^2/8S$. Then,

$$z = -\frac{q}{2S}x^2 + \frac{qh^2}{8S} \qquad (7.54)$$

Thus, at $x = h/2$, we have

$$-\frac{dz}{dx}\Big|_{max} = \pm\frac{qh}{2S} \qquad (7.55)$$

Denoting the volume occupied by the membrane by V,

$$V \cong 2\int_0^{h/2} bzdx = \frac{qbh^3}{12S} \qquad (7.56)$$

From the correspondences between $2G\theta_0$ and q/S,

$$T = 2V = \frac{bh^3}{3}G\theta_0 \qquad (7.57)$$

$$\tau = 2G\theta_0 x \qquad (7.58)$$

$$\tau_{max} = hG\theta_0 \qquad (7.59)$$

$$\theta_0 = \frac{3T}{bh^3G} \qquad (7.60)$$

$$\tau_{max} = \frac{3T}{bh^2} \qquad (7.61)$$

Example 2   Torsion of a thin tube having a longitudinal slit
In the case of a thin tube having a longitudinal slit and with $h \ll D$ like Figure 7.13, the dimension corresponding to b in Example 1 is πD. By this replacement, the solution of Example 1 can be applied automatically.

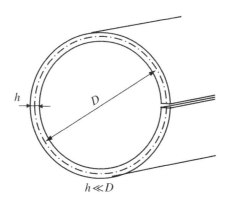

Figure 7.13   Thin tube having a longitudinal slit

Example 3   Torsion of bars having various unclosed cross sections
The solution of Example 1 can be applied also to the bars having cross sections like
Figure 7.14a–c. The dimension corresponding to b in Example 1 is $b = \sim 2a$ for part (a) of
Figure 7.14, $b \sim b_1 + 2b_2$ for part (b) and $b \sim b_1 + 2b_2$ for part (c), respectively.

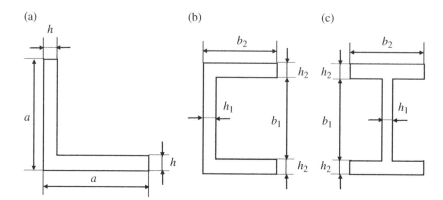

Figure 7.14

Example 4   Torsion of a bar having a rectangular cross section
The torsion problem of a bar having a rectangular cross section [1], as in Figure 7.15, is different from bending problems and must not be solved by a simple application of the polar
moment of inertia which results in a completely wrong solution.

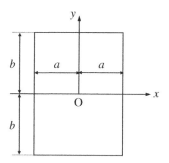

Figure 7.15

The basic equation of membrane analogy for this problem is

$$\frac{\partial^2 z}{\partial x^2} + \frac{\partial^2 z}{\partial y^2} = -\frac{q}{S}$$ (7.62)

Imagining the deformation as that of a soap membrane, we assume the form of z as follows:

$$z = \sum_{n=1,3,5}^{\infty} \beta_n \cos\frac{n\pi x}{2a} Y_n, \quad \beta_n : \text{constant}, \quad Y_n = Y_n(y) \tag{7.63}$$

Applying the Fourier expansion to the right hand of Equation 7.62, $-q/S$, within $-a < x < a$,

$$-\frac{q}{S} = -\sum_{n=1,3,5}^{\infty} \frac{q}{S} \cdot \frac{4}{n\pi} (-1)^{\frac{n-1}{2}} \cos\frac{n\pi x}{2a} \tag{7.64}$$

Substituting Equations 7.63 and 7.64 into Equation 7.62, we have

$$Y_n'' - \frac{n^2\pi^2}{4a^2} Y_n = -\frac{q}{S} \cdot \frac{4}{n\pi\beta_n} (-1)^{(n-1)/2} \tag{7.65}$$

Solving this equation, we have

$$Y_n = A\sinh\frac{n\pi y}{2a} + B\cosh\frac{n\pi y}{2a} + \frac{16qa^2}{Sn^3\pi^3\beta_n}(-1)^{(n-1)/2} \tag{7.66}$$

From the symmetry of the membrane deformation, $A = 0$. B can be determined from the condition $Y_n(\pm b) = 0$ due to $z = 0$ at $y = \pm b$. Consequently,

$$Y_n = \frac{16qa^2}{Sn^3\pi^3\beta_n}(-1)^{(n-1)/2}\left[1 - \frac{\cosh(n\pi y/2a)}{\cosh(n\pi b/2a)}\right] \tag{7.67}$$

z can be obtained by inserting Y_n of Equation 7.67 into Equation 7.63.

Finally, ϕ is determined by the one to one correspondence between q/S and $2G\theta_0$ and also between z and ϕ. Once the form of ϕ is determined, the shear stresses (τ_{zx} or τ_{yz}) and the equation of twisting angle can be determined after some numerical calculations on the equations expanded by series.

Table 7.2 shows the result of the numerical calculations k, k_1 and k_2 for the expressions of $\tau_{max} = k_2 G\theta_0 a$, $T = k_1 G\theta_0(2a)^3(2b)$, $\tau_{max} = T/k_2(2a)^2(2b)$. It must be noted that the maximum shear stress occurs at the midpoint of the longer side of the rectangular section. This nature of shear stress distribution in rectangular section is very often misunderstood.

Table 7.2

b/a	k	k_1	k_2
1.0	0.675	0.1406	0.208
1.2	0.759	0.166	0.219
1.5	0.847	0.196	0.231
2.0	0.930	0.229	0.246
2.5	0.968	0.249	0.258
3	0.985	0.263	0.267
4	0.997	0.281	0.282
5	0.999	0.291	0.291
10	1.0	0.312	0.312
∞	1.0	0.333	0.333

7.7 Comparison of Torsional Rigidity between a Bar with an Open Section and a Bar with a Closed Section

It must be noted that the reason why the torsional rigidity of a bar with a closed section is much larger than that of a bar with an open section is due to the contribution of the large area inside the boundary, as shown in Equations 7.14 and 7.15. A few examples will be shown in the following.

Example 1   A thin rectangular section plate and a thin circular pipe (Figure 7.16)
The twisting angles per unit length, θ_{01} and θ_{02}, for Figure 7.16a and b are

$$\theta_{01} = \frac{3T_1}{Gbh^3} \tag{7.68}$$

$$\theta_{02} = \frac{T_2 s_0}{4A^2 Gh} = \frac{4T_2}{G\pi b^3 h} \tag{7.69}$$

Therefore, if we compare the twisting angles for the case of $T_1 = T_2$, we have

$$\frac{\theta_{01}}{\theta_{02}} = \frac{3\pi b^3 h}{4bh^3} = \frac{3\pi}{4}\left(\frac{b}{h}\right)^2 \gg 1 \tag{7.70}$$

If we compare the twisting moments for the case of $\theta_{01} = \theta_{02}$ (= the same twisting angle), we have

$$\frac{T_1}{T_2} = \frac{4}{3\pi}\left(\frac{h}{b}\right)^2 \tag{7.71}$$

and $T_1/T_2 \sim 4 \times 10^{-3}$ for $h/b = 0.1$. It follows that welding (a) of Figure 7.16 to the diameter section of (b) of Figure 7.16 does not contribute to an increase in the torsional rigidity of the pipe.

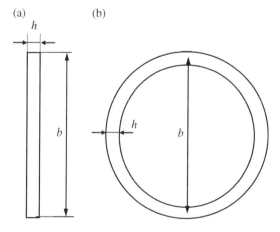

Figure 7.16

Example 2 A thin pipe having a slit and a closed thin pipe
If we compare the twisting moments for Figure 7.17a and b with the identical twisting angle,
we have

$$\frac{T_1}{T_2} = \frac{4}{3}\left(\frac{h}{d}\right)^2 \tag{7.72}$$

This result means a large difference in the twisting rigidity of the two pipes (a) and (b) in
Figure 7.17 as $T_1/T_2 \sim 0.01$ for $h/d = 0.1$ and $T_1/T_2 \sim 4 \times 10^{-4}$ for $h/d = 0.01$.

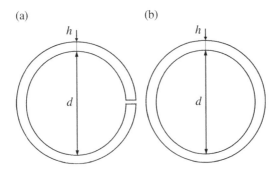

Figure 7.17

Example 3 A thin rectangular pipe with a closed section having a rib (Figure 7.18)
Applying the equations obtained so far to the problem of Figure 7.18, we can obtain the following equations.

$$\tau_1 h_1 = \tau_2 h_2 + \tau_3 h_3 \tag{7.73}$$

$$T = 2A_1\tau_1 h_1 + 2A_2\tau_2 h_2 \tag{7.74}$$

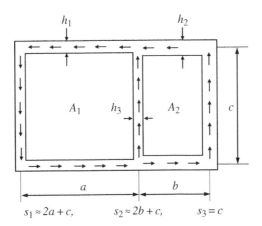

$s_1 \approx 2a + c, \qquad s_2 \approx 2b + c, \qquad s_3 = c$

Figure 7.18

Taking Equation 7.49 into consideration, we have

$$\tau_1 s_1 + \tau_3 s_3 = 2G\theta_0 A_1 \tag{7.75}$$

$$\tau_2 s_2 - \tau_3 s_3 = 2G\theta_0 A_2 \tag{7.76}$$

From the above equations, we can obtain τ_1, τ_2 and τ_3. Only the expression for τ_1 is shown in the following.

$$\tau_1 = \frac{T\left[s_2 h_3 A_1 + s_2 h_2 (A_1 + A_2)\right]}{2\left[s_2 h_1 h_3 A_1{}^2 + s_1 h_2 h_3 A_2{}^2 + s_3 h_1 h_2 (A_1 + A_2)^2\right]} \tag{7.77}$$

Problems of Chapter 7

1. Determine the twisting angle φ for a square sectioned thin tube of Figure 7.19 which is fixed at one end and subjected to a twisting moment T at the other, open end. The shear modulus of the material is G.

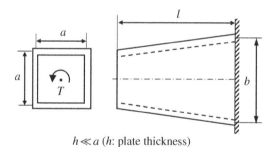

$h \ll a$ (h: plate thickness)

Figure 7.19

2. Determine the twisting angle φ for a thin cantilever plate of Figure 7.20 which is fixed at one end and subjected to a twisting moment T at the free end. Assume that the thickness of the plate t = constant and the height of the plate linearly varies from a_1 to a_2. The Young's modulus of the material is E and the Poisson's ratio is ν.
3. Determine the twisting angle φ at the central position of the thin tube of Figure 7.21 with mean diameter d and thickness h which has a slit for half the length of the tube. The tube is fixed at both ends and is subjected to a twisting moment T at the the central position of the tube. The shear modulus of the material is G.
4. Determine the twisting angle θ_0 per unit length when a pipe having a cross section like Figure 7.22 is subjected to a twisting moment T. Assume the thickness of the pipe is h and $h \ll D$. The shear modulus of the material is G.

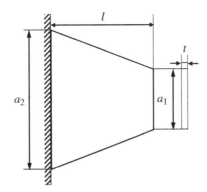

Figure 7.20

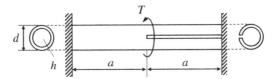

Figure 7.21

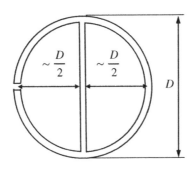

Figure 7.22

Reference

[1] S. P. Timoshenko and J. N. Goodier (1982) *Theory of Elasticity*, *3rd edn*, McGraw–Hill International, New York, p. 309.

8

Energy Principles

The several basic principles of the theory of elasticity have been obtained by paying attention to the energy of a system. The typical principles of these will be explained in this chapter.

8.1   Strain Energy

The historical sequence of describing the change of internal state of an elastic solid under the application of external forces was to pay attention to the change in location of atoms and molecules. However, the trial was not successful, though recent progress in computer analysis is making a new trend. The work done by the external forces results in a change in the location of atoms, namely stored as the strain energy. The expression of strain energy in the present form has been made possible by a simple and basic definition of stress and strain by Cauchy.

The definition of the concept of *stress* and *strain* which is used currently in general was given by Cauchy (1868)[1]. Without getting into the details of the microstructural conditions of solids, Cauchy thought the physical quantities of stress and strain should be related to deformation and failure. This simple idea became the beginning of the theory of elasticity.

Equations 8.1–8.4 were all derived by Cauchy. p_x, p_y and p_z in Equation 8.1 are the x, y and z components of the resultant force per unit area acting in the x, y and z directions on the plane having the direction cosines l, m, n (Figure 8.1; see the explanation related to Figure 1.13). Equations 8.2–8.4 are already explained in earlier chapters.

$$\left.\begin{array}{l} p_x = \sigma_x l + \tau_{yx} m + \tau_{zx} n \\ p_y = \tau_{xy} l + \sigma_y m + \tau_{zy} n \\ p_z = \tau_{xz} l + \tau_{yz} m + \sigma_z n \end{array}\right\} \tag{8.1}$$

---

[1] Regarding the details, see Reference [1].

Theory of Elasticity and Stress Concentration, First Edition. Yukitaka Murakami.
© 2017 John Wiley & Sons, Ltd. Published 2017 by John Wiley & Sons, Ltd.

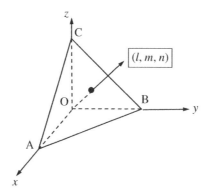

Figure 8.1

$$\tau_{xy} = \tau_{yx}, \quad \tau_{zx} = \tau_{xz}, \quad \tau_{yz} = \tau_{zy} \tag{8.2}$$

$$\left.\begin{array}{l} \dfrac{\partial \sigma_x}{\partial x} + \dfrac{\partial \tau_{yx}}{\partial y} + \dfrac{\partial \tau_{zx}}{\partial z} + X = 0 \\[3mm] \dfrac{\partial \tau_{xy}}{\partial x} + \dfrac{\partial \sigma_y}{\partial y} + \dfrac{\partial \tau_{zy}}{\partial z} + Y = 0 \\[3mm] \dfrac{\partial \tau_{xz}}{\partial x} + \dfrac{\partial \tau_{yz}}{\partial y} + \dfrac{\partial \sigma_z}{\partial z} + Z = 0 \end{array}\right\} \text{(Equilibrium equation)} \tag{8.3}$$

$$\left.\begin{array}{l} \varepsilon_x = \dfrac{\partial u}{\partial x}, \quad \varepsilon_y = \dfrac{\partial v}{\partial y}, \quad \varepsilon_z = \dfrac{\partial w}{\partial z} \\[3mm] \gamma_{xy} = \dfrac{\partial u}{\partial y} + \dfrac{\partial v}{\partial x}, \quad \gamma_{yz} = \dfrac{\partial v}{\partial z} + \dfrac{\partial w}{\partial y}, \quad \gamma_{zx} = \dfrac{\partial w}{\partial x} + \dfrac{\partial u}{\partial z} \end{array}\right\} \tag{8.4}$$

Clapeyron (1799–1864) pointed out that the work done by external forces under a constant temperature and iso-entropy process is equal to the *strain energy*.

The initial state in which no external forces are applied is a thermodynamically equilibrium condition and is called a *natural state*. The energy in the natural state is termed E_0. If external forces are applied to a body, the energy state E_1 of the body in question is increased due to the disturbance from the natural state. If the external forces are unloaded, the natural state is recovered and the difference ΔE between E_1 and E_0 is written as follows:

$$\Delta E = E_1 - E_0$$

We can regard $\Delta E > 0$. We call ΔE the strain energy, and the nature of $\Delta E > 0$ is called the *positive form*.

We express the strain energy with stresses and strains in a body as follows.

Let us imagine an infinitesimal rectangular solid in a body, as in Figure 8.2, and define the stresses by σ_x, σ_y, σ_z, τ_{xy} and so on, the strains by ε_x, ε_y, ε_z, γ_{xy} and so on and the body forces by X, Y, Z.

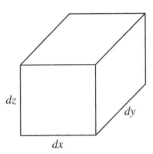

Figure 8.2

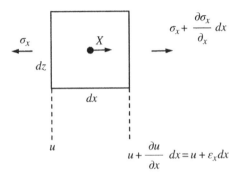

Figure 8.3

Considering the contribution of σ_x, ε_x, and X to strain energy in terms of Figure 8.3:

$$\frac{1}{2}\left(\sigma_x+\frac{\partial\sigma_x}{\partial x}dx\right)dy\cdot dz(u+\varepsilon_x dx)-\frac{1}{2}\sigma_x dy\cdot dz\cdot u$$

$$\cong\frac{1}{2}\sigma_x\varepsilon_x dx\cdot dy\cdot dz+\frac{1}{2}u\frac{\partial\sigma_x}{\partial x}dx\cdot dy\cdot dz \qquad (8.5)$$

$$=\frac{1}{2}\left(\sigma_x\varepsilon_x+u\frac{\partial\sigma_x}{\partial x}\right)dx\cdot dy\cdot dz$$

$$\frac{1}{2}X\left(u+\frac{1}{2}\varepsilon_x dx\right)dx\cdot dy\cdot dz\cong\frac{1}{2}Xudx\cdot dy\cdot dz \qquad (8.6)$$

Referring Figure 8.4, the contribution of τ_{xy} to strain energy can be expressed as follows:

$$\frac{1}{2}\left(\tau_{xz}+\frac{\partial\tau_{xz}}{\partial x}dx\right)dy\cdot dz\left(w+\frac{\partial w}{\partial x}dx\right)-\frac{1}{2}\tau_{xz}dy\cdot dz\cdot w$$

$$\cong\frac{1}{2}\left(\tau_{xz}\frac{\partial w}{\partial x}+w\frac{\partial\tau_{xz}}{\partial x}\right)dx\cdot dy\cdot dz \qquad (8.7)$$

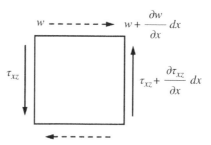

Figure 8.4

Likewise, the contribution of other stresses and strains can be expressed as follows:

$$\frac{1}{2}\left(\sigma_y \varepsilon_y + v\frac{\partial \sigma_y}{\partial y}\right)dx \cdot dy \cdot dz, \quad \frac{1}{2}\left(\sigma_z \varepsilon_z + w\frac{\partial \sigma_z}{\partial z}\right)dx \cdot dy \cdot dz,$$

$$\frac{1}{2}\left(\tau_{xy}\frac{\partial v}{\partial x} + v\frac{\partial \tau_{xy}}{\partial x}\right)dx \cdot dy \cdot dz, \quad \frac{1}{2}\left(\tau_{xy}\frac{\partial u}{\partial y} + u\frac{\partial \tau_{xy}}{\partial y}\right)dx \cdot dy \cdot dz,$$

$$\frac{1}{2}\left(\tau_{yz}\frac{\partial w}{\partial y} + w\frac{\partial \tau_{yz}}{\partial y}\right)dx \cdot dy \cdot dz, \quad \frac{1}{2}\left(\tau_{yz}\frac{\partial v}{\partial x} + v\frac{\partial \tau_{yz}}{\partial z}\right)dx \cdot dy \cdot dz,$$

$$\frac{1}{2}\left(\tau_{xz}\frac{\partial u}{\partial z} + u\frac{\partial \tau_{xz}}{\partial z}\right)dx \cdot dy \cdot dz$$

The contribution of the body forces Y, Z to strain energy can also be expressed as

$$\frac{1}{2}Yvdx \cdot dy \cdot dz, \quad \frac{1}{2}Zwdx \cdot dy \cdot dz.$$

Denoting the sum of all the contributions listed above by dU, and considering Equation 8.2 and Equation 8.4, we have

$$dU = \frac{1}{2}\left[\left(\sigma_x \varepsilon_x + \sigma_y \varepsilon_y + \sigma_z \varepsilon_z + \tau_{xy}\gamma_{xy} + \tau_{yz}\gamma_{yz} + \tau_{zx}\gamma_{zx}\right)\right.$$

$$+ u\left(\frac{\partial \sigma_x}{\partial x} + \frac{\partial \tau_{yx}}{\partial y} + \frac{\partial \tau_{zx}}{\partial z} + X\right) + v\left(\frac{\partial \tau_{xy}}{\partial x} + \frac{\partial \sigma_y}{\partial y} + \frac{\partial \tau_{zy}}{\partial z} + Y\right)$$

$$\left.+ w\left(\frac{\partial \tau_{xz}}{\partial x} + \frac{\partial \tau_{yz}}{\partial y} + \frac{\partial \sigma_z}{\partial z} + Z\right)\right]dx \cdot dy \cdot dz \qquad (8.8)$$

The second, third and fourth terms in [ ] are reduced to zero, because they include the equilibrium conditions itself. Therefore,

$$dU = \frac{1}{2}\left(\sigma_x \varepsilon_x + \sigma_y \varepsilon_y + \sigma_z \varepsilon_z + \tau_{xy}\gamma_{xy} + \tau_{yz}\gamma_{yz} + \tau_{zx}\gamma_{zx}\right)dx \cdot dy \cdot dz \qquad (8.9)$$

Denoting the strain energy per unit volume by U_0 and using the relationship $dU = U_0 dx \cdot dy \cdot dz$:

$$U_0 = \frac{1}{2}\left(\sigma_x \varepsilon_x + \sigma_y \varepsilon_y + \sigma_z \varepsilon_z + \tau_{xy}\gamma_{xy} + \tau_{yz}\gamma_{yz} + \tau_{zx}\gamma_{zx}\right) \tag{8.10}$$

U_0 is called the *strain energy density function*.
Using Hooke's law, U_0 can be expressed only by stresses or strains as follows:

$$U_0 = \frac{1}{2E}\left\{\left(\sigma_x^2 + \sigma_y^2 + \sigma_z^2\right) - 2\nu\left(\sigma_x\sigma_y + \sigma_y\sigma_z + \sigma_z\sigma_x\right)\right\} + \frac{1}{2G}\left(\tau_{xy}^2 + \tau_{yz}^2 + \tau_{zx}^2\right) \tag{8.11}$$

$$U_0 = \frac{E\nu}{2(1+\nu)\cdot(1-2\nu)}\left(\varepsilon_x + \varepsilon_y + \varepsilon_z\right)^2 + G\left\{\left(\varepsilon_x^2 + \varepsilon_y^2 + \varepsilon_z^2\right) + \frac{1}{2}\left(\gamma_{xy}^2 + \gamma_{yz}^2 + \gamma_{zx}^2\right)\right\} \tag{8.12}$$

If U_0 expressed by the polar coordinate is prepared, many applications in polar coordinate problems will be possible.

Since $U_0 \geq 0$ (positive form), the possible range of Poisson's ratio ν can be determined as follows:

$$-1 < \nu < 1/2 \tag{8.13}$$

Poisson's ratios for most metallic materials are $\nu = 0.25$–0.33; and especially for steels, $\nu \sim 0.3$. According to Equation 8.13, Poisson's ratio can have a negative value, though no material having $\nu < 0$ is known.

Problem 8.1.1   Derive Equations 8.11 and 8.12 from Equation 8.10.

Problem 8.1.2   Derive Equation 8.13.

Problem 8.1.3   Verify the following relationships.

$$\varepsilon_x = \frac{\partial U_0}{\partial \sigma_x}, \quad \varepsilon_y = \frac{\partial U_0}{\partial \sigma_y}, \quad \varepsilon_z = \frac{\partial U_0}{\partial \sigma_z}, \quad \gamma_{xy} = \frac{\partial U_0}{\partial \tau_{xy}}, \quad \cdots \tag{8.14}$$

$$\sigma_x = \frac{\partial U_0}{\partial \varepsilon_x}, \quad \sigma_y = \frac{\partial U_0}{\partial \varepsilon_y}, \quad \sigma_z = \frac{\partial U_0}{\partial \varepsilon_z}, \quad \tau_{xy} = \frac{\partial U_0}{\partial \gamma_{xy}}, \quad \cdots \tag{8.15}$$

Problem 8.1.4   Figure 8.5a, b shows a wide plate with unit thickness under a uniform remote tensile biaxial stress σ_0. The plate of Figure 8.5a does not contain a hole. The plate of Figure 8.5b contains a circular hole with radius a.

Answer the following questions.

1. Calculate the strain energy U_1 for the plate of Figure 8.5a and U_2 for the plate of Figure 8.5b inside the region included by the dotted line. The Young's modulus and the Poisson's ratio of the plate are E and ν, respectively.
2. When the radius a is reduced to 0, $a \to 0$, which one of the following relationships is correct.

$$U_1 = U_2 \quad U_1 > U_2 \quad U_1 < U_2$$

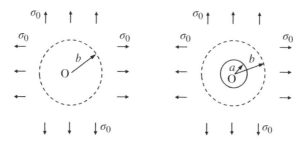

Figure 8.5

8.2 Uniqueness of the Solutions of Elasticity Problems

The uniqueness of the solution of elasticity problems can be solved in the following. Assuming two solutions for Problem A, we denote the solutions by Solution 1 and Solution 2, namely:

$$\text{Solution 1: } \sigma_x', \sigma_y', \sigma_z', \tau_{xy}' \ldots, \varepsilon_x', \varepsilon_y', \varepsilon_z', \gamma_{xy}' \ldots$$

$$\text{Solution 2: } \sigma_x'', \sigma_y'', \sigma_z'', \tau_{xy}'' \ldots, \varepsilon_x'', \varepsilon_y'', \varepsilon_z'', \gamma_{xy}'' \ldots$$

Since Solutions 1 and 2 both satisfy the compatibility equation, the equilibrium condition and the boundary condition, first of all we write the equilibrium condition with Solution 1 as:

$$\left.\begin{aligned}
\frac{\partial \sigma_x'}{\partial x} + \frac{\partial \tau_{xy}'}{\partial y} + \frac{\partial \tau_{zx}'}{\partial z} + X = 0 \\
\frac{\partial \tau_{xy}'}{\partial x} + \frac{\partial \sigma_y'}{\partial y} + \frac{\partial \tau_{yz}'}{\partial z} + Y = 0 \\
\frac{\partial \tau_{zx}'}{\partial x} + \frac{\partial \tau_{yz}'}{\partial y} + \frac{\partial \sigma_z'}{\partial z} + Z = 0
\end{aligned}\right\}
\qquad (8.16)$$

denoting the resultant forces per unit area in the x, y and z directions by p_x, p_y and p_z and the direction cosines with respect to the x, y and z axes by (l, m, n):

$$\left.\begin{aligned}
\sigma_x' l + \tau_{yx}' m + \tau_{zx}' n = p_x \\
\tau_{yx}' l + \sigma_y' m + \tau_{zy}' n = p_y \\
\tau_{xz}' l + \tau_{yz}' m + \sigma_z' n = p_z
\end{aligned}\right\}
\qquad (8.17)$$

Likewise regarding Solution 2, we can obtain equations with the same form as Equations 8.16 and 8.17. We denote these equations as Equations 8.18 and 8.19.

$$\text{The equilibrium condition for Solution 2} \qquad (8.18)$$

$$\text{The boundary condition for Solution 2} \qquad (8.19)$$

If we denote the subtraction between Solution 1 and Solution 2 by $\sigma_x = \sigma_x' - \sigma_x''$, $\sigma_y = \sigma_y' - \sigma_y''$, ...

$$\left.\begin{array}{l} \dfrac{\partial \sigma_x}{\partial x} + \dfrac{\partial \tau_{xy}}{\partial y} + \dfrac{\partial \tau_{zx}}{\partial z} = 0 \\[2mm] \dfrac{\partial \tau_{xy}}{\partial x} + \dfrac{\partial \sigma_y}{\partial y} + \dfrac{\partial \tau_{yz}}{\partial z} = 0 \\[2mm] \dfrac{\partial \tau_{zx}}{\partial x} + \dfrac{\partial \tau_{yz}}{\partial y} + \dfrac{\partial \sigma_z}{\partial z} = 0 \end{array}\right\} \tag{8.20}$$

$$\left.\begin{array}{l} \sigma_x l + \tau_{yx} m + \tau_{zx} n = 0 \\[1mm] \tau_{xy} l + \sigma_y m + \tau_{zy} n = 0 \\[1mm] \tau_{xz} l + \tau_{yz} m + \sigma_z n = 0 \end{array}\right\} \tag{8.21}$$

Equations 8.20 and 8.21 mean that the subtraction between Solution 1 and Solution 2 is the solution for a problem with no body force and no external force, namely the solution for a problem of natural state. It can be easily confirmed that the solution satisfies the compatibility condition. Considering that the strain energy of the natural state is zero and U_0 has the positive form, from Equations 8.11 and 8.12 we can derive the following result:

$$\sigma_x = \sigma_y = \sigma_z = \cdots = 0 \tag{8.22}$$

It follows that

$$\sigma_x' = \sigma_x'', \quad \sigma_y' = \sigma_y'', \cdots \tag{8.23}$$

Thus, the solution for a linear elastic problem is unique. The above discussion is the natural conclusion of the linearity of the basic equations and as the result the *superposition principle* can be understood, based on the nature of the governing equations. Therefore, we can write:

The solution for Problem A + the solution for Problem B = the solution for a problem which has superimposed boundary conditions for Problems A and B.

On the other hand, the superposition principle does not hold in general for nonlinear material problems and geometrically nonlinear problems.

8.3   Principle of Virtual Work

If external forces applied on a rigid body are in equilibrium condition and virtual displacements are applied to the rigid body, the work done by the external forces (surface forces and body forces) with the virtual displacement is calculated to be zero. This nature is called the *principle of virtual* **displacement**.

On the other hand, if external forces applied on an elastic body are in equilibrium condition and virtual displacements[2] are applied to the elastic body, the work done by the external forces

---

[2] We assume the virtual displacements are continuous functions of x, y and z and sufficiently small compared to the actual displacements and do not break the original displacement boundary conditions.

with the virtual displacement is not zero. The work is equal to that done by stresses with strains, that is the variation of strain energy. This is termed the *principle of virtual work*. This principle is verified as follows.

The necessary terms for the verification are defined in the following:

- S: Surface of elastic body.
- V: Domain included by the surface S.
- S_u: Part of S where the displacement boundary conditions are given.
- S_σ: Part of S where the force or stress boundary conditions are given.
- $p(p_x, p_y, p_z)$: The boundary conditions on S_σ.
- $F(X, Y, Z)$: Body forces.
- $\delta u, \delta v, \delta w$: The virtual displacements on S excluding S_u.

Denoting the work due to the surface forces by U_s,

$$U_s = \int_S (p_x \delta u + p_y \delta v + p_z \delta w) dS \tag{8.24}$$

Here, since $p = 0$ on S excluding S_σ, we may replace S by S_σ in Equation 8.24. Using Equation 8.1,

$$U_s = \int_S [(\sigma_x l + \tau_{xy} m + \tau_{zx} n)\delta u + (\tau_{xy} l + \sigma_y m + \tau_{yz} n)\delta v + (\tau_{zx} l + \tau_{yz} m + \sigma_z n)\delta w] dS$$

$$= \int_S [(\sigma_x \delta u + \tau_{xy} \delta v + \tau_{zx} \delta w)l + (\tau_{xy} \delta u + \sigma_y \delta v + \tau_{yz} \delta w)m + (\tau_{zx} \delta u + \tau_{yz} \delta v + \sigma_z \delta w)n] dS \tag{8.25}$$

The above equation can be reduced by the application of the Gauss diversion theory (see Appendix A.2 of Part I) as follows.

$$U_s = \int_V \left[\frac{\partial}{\partial x}(\sigma_x \delta u + \tau_{xy} \delta v + \tau_{zx} \delta w) + \frac{\partial}{\partial y}(\tau_{xy} \delta u + \sigma_y \delta v + \tau_{yz} \delta w) \right.$$

$$\left. + \frac{\partial}{\partial z}(\tau_{zx} \delta u + \tau_{yz} \delta v + \sigma_z \delta w) \right] dV$$

$$= \int_V \left[\left(\frac{\partial \sigma_x}{\partial x} + \frac{\partial \tau_{xy}}{\partial y} + \frac{\partial \tau_{zx}}{\partial z} \right) \delta u + \left(\frac{\partial \tau_{xy}}{\partial x} + \frac{\partial \sigma_y}{\partial y} + \frac{\partial \tau_{yz}}{\partial z} \right) \delta v \right. \tag{8.26}$$

$$+ \left(\frac{\partial \tau_{zx}}{\partial x} + \frac{\partial \tau_{yz}}{\partial y} + \frac{\partial \sigma_z}{\partial z} \right) \delta w + \sigma_x \frac{\partial \delta u}{\partial x} + \sigma_y \frac{\partial \delta v}{\partial y} + \sigma_z \frac{\partial \delta w}{\partial z}$$

$$\left. + \tau_{xy} \left(\frac{\partial \delta v}{\partial x} + \frac{\partial \delta u}{\partial y} \right) + \tau_{yz} \left(\frac{\partial \delta w}{\partial y} + \frac{\partial \delta v}{\partial z} \right) + \tau_{zx} \left(\frac{\partial \delta u}{\partial z} + \frac{\partial \delta w}{\partial x} \right) \right] dV$$

Considering the equilibrium equation (Equation 8.3),

$$\int_{S_\sigma} (p_x \delta u + p_y \delta v + p_z \delta w) dS_\sigma + \int_V (X \delta u + Y \delta v + Z \delta w) dV$$

$$= \int_V (\sigma_x \delta \varepsilon_x + \sigma_y \delta \varepsilon_y + \sigma_z \delta \varepsilon_z + \tau_{xy} \delta \gamma_{xy} + \tau_{yz} \delta \gamma_{yz} + \tau_{zx} \delta \gamma_{zx}) dV$$

(8.27)

The virtual displacements defined in the above discussion are only assumed quantities in the equations and have no actual influence on the external forces or stresses in question. As the equilibrium equations are used in the verification, the principle of virtual work does not hold for the system in which the equilibrium condition does not hold. *In other words, the principle of virtual work is an alternative expression for the equilibrium equations.* Namely, the principle of virtual work is equivalent to the equilibrium equations. Only the usage or applicability of the principle is different from the equilibrium equations.

From the above discussion, when arbitrary small virtual displacements which do not break the displacement boundary condition of a problem are given, the stresses which can satisfy the principle of virtual work are only those which satisfy the equilibrium equations. It should be noted that, since only the equilibrium equations are used for the derivation of the principle of virtual work, the principle can be applied not only to linear elastic problems but also to non-linear elastic problems and elasto-plastic problems.

Example problem 8.1

As shown in Figure 8.6, two elastic rods are connected with a pin. P_x and P_y are the external loads acting to the connecting pin and Q and R are the forces assumed to act in rod 1 and rod 2, respectively. When the virtual displacement δ is given at the pin in the direction θ with respect to the horizontal direction, the principle of virtual work can be written as follows.

The principle of virtual work

$$P_x \delta \cos\theta + P_y \delta \sin\theta = Q\delta \cos(\pi/2 - \theta + \alpha) + R\delta \cos(\pi/2 - \theta - \beta)$$

Reforming this equation,

$$P_x \cos\theta \cdot \delta + P_y \sin\theta \cdot \delta$$
$$= (-Q\sin\alpha + R\sin\beta)\cos\theta \cdot \delta + (Q\cos\alpha + R\cos\beta)\sin\theta \cdot \delta$$

Since the above equation must be satisfied for arbitrary δ, the following equation must hold:

$$P_x = -Q\sin\alpha + R\sin\beta$$
$$P_y = Q\cos\alpha + R\cos\beta$$

These equations are exactly the same as the equilibrium equations for the connecting pin.

The equilibrium equations

In the above example, the principle of virtual work expresses two equilibrium equations by one equation and the meaning of the principle of virtual work is equivalent to the equilibrium equations.

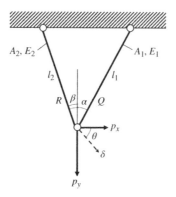

Figure 8.6

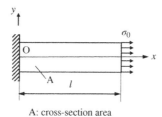

A: cross-section area

Figure 8.7

Problem 8.3.1   Figure 8.7 shows the tension problem of a bar in which the normal stress σ_0 is applied at $x = l$. Assume normal stress at $x = x$ in the x direction by σ_x (unknown) and virtual displacement by $\delta u = \alpha x^2 (\alpha = \text{constant})$ at $x = x$. Applying the principle of the virtual work, verify $\sigma_x = \sigma_0$.

Although it is evident from the equilibrium condition that $\sigma_x = \sigma_0$, this problem is an example to understand the derivation of the result $\sigma_x = \sigma_0$ assuming that we do not have knowledge of the equilibrium equations.

8.4 Principle of Minimum Potential Energy

Let us investigate the physical meaning of the subtraction between the right hand equation and the left hand equation for the principle of the virtual work, Equation 8.27.

Considering Equation 8.15, we have

$$
\int_V \left(\sigma_x \delta\varepsilon_x + \sigma_y \delta\varepsilon_y + \sigma_z \delta\varepsilon_z + \tau_{xy}\delta\gamma_{xy} + \tau_{yz}\delta\gamma_{yz} + \tau_{zx}\delta\gamma_{zx} \right) dV
$$

$$
= \int_V \left(\frac{\partial U_0}{\partial \varepsilon_x}\delta\varepsilon_x + \frac{\partial U_0}{\partial \varepsilon_y}\delta\varepsilon_y + \frac{\partial U_0}{\partial \varepsilon_z}\delta\varepsilon_z + \frac{\partial U_0}{\partial \gamma_{xy}}\delta\gamma_{xy} + \frac{\partial U_0}{\partial \gamma_{yz}}\delta\gamma_{yz} + \frac{\partial U_0}{\partial \gamma_{zx}}\delta\gamma_{zx} \right) dV \qquad (8.28)
$$

$$
= \int_V \delta U_0 \, dV
$$

Denoting the subtraction between the right hand equation and the left hand equation of Equation 8.27 by $\delta\Pi$,

$$\delta\Pi = \int_V \delta U_0 dV - \int_{S_\sigma} \left(p_x\delta u + p_y\delta v + p_z\delta w\right)dS - \int_V (X\delta u + Y\delta v + Z\delta w)dV \tag{8.29}$$

Adding the constraint that $p(p_x, p_y, p_z)$ and $F(X, Y, Z)$ are constant,

$$\begin{aligned}\delta\Pi = \delta\int_V U_0 dV - \delta\int_{S_\sigma} \left(p_x u + p_y v + p_z w\right)dS \\ -\delta\int_V (Xu + Yv + Zw)dV\end{aligned} \tag{8.30}$$

Thus,

$$\begin{aligned}\Pi = \int_V U_0 dV - \int_{S_\sigma} \left(p_x u + p_y v + p_z w\right)dS \\ -\int_V (Xu + Yv + Zw)dV\end{aligned} \tag{8.31}$$

The quantity Π defined by Equation 8.31 is called the *potential energy* of the system. The first term of the right hand equation of Equation 8.31 is called the strain energy, the second term is called the potential energy of the external forces and the third term is the so-called potential energy of body forces.

If we add the condition of Equation 8.15 and the constraint that the external forces are constant to the principle of virtual work, we can derive the following relationship.

The principle of virtual work $\rightarrow \delta\Pi = 0$. (*The potential energy has a stationary value.*)

Since this equation was derived after adding some constraints to the principle of virtual work, it loses generality unlike the principle of virtual work. However, since most of actual problems satisfy these constraints, the form $\delta\Pi = 0$ has many applications.

Problem 8.4.1 Verify that Π takes the minimum value for $\delta\Pi = 0$. This nature is called the *principle of minimum potential energy*.

Hint for the solution

Verify $\delta\Pi = \Pi(u + \delta u, v + \delta v, w + \delta w) - \Pi(u, v, w) \geq 0$.

$$U_0\left(\varepsilon_x + \delta\varepsilon_x, \varepsilon_y + \delta\varepsilon_y, \cdots\right) = U_0\left(\varepsilon_x, \varepsilon_y, \cdots\right) + \frac{\partial U_0}{\partial\varepsilon_x}\delta\varepsilon_x + \frac{\partial U_0}{\partial\varepsilon_y}\delta\varepsilon_y + \cdots$$

$$+ \frac{1}{2}\left\{\frac{\partial^2 U_0}{\partial\varepsilon_x{}^2}(\delta\varepsilon_x)^2 + \frac{\partial^2 U_0}{\partial\varepsilon_y{}^2}(\delta\varepsilon_y)^2 + \cdots + \frac{\partial^2 U_0}{\partial\varepsilon_x\partial\varepsilon_y}\delta\varepsilon_x\delta\varepsilon_y + \cdots\right\} + \cdots$$

At natural state, $\varepsilon_x = \varepsilon_y = \cdots = 0$, $\sigma_x = \sigma_y = \cdots = 0$, $U_0 = 0$

From the nature of the positive form of strain energy, $U_0\left(\delta\varepsilon_x, \delta\varepsilon_y, \ldots\right) > 0$.

Then, we can reduce

$$\frac{1}{2}\left\{\frac{\partial^2 U_0}{\partial \varepsilon_x^2}(\partial \varepsilon_x)^2 + \cdots\right\} > 0$$

Example problem 8.2

Let us determine the displacement of the cantilever of Figure 8.8 by the minimum potential energy.

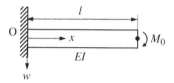

Figure 8.8

We assume the displacement in the cantilever as

$$w = a_1 + a_2 x + a_3 x^2 + a_4 x^3$$

The boundary condition:
From $w|_{x=0} = 0$, $a_1 = 0$, and from $w'|_{x=0} = 0$, $a_2 = 0$.
Therefore, we have

$$w = a_3 x^2 + a_4 x^3,\ w' = 2a_3 x + 3a_4 x^2,\ w'' = 2a_3 + 6a_4 x.$$

Since the strain energy for bending of cantilever is expressed as

$$\int_0^l (M^2/2EI)\,dx \quad \text{or} \quad \frac{1}{2}EI\int_0^l (w'')^2\,dx$$

$$\Pi = \frac{1}{2}EI\int_0^l (w'')^2\,dx - M_0\theta_0$$

Now, considering $\theta_0 = w'|_{x=l}$,

$$\Pi = \frac{1}{2}EI\int_0^l (w'')^2\,dx - M_0(w')_{x=l}$$

$$= \frac{1}{2}EI\int_0^l (2a_3 + 6a_4 x)^2\,dx - M_0(2a_3 l + 3a_4 l^2)$$

$$= \frac{1}{2}EI(4a_3^2 l + 12a_3 a_4 l^2 + 12a_4^2 l^3) - M_0(2a_3 l + 3a_4 l^2)$$

Since $\delta\Pi = 0$ is required for the exact solution, we operate the following differentiation: $\dfrac{\partial \Pi}{\partial a_3} = 0,\ \dfrac{\partial \Pi}{\partial a_4} = 0$

And we have

$$\left. \begin{array}{l} \dfrac{1}{2}EI\left(8la_3 + 12l^2a_4\right) - 2M_0l = 0 \\[2mm] \dfrac{1}{2}EI\left(12l^2a_3 + 24l^3a_4\right) - 3M_0l = 0 \end{array} \right\}$$

Solving these equations, we obtain

$$a_3 = \frac{M_0}{2EI}, \quad a_4 = 0$$

Then,

$$w = \frac{M_0}{2EI}x^2$$

The above solution method in which the displacements are first assumed and next the unknown constants are determined by the application of the minimum potential energy is called the *Rayleigh–Ritz method*.

Problem 8.4.2   Determine the displacement of the cantilever of Figure 8.9 by the Rayleigh–Ritz method.

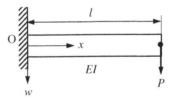

Figure 8.9

8.5   Castigliano's Theorem

Let us investigate the relationship among the strain energy U, the load P and the displacement λ at the locations of applied forces and their incremental values in the problem of Figure 8.10. The relationship between P and λ in an elastic body is expressed as Figure 8.10.

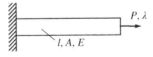

Figure 8.10

The quantity corresponding to $\triangle OBC$ is denoted by U_c and is called the *complementary strain energy*. Namely, the following relationship holds in the problem of Figure 8.11:

$$U + U_c = P\lambda \tag{8.32}$$

From Figure 8.10, it follows:

$$\Delta U = P \cdot \delta\lambda, \quad \text{or} \quad P = \frac{\partial U}{\partial \lambda} \tag{8.33}$$

$$\delta U_c = \lambda \cdot \delta P, \quad \text{or} \quad \lambda = \frac{\partial U_c}{\partial P} \tag{8.34}$$

On the other hand, we have the following relationship in the strength of material.

$$\lambda = \frac{Pl}{EA} \tag{8.35}$$

$$U = U_c = \frac{1}{2}P\lambda = \frac{P^2 l}{2EA} = \frac{EA\lambda^2}{2l} \tag{8.36}$$

If tensile load P is replaced by bending moment M and displacement λ is replaced by angular displacement (or rotation angle) θ, a similar relationship holds.

Although for a problem in which Equation 8.36 holds, λ can be obtained by $\partial U/\partial P$, λ essentially should be calculated by Equation 8.34. Figure 8.12a, b shows the cases for $U \neq U_c$ in the relationship between P and λ in which the wrong value of λ is obtained by $\partial U/\partial P$.

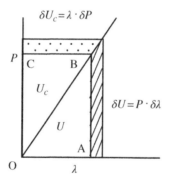

Figure 8.11

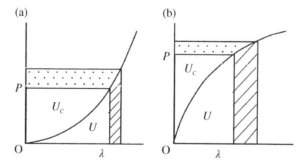

Figure 8.12

Problem 8.5.1   Figure 8.13 shows two bars connected horizontally with a pin. The two bars have length l, cross sectional area A and Young's modulus E. Calculate U and U_c of this problem, and show that $U \neq U_c$. Explain whether the relationship between P and λ corresponds to Figure 8.12 part (a) or part (b).

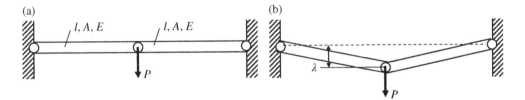

Figure 8.13

In order to generalize the above mentioned relationship, a variation[3] of continuous and infinitesimal stresses (external forces) is given to an elastic body *without breaking the equilibrium condition*. Therefore, it follows that:

$$
\left.
\begin{aligned}
\frac{\partial(\sigma_x + \delta\sigma_x)}{\partial x} + \frac{\partial(\tau_{xy} + \delta\tau_{xy})}{\partial y} + \frac{\partial(\tau_{zx} + \partial\tau_{zx})}{\partial z} + X + \delta X &= 0 \\
\cdots\cdots\cdots\cdots\cdots\cdots\cdots\cdots\cdots\cdots + Y + \delta Y &= 0 \\
\cdots\cdots\cdots\cdots\cdots\cdots\cdots\cdots\cdots\cdots + Z + \delta Z &= 0
\end{aligned}
\right\}
\qquad (8.37)
$$

Now, the following relationship holds between the variation of the complementary strain energy and the external forces. This relationship is called *Castigliano's theorem*.

$$
\delta U_c = \int_V \left(\frac{\partial U_c}{\partial \sigma_x}\delta\sigma_x + \frac{\partial U_c}{\partial \sigma_y}\delta\sigma_y + \frac{\partial U_c}{\partial \sigma_z}\delta\sigma_z + \frac{\partial U_c}{\partial \tau_{xy}}\delta\tau_{xy} + \frac{\partial U_c}{\partial \tau_{yz}}\delta\tau_{yz} + \frac{\partial U_c}{\partial \tau_{zx}}\delta\tau_{zx} \right) dV
\qquad (8.38)
$$

Using the relationship $(\partial U_c/\partial \sigma_x) = \varepsilon_x$, $(\partial U_c/\partial \sigma_y) = \varepsilon_y$, $\cdots$

$$
\delta U_c = \int_V \left(\varepsilon_x\delta\sigma_x + \varepsilon_y\delta\sigma_y + \varepsilon_z\delta\sigma_z + \gamma_{xy}\delta\tau_{xy} + \gamma_{yz}\delta\tau_{yz} + \gamma_{zx}\delta\tau_{zx} \right) dV
\qquad (8.39)
$$

Equation 8.38 holds for the problems in which Hooke's law does not hold, because Equation 8.38 describes only the variation of U_c, and U_c does not necessarily mean $U_c = 1/2\sigma_x\varepsilon_x + 1/2\sigma_y\varepsilon_y + 1/2\sigma_z\varepsilon_z + \cdots\cdots\cdots$

---

[3] Note that the boundary conditions were kept unchanged in the principle of minimum potential energy.

Thus,

$$\delta U_c = \int_V \left[\left(\frac{\partial u}{\partial x} \delta\sigma_x + \frac{\partial v}{\partial y} \delta\sigma_y + \frac{\partial w}{\partial z} \delta\sigma_z \right) + \left(\frac{\partial u}{\partial y} + \frac{\partial v}{\partial x} \right) \delta\tau_{xy} + \left(\frac{\partial v}{\partial z} + \frac{\partial w}{\partial y} \right) \delta\tau_{yz} + \left(\frac{\partial w}{\partial x} + \frac{\partial u}{\partial z} \right) \delta\tau_{zx} \right] dV$$

$$= \int_V \left[\left\{ \frac{\partial(u\delta\sigma_x)}{\partial x} - u\frac{\partial(\delta\sigma_x)}{\partial x} \right\} + \left\{ \frac{\partial(v\delta\sigma_y)}{\partial y} - v\frac{\partial(\delta\sigma_y)}{\partial y} \right\} + \left\{ \frac{\partial(w\delta\sigma_z)}{\partial z} - w\frac{\partial(\delta\sigma_z)}{\partial z} \right\} \right.$$

$$+ \left\{ \frac{\partial(u\delta\tau_{xy})}{\partial y} - u\frac{\partial(\delta\tau_{xy})}{\partial y} \right\} + \left\{ \frac{\partial(v\delta\tau_{xy})}{\partial x} - v\frac{\partial(\delta\tau_{xy})}{\partial x} \right\} + \left\{ \frac{\partial(v\delta\tau_{yz})}{\partial z} - v\frac{\partial(\delta\tau_{yz})}{\partial z} \right\} \qquad (8.40)$$

$$\left. + \left\{ \frac{\partial(w\delta\tau_{yz})}{\partial y} - w\frac{\partial(\delta\tau_{yz})}{\partial y} \right\} + \left\{ \frac{\partial(w\delta\tau_{zx})}{\partial x} - w\frac{\partial(\delta\tau_{zx})}{\partial x} \right\} + \left\{ \frac{\partial(u\delta\tau_{zx})}{\partial z} - u\frac{\partial(\delta\tau_{zx})}{\partial z} \right\} \right] dV$$

Taking Equation 8.37 into consideration,

$$\delta U_c = \int_V \left[\left\{ \frac{\partial(u\delta\sigma_x)}{\partial x} + \frac{\partial(u\delta\tau_{xy})}{\partial y} + \frac{\partial(u\delta\tau_{zx})}{\partial z} \right\} + \left\{ \frac{\partial(v\delta\tau_{xy})}{\partial x} + \frac{\partial(v\delta\sigma_y)}{\partial y} + \frac{\partial(v\delta\tau_{yz})}{\partial z} \right\} \right.$$

$$\left. + \left\{ \frac{\partial(w\delta\tau_{zx})}{\partial x} + \frac{\partial(w\delta\tau_{yz})}{\partial y} + \frac{\partial(w\delta\sigma_z)}{\partial z} \right\} + (u\delta X + v\delta Y + w\delta Z) \right] dV.$$

$$= \int_S \left[(lu\delta\sigma_x + mu\delta\tau_{xy} + nu\delta\tau_{zx}) + (lv\delta\tau_{xy} + mv\delta\sigma_y + nv\delta\tau_{yz}) + (lw\delta\tau_{zx} + mw\delta\tau_{yz} + nw\delta\sigma_z) \right] dS$$

$$+ \int_V (u\delta X + v\delta Y + w\delta Z) dV \qquad (8.41)$$

$$= \int_S (u\delta p_x + v\delta p_y + w\delta p_z) dS + \int_{S_u} (u\delta X + v\delta Y + w\delta Z) dV \qquad (8.42)$$

$$= \int_{S_\sigma} (u\delta p_{x0} + v\delta p_{y0} + w\delta p_{z0}) dS + \int_{S_u} (u_0\delta p_x + v_0\delta p_y + w_0\delta p_z) dS$$

$$+ \int_V (u\delta X + v\delta Y + w\delta Z) dV \qquad (8.43)$$

Now, in the case where $\delta X = \delta Y = \delta Z = 0$, $u_0 = v_0 = w_0 = 0$ (the fixed displacement boundary conditions) and $p_{x0}dS_\sigma = P_x$, $p_{y0}dS_\sigma = P_y$, $p_{z0}dS_\sigma = P_z$ (concentrated external forces),

$$\delta U_c = \sum_{i=1}^{n} (u\delta P_x + v\delta P_y + w\delta P_z)_i \qquad (8.44)$$

From these relationships, we can deduce the following equations.

$$u_i = \frac{\partial U_c}{\partial P_{xi}}, \quad v_i = \frac{\partial U_c}{\partial P_{yi}}, \quad w_i = \frac{\partial U_c}{\partial P_{zi}} \tag{8.45}$$

Therefore, regardless of the x, y, and z directions, in general, the displacement λ_i in the direction of a concentrated force P_i can be given by the following equation.

$$\lambda_i = \frac{\partial U_c}{\partial P_i} \tag{8.46}$$

On the other hand, if we want to know the external force or reaction arose by a given displacement, we only have to pay attention to U instead of U_c and follow the same deriving procedure as in the above equations. Thus,

$$P_i = \frac{\partial U}{\partial \lambda_i} \tag{8.47}$$

8.6   The Reciprocal Theorem

If we compare the displacement δ_A at point A in Figure 8.14a and the displacement δ_B at point B in Figure 8.14b, we can show that $\delta_A = \delta_B$. In the same way, if we compare the displacement δ_B at point B in Figure 8.14c and the angular displacement θ_A at point A in Figure 8.14d, we can show that $\delta_B/M = \theta_A/P$. These relationships are not coincidence by chance. These relationships hold necessarily. These relationships come from the *reciprocal theorem*. The reciprocal theorem is useful when the solution of a problem can be obtained by replacing the original problem by an alternative problem and also for examining the validity of solutions.

The reciprocal theorem is derived as follows. The influence coefficients for a linear elastic body in equilibrium condition as Figure 8.15 are defined as follows.

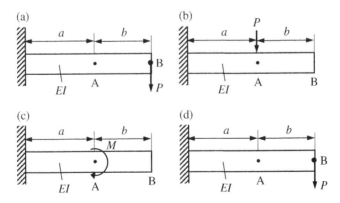

Figure 8.14

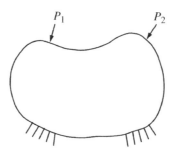

Figure 8.15

a_{ij}: The displacement at point i for a unit force in P_i direction when a unit force is applied at point j in P_j direction. Forces and displacements defined here have a general meaning such as moment and angular displacement.

Referring to Figure 8.15, let us make clear the nature of the influence coefficients by changing the order of the application of forces P_1 and P_2.

1. Define the strain energy by U_1 for the case that P_1 is first applied and followed by P_2. U_1 is expressed as follows:

$$U_1 = \frac{1}{2}P_1 P_1 a_{11} + \frac{1}{2}P_2 P_2 a_{22} + P_1 P_2 a_{12} \tag{8.48}$$

where the reason why 1/2 is not multiplied for the third term of the right hand side of Equation 8.48 is that P_1 is already applied when P_2 is applied.

2. Define the strain energy by U_2 for the case that P_2 is first applied and followed by P_1. U_2 is expressed as follows:

$$U_2 = \frac{1}{2}P_2 P_2 a_{22} + \frac{1}{2}P_1 P_1 a_{11} + P_2 P_1 a_{21} \tag{8.49}$$

Since the final conditions of (1) and (2) are the same, $U_1 = U_2$. Therefore,

$$a_{12} = a_{21} \tag{8.50}$$

The same verification is possible for the case of the increased number of forces P_i. Namely, we have in general (Maxwell, 1831–1879):

$$a_{ij} = a_{ji} \tag{8.51}$$

The relationship of Equation 8.50 for Figure 8.14 is a special case of Equation 8.51.

A similar relationship can be derived for the case where the number of applied forces is more than two (Betti, 1872). First, a group of forces $P_k (k = 1 \sim K)$ are applied[4] followed by the application of another group of forces $Q_j (j = 1 \sim J)$. We denote by U_{12} the work done by the group of

---

[4] It may be assumed that all the forces $P_k (k = 1 \sim k)$ are applied at the same time.

P_k by the application of the group of Q_j, and by U_{21} vice versa. The relationship $U_{12} = U_{21}$ is verified as follows. First, define the displacement as follows.

δ_{PQk}: The displacement at point k in the direction of P_k at the P_k group by the application of the Q_j group. P_k group is applied first followed by the Q_j group.
δ_{QPj}: The displacement at point j in the direction of Q_j at the Q_j group by the application of the P_k group. Q_j group is applied first followed by the P_k group.
δ_{PPk}: The displacement at point k in the direction of P_k by the application of the P_k group.
δ_{QQj}: The displacement at point j in the direction of Q_j by the application of the Q_j group.

Denoting the strain energy produced by applying the P_k group first followed by the Q_j group by U_1 and the strain energy produced by applying the Q_j group first followed by the P_k group by U_2, the following relationships are obtained.

$$U_1 = \frac{1}{2}\sum_{k=1}^{K} P_k \delta_{PPk} + \sum_{k=1}^{K} P_k \delta_{PQk} + \frac{1}{2}\sum_{j=1}^{J} Q_j \delta_{QQj} \tag{8.52}$$

$$U_2 = \frac{1}{2}\sum_{j=1}^{J} Q_j \delta_{QQj} + \sum_{j=1}^{J} Q_j \delta_{QPj} + \frac{1}{2}\sum_{k=1}^{K} P_k \delta_{PPk} \tag{8.53}$$

The above relationship can be easily understood by solving Problem 8.6.1.

Denoting the second term of Equations 8.52 and 8.53 by U_{12} and U_{21} respectively, and considering $U_1 = U_2$, we have

$$U_{12} = U_{21}, \text{ that is } \sum_{k=1}^{K} P_k \delta_{PQk} = \sum_{j=1}^{J} Q_j \delta_{QPj} \tag{8.54}$$

Problem 8.6.1   We denote the strain energy by U_1 for the case of Figure 8.16 in which two forces P_1 and P_2 are applied simultaneously and U_2 for the case in which two forces P_1 and P_2 are applied in series. Verify that $U_1 = U_2$. (This problem is not for the application of the reciprocal theorem but for understanding the relationship between the order of application of forces and the stored energy or work done by forces.)

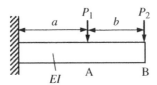

Figure 8.16

Problem 8.6.2   Figure 8.17a shows a cantilever with forces P at point A and Q at point C applied simultaneously. In this condition, we assume the displacement at point B is δ_B. In Figure 8.17b when force R is applied only at point B, the displacement measured at point C is δ_C. On this occasion (Figure 8.17b), determine the displacement δ_A at point A.

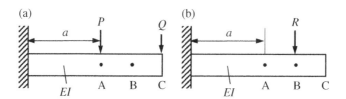

Figure 8.17

Problems of Chapter 8

1.   Determine the displacement δ_0 at the central point of a simply supported lever by approximating the displacement of a lever $u = \delta_0 \cos(\pi x/2l)$ and using the Rayleigh–Ritz method. The lever is simply supported at both ends. Compare the result with the value obtained by the elementary theory of the strength of materials and calculate the difference as a percentage. Here, x is the distance from the point of application of force; assume the bending rigidity as EI.
2.   Define the strain energy in the final condition by U_1 for a linear elastic body subjected to forces $P_1, P_2, \ldots, P_i, \ldots, P_n$ which are applied in series.

Define the strain energy in the final condition by U_2 for the case in which the forces are applied in reverse order, $P_n, P_{n-1}, \ldots, P_i, \ldots, P_2, P_1$. Assume the displacements produced by individual forces are linearly proportional to the forces.

Verify that $U_1 = U_2$ and the influence factors satisfy the relationship of Equation 8.51, $a_{ij} = a_{ji}$.

Verify the displacement δ_i at the point of application of P_i in the direction of P_i is given by the following equation:

$$\delta_i = \frac{\partial U}{\partial P_i}, \quad (U = U_1 = U_2)$$

Reference

[1] S. P. Timoshenko (1953) *History of Strength of Materials*, McGraw Hill, New York.

9

Finite Element Method

As explained in earlier chapters, the problems of elasticity can be solved by satisfying three conditions, that is equilibrium conditions, compatibility conditions and boundary conditions. However, except for simple problems, it is not easy to find the solutions in a closed form for general problems by the stress function method.

The *finite element method* (FEM) is nowadays the numerical analysis used most commonly in structural engineering fields. FEM is a computational numerical method in which, once the computer software is completed, the difference of problems is treated as the difference of the input data for boundary conditions. The advantage of FEM is that approximate solutions for most practical problems can always be obtained without advanced knowledge of stress analysis.

A basic knowledge of the strength of materials or the theory of elasticity may be sufficient for using a FEM program alone. However, for obtaining solutions with higher accuracy and for more extended applications of the method without mistakes, it is necessary to understand sufficiently the basic principle of FEM.

In the early stage of the progress of FEM in the 1960s, researchers and engineers involved in structural analysis developed so-called handmade computer software for themselves. Since the capacity of computers was limited, the number of elements and nodes was limited within several hundreds. Those researchers and engineers naturally understood the basic theory of elasticity and the programming necessary for FEM. They developed new elements and methods to improve the accuracy of their results using a small number of elements. Nowadays, it is not rare to use millions of elements in automatic meshing software. However, most users of this modern commercial FEM software do not necessarily understand the basic theory installed in the software. It is not surprising for many users to make mistakes in using the commercial software as a black box. Most mistakes come from a lack of knowledge of the basic theory of elasticity, especially by applying incorrect boundary conditions and misunderstanding the concept of stress concentration.

Theory of Elasticity and Stress Concentration, First Edition. Yukitaka Murakami.
© 2017 John Wiley & Sons, Ltd. Published 2017 by John Wiley & Sons, Ltd.

Most FEM theories have been treated mathematically and directly put into computer software schemes so that users of FEM software cannot touch the inside of the software. In the following sections, the most simple and basic concept of FEM for 2D problems will be explained using several examples which will help FEM users to understand the basic theoretical principle of modern commercial FEM software.

9.1 FEM for One Dimensional Problems

Example 1   Two connected springs A and B under the application of forces or displacements
Basic equation (Hooke's law).
$F = ku$, (F: external force, k: spring constant, u: elongation)
The situation of Figure 9.1 can be expressed using Hooke's law as follows. Regarding the element A,

$$F_{A1} = k_{A1}(u_{A1} - u_{A2}), \quad F_{A2} = k_A(u_{A2} - u_{A1}) \tag{9.1}$$

Expressing these equations by matrix form, we have

$$\begin{bmatrix} k_A & -k_A \\ -k_A & k_A \end{bmatrix} \begin{Bmatrix} u_{A1} \\ u_{A2} \end{Bmatrix} = \begin{Bmatrix} F_{A1} \\ F_{A2} \end{Bmatrix} \tag{9.2}$$

Equation 9.2 is rewritten in general as follows:

$$\begin{bmatrix} k_{11}^A & k_{12}^A \\ k_{21}^A & k_{22}^A \end{bmatrix} \begin{Bmatrix} u_{A1} \\ u_{A2} \end{Bmatrix} = \begin{Bmatrix} F_{A1} \\ F_{A2} \end{Bmatrix} \tag{9.3}$$

In the case of 2D problems which will be treated later, not only the sign of $k^A{}_{11}$ is different from that of $k^A{}_{12}$ but also the absolute values are different, and in general the values of k are different except for $k_{ij} = k_{ji}$.

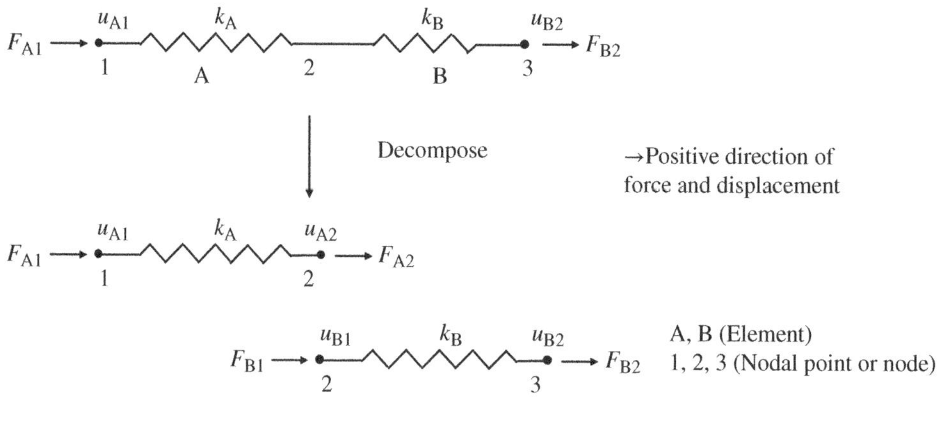

Figure 9.1

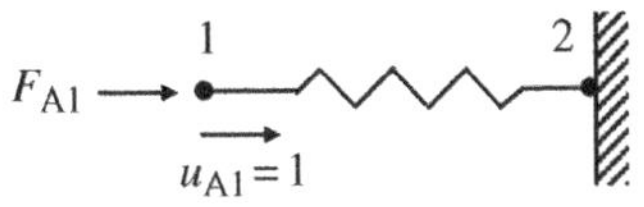

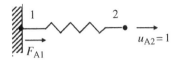

Figure 9.2

Figure 9.3

Now, let us think about the physical meaning of k_{ij}.

$k^A{}_{11}$ means the external force to be applied at node 1 when the unit displacement $u_{A1} = 1$ is given at node 1 by fixing node 2 (Figure 9.2).

$k^A{}_{12}$ similarly means the reaction to be produced at node 1 when the unit displacement $u_{A2} = 1$ is given at node 2 by fixing node 1 (Figure 9.3).

$k^A{}_{21}$ and $k^A{}_{22}$ have similar meanings to $k^A{}_{11}$ and $k^A{}_{12}$.

In the case of a one dimensional spring, as in Figure 9.1, the following equations hold:

$$k^A_{11} = k_A, \quad k^A_{12} = -k_A, \quad k^A_{21} = -k_A, \quad k^A_{22} = k_A$$

Equation 9.3 is in some cases expressed by the following form for simplicity.

$$[k^A]\{u_A\} = \{f_A\} \tag{9.4}$$

Regarding element B, we can express the relationship between the spring constants, displacements and forces as follows:

$$\begin{bmatrix} k^B_{11} & k^B_{12} \\ k^B_{21} & k^B_{22} \end{bmatrix} \begin{Bmatrix} u_{B1} \\ u_{B2} \end{Bmatrix} = \begin{Bmatrix} F_{B1} \\ F_{B2} \end{Bmatrix} \tag{9.5}$$

$$[k^B]\{u_B\} = \{f_B\} \tag{9.6}$$

$$k^B_{11} = k_B, \quad k^B_{12} = -k_B, \quad k^B_{21} = -k_B, \quad k^B_{22} = k_B$$

Considering $u_{A2} = u_{B1}$ for the two connected springs of the top figure of Figure 9.1, Equations 9.3 and 9.5 can be put into one matrix as follows.

$$\begin{bmatrix} k^A_{11} & k^A_{12} & 0 \\ k^A_{21} & k^A_{22} + k^B_{11} & k^B_{12} \\ 0 & k^B_{21} & k^B_{22} \end{bmatrix} \begin{Bmatrix} u_1 \\ u_2 \\ u_3 \end{Bmatrix} = \begin{Bmatrix} F_{A1} \\ F_{A2} + F_{B1} \\ F_{B2} \end{Bmatrix} \tag{9.7}$$

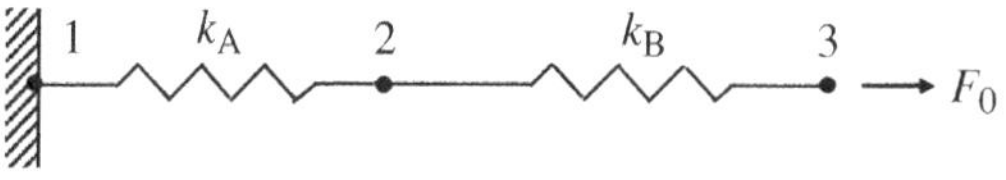

Figure 9.4

where, $u_1 = u_{A1}$, $u_2 = u_{A2} = u_{B1}$, $u_3 = u_{B2}$.

Now, let us solve the problem of Figure 9.1 under the following boundary conditions:

Boundary condition (Figure 9.4) $\begin{cases} u_1 = 0 \\ F_{B2} = F_0 \\ F_{A2} + F_{B1} = 0 \end{cases}$

(No external force is applied at node 2)

Replacing as $k_{11}^A = k_A$, $k_{12}^A = -k_A$, ..., Equation 9.7 is expressed as

$$\begin{bmatrix} k_A & -k_A & 0 \\ -k_A & (k_A + k_B) & -k_B \\ 0 & -k_B & k_B \end{bmatrix} \begin{Bmatrix} 0 \\ u_2 \\ u_3 \end{Bmatrix} = \begin{Bmatrix} F_{A1} \\ 0 \\ F_0 \end{Bmatrix} \tag{9.8}$$

Unknowns of Equation 9.8 are u_2, u_3 and F_{A1}. This means that the reaction is unknown at the node where displacement is given and the displacement is unknown at the node where the external force is given.

Expanding Equation 9.8, we have the following linear simultaneous equation.

$$\left. \begin{aligned} -k_A u_2 &= F_{A1} \\ (k_A + k_B)u_2 - k_B u_3 &= 0 \\ -k_B u_2 + k_B u_3 &= F_0 \end{aligned} \right\} \tag{9.9}$$

In order to determine u_2 and u_3, we use the second and the third equations of Equation 9.9. The first equation of Equation 9.9 is related to node 1 and it cannot be used to find the displacement u_2. The first equation is used to determine the reaction force F_{A1} after the displacement u_2 is determined. Thus, the solutions of Equation 9.9 are obtained as follows.

$$u_2 = \frac{F_0}{k_A}, u_3 = \frac{F_0}{k_A} + \frac{F_0}{k_B}, F_{A1} - -F_0 \text{ (pay attention to sign)} \tag{9.10}$$

The above explanation is an example of two connected springs expressed in FEM style. According to this example, any problem of springs can be solved completely in the same way regardless of the number of springs and different spring constants. Namely, the *total stiffness matrix* corresponding to Equation 9.7 is composed by including all the *stiffness matrices* of individual springs (*elements*). The solution is obtained by solving linear simultaneous equations similar to Equation 9.8.

Example 2   Tension of bars

For element A, an expression similar to Equation 9.1 of Example 1 can be made as follows (Figure 9.5). The meaning of k is the same as the previous example.

$$\begin{bmatrix} k_{11}^A & k_{12}^A \\ k_{21}^A & k_{22}^A \end{bmatrix} \begin{Bmatrix} u_{A1} \\ u_{A2} \end{Bmatrix} = \begin{Bmatrix} F_{A1} \\ F_{A2} \end{Bmatrix} \tag{9.11}$$

Here, considering Hooke's law of the strength of materials $\lambda = Pl/ES$ and $P = k\lambda$,

$$k_{11}^A = \frac{S_A E_A}{l_A}, k_{12}^A = -\frac{S_A E_A}{l_A}, k_{21}^A = -\frac{S_A E_A}{l_A}, k_{22}^A = \frac{S_A E_A}{l_A} \tag{9.12}$$

For element B, we can write

$$\begin{bmatrix} k_{11}^B & k_{12}^B \\ k_{21}^B & k_{22}^B \end{bmatrix} \begin{Bmatrix} u_{B1} \\ u_{B2} \end{Bmatrix} = \begin{Bmatrix} F_{B1} \\ F_{B2} \end{Bmatrix} \tag{9.13}$$

Here,

$$k_{11}^B = \frac{S_B E_B}{l_B}, k_{12}^B = -\frac{S_B E_B}{l_B}, k_{21}^B = -\frac{S_B E_B}{l_B}, k_{22}^B = \frac{S_B E_B}{l_B} \tag{9.14}$$

Combining Equations 9.11 and 9.13 into one matrix, we have the equation as Example 1 as

$$\begin{bmatrix} k_{11}^A & k_{12}^A & 0 \\ k_{21}^A & k_{22}^A + k_{11}^B & k_{12}^B \\ 0 & k_{21}^B & k_{22}^B \end{bmatrix} \begin{Bmatrix} u_1 \\ u_2 \\ u_3 \end{Bmatrix} = \begin{Bmatrix} F_{A1} \\ F_{A2} + F_{B1} \\ F_{B2} \end{Bmatrix} \tag{9.15}$$

Here, $u_1 = u_{A1}$, $u_2 = u_{A2} = u_{B1}$, $u_3 = u_{B2}$.

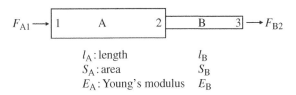

l_A : length l_B
S_A : area S_B
E_A : Young's modulus E_B

Figure 9.5

The structure of Figure 9.5 is written more concretely as follows:

$$
\begin{bmatrix}
\dfrac{S_A E_A}{l_A} & -\dfrac{S_A E_A}{l_A} & 0 \\[2ex]
-\dfrac{S_A E_A}{l_A} & \left(\dfrac{S_A E_A}{l_A} + \dfrac{S_B E_B}{l_B}\right) & -\dfrac{S_B E_B}{l_B} \\[2ex]
0 & -\dfrac{S_B E_B}{l_B} & \dfrac{S_B E_B}{l_B}
\end{bmatrix}
\begin{Bmatrix} u_1 \\ u_2 \\ u_3 \end{Bmatrix}
=
\begin{Bmatrix} F_{A1} \\ F_{A2} + F_{B1} \\ F_{B2} \end{Bmatrix}
\tag{9.16}
$$

Equation 9.16 can be easily solved by giving the boundary conditions of displacement and force.

Problem 9.1.1   Write the expression corresponding to Equation 9.16 for the case of three bars.

9.2   Analysis of Plane Stress Problems by the Finite Element Method

In the previous section, the basic concept of FEM was explained by a one dimensional problem. The advantage of FEM is more extensively shown in 2D and 3D problems rather than 1D problems. As already studied in Chapter 6 of Part I, a simple problem such as a circular hole in an infinite plate under uniform tension can be solved by the stress function method. However, most problems encountered in practical situations, having complex shapes and boundary conditions, cannot be solved in a closed form by the stress function method. Even for such cases, FEM enables one to obtain an approximate solution. Using FEM for solutions, it is not necessary to change the computer program for solving individual problems. Once the program is completed, the change in problems can be treated as the difference of input data for the shape of structures and boundary conditions.

Thus, nowadays major commercial computer softwares such as ABAQUS and ANSYS are used worldwide for design and research.

As described above, although the basic theory of FEM is already established, the most simple principle of 2D FEM is explained in this section.

9.2.1   Approximation of 2D Plate Problems by a Set of Triangular Elements

Figure 9.6a shows the practical plate in question and Figure 9.6b shows the plate approximated by a set of triangular plates. The individual triangular plate is called a *triangular plate element*, as shown in Figure 9.6c.

In general, the stress and strain inside the actual plate of Figure 9.6a are complex, in the model of Figure 9.6b a relatively simple state of stress and strain is assumed to obtain an approximate solution.

For understanding the concept of FEM for 2D problems, triangular elements are treated as a kind of springs as an extension of the one dimensional problem explained in Section 9.1.

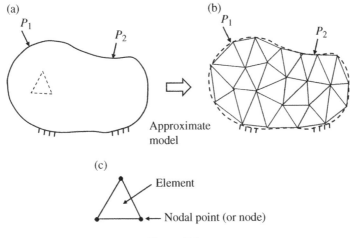

(a)

(b)

Approximate model

(c) Element — Nodal point (or node)

Figure 9.6

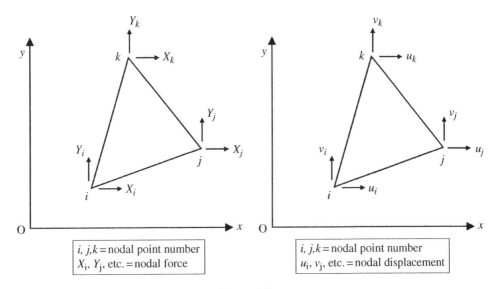

| i, j,k = nodal point number |
| X_i, Y_j, etc. = nodal force |

| i, j,k = nodal point number |
| u_i, v_j, etc. = nodal displacement |

Figure 9.7

However, since one plate element has three nodes as shown in Figure 9.7, one node has two degrees of freedom with respect to forces and displacements, resulting in 6 degrees of freedom.

Considering this nature of a plate element in a similar way to Examples 1 and 2, we can express the relationship between forces and displacements at nodes as follows.

$$[k]\{u\} = \{F\} \qquad (9.17)$$

$$[k] = \begin{bmatrix} k_{11} & k_{12} & k_{13} & k_{14} & k_{15} & k_{16} \\ k_{21} & k_{22} & k_{23} & k_{24} & k_{25} & k_{26} \\ k_{31} & k_{32} & k_{33} & k_{34} & k_{35} & k_{36} \\ k_{41} & k_{42} & k_{43} & k_{44} & k_{45} & k_{46} \\ k_{51} & k_{52} & k_{53} & k_{54} & k_{55} & k_{56} \\ k_{61} & k_{62} & k_{63} & k_{64} & k_{65} & k_{66} \end{bmatrix} \tag{9.18}$$

$$(6 \times 6)$$

$$\{u\} = \begin{Bmatrix} u_i \\ v_i \\ u_j \\ v_j \\ u_k \\ v_k \end{Bmatrix}, \quad \{f\} = \begin{Bmatrix} X_i \\ Y_i \\ X_j \\ Y_j \\ X_k \\ Y_k \end{Bmatrix} \tag{9.19}$$

$$(6 \times 1)(6 \times 1)$$

Here, the meaning of an element of the stiffness matrix k_{23} for example is interpreted as follows.

k_{23}: The reaction force which the node i is subjected to in the y direction for $u_j = 1$ and when the displacements of other nodes are fixed $(u_i = v_i = v_j = u_k = v_k = 0)$.

In Equation 9.18, it is known that the relationship $k_{lm} = k_{ml}$ holds (see Problem 1 at the end of this chapter).

In order to solve 2D problems, it is necessary to determine $k_{lm}(l = 1 \sim 6, m = 1 \sim 6)$. Once the values of all the elements k_{lm} are determined, all the stiffness matrices are combined into the form of a simultaneous equation as in Examples 1 and 2 for a one dimensional problem.

The following factors influence the value of k_{lm}.

• Elastic constant of material (E, ν)
• Shape and size of triangular element, that is (x_i, y_i), (x_j, y_j), (x_k, y_k)
• Thickness t of plate.

If we try to find an exact solution, we need to consider the variations of stress and strain inside an element depending on the local coordinate. However, such an approach makes the problem very complicated[1]. Therefore, a more efficient method is to make some assumption regarding the variation of stress and strain inside the element. For example, the simplest way is to assume the stress and strain inside an element is constant, though stresses and strains in individual elements are naturally different. Meshing the plate in question into sufficiently small elements, it is expected that practically sufficient accuracy is guaranteed by this assumption.

---

[1] Even if the exact value of k_{lm} is determined, using such value is not necessarily appropriate from the viewpoint of the compatibility equation.

One of the methods to determine k_{lm} is to apply the principle of virtual work which was explained in Chapter 8 of Part I. For the preparation of this calculation, we express the stress–strain relationship (Hooke's law) in the style of a FEM matrix.

9.2.2   Relationship between Stress and Strain in Plane Stress Problem

The relationship between stress and strain for a plane stress problem is expressed as follows:

$$\left.\begin{array}{c} \sigma_x = \dfrac{E}{1-\nu^2}\left(\varepsilon_x + \nu\varepsilon_y\right), \sigma_y = \dfrac{E}{1-\nu^2}\left(\varepsilon_y + \nu\varepsilon_x\right), \sigma_z = 0 \\[2mm] \tau_{xy} = G\gamma_{xy} = \dfrac{E}{2(1+\nu)}\gamma_{xy}, \tau_{yz} = 0, \tau_{zx} = 0 \end{array}\right\} \tag{9.20}$$

Expressing Equation 9.20 in the form of a matrix:

$$\{\sigma\} = [D]\{\varepsilon\} \tag{9.21}$$

where,

$$\{\sigma\} = \left\{\begin{array}{c} \sigma_x \\ \sigma_y \\ \tau_{xy} \end{array}\right\}, \{\varepsilon\} = \left\{\begin{array}{c} \varepsilon_x \\ \varepsilon_y \\ \gamma_{xy} \end{array}\right\}, [D] = \frac{E}{1-\nu^2}\begin{bmatrix} 1 & \nu & 0 \\ \nu & 1 & 0 \\ 0 & 0 & (1-\nu)/2 \end{bmatrix} \tag{9.22}$$

Namely,

$$\left\{\begin{array}{c} \sigma_x \\ \sigma_y \\ \tau_{xy} \end{array}\right\} = \frac{E}{1-\nu^2}\begin{bmatrix} 1 & \nu & 0 \\ \nu & 1 & 0 \\ 0 & 0 & (1-\nu)/2 \end{bmatrix}\left\{\begin{array}{c} \varepsilon_x \\ \varepsilon_y \\ \gamma_{xy} \end{array}\right\} \tag{9.23}$$

9.2.3   Stiffness Matrix of a Triangular Plate Element

Let us pay attention to a triangular domain imagined in Figure 9.6a (illustrated by dotted lines). If the domain is sufficiently small, the stress and strain in this domain may be regarded almost constant. Thus, assuming the stress inside the domain is constant, we imagine other triangular domains orderly in the neighborhood of the domain and assume the stresses within those domains are constant, though the constant values naturally differ from domain to domain. In this way, we fill the plate in question with those triangular domains. Based on this assumption, if we can determine the stiffness matrix of an individual element, we can easily determine the stiffness matrix of the total system.

Although the above assumption is the simplest one, it is possible to make other assumptions for elements in which the values of stress and strain vary linearly or in a more complex way. In this way, we may be able to expect an improvement in the accuracy of solutions, though the determination of k becomes complicated with increasing degrees of freedom of stress and strain distribution in an element. In recent FEM commercial software, all these difficulties are solved internally, as in a black box, and no additional knowledge is required for software users. However, using commercial software as a black box is always a concern, for users can make simple mistakes. To avoid this kind of uncertainty and mistakes, it will be useful to experience the derivation of a stiffness matrix for the simplest case, as follows.

Since we assumed the strains in an element are constant, the displacements in the element are expressed by

$$u = \alpha_1 + \alpha_2 x + \alpha_3 y, \quad v = \alpha_4 + \alpha_5 x + \alpha_6 y \tag{9.24}$$

The importance of the assumption of Equation 9.24 is that a straight line of a triangular element before deformation keeps a straight line after deformation and accordingly the compatibility condition (see Chapter 2 of Part I) is satisfied. In other words, elements after deformation do not interfere like Figure 9.8 or produce clearance like Figure 9.9. If the displacements inside an element which do not satisfy the compatibility condition are assumed, the exact solution cannot be obtained even with precise meshing.

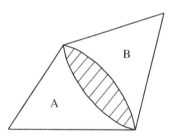

Figure 9.8

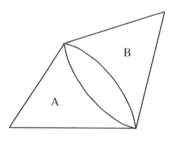

Figure 9.9

The displacements at the nodes of an element are expressed by inputting nodal coordinates (x, y) into Equation 9.24 as follows.

$$\left.\begin{array}{ll} u_i = \alpha_1 + \alpha_2 x_i + \alpha_3 y_i, & v_i = \alpha_4 + \alpha_5 x_i + \alpha_6 y_i \\[2mm] u_j = \alpha_1 + \alpha_2 x_j + \alpha_3 y_j, & v_j = \alpha_4 + \alpha_5 x_j + \alpha_6 y_j \\[2mm] u_k = \alpha_1 + \alpha_2 x_k + \alpha_3 y_k, & v_k = \alpha_4 + \alpha_5 x_k + \alpha_6 y_k \end{array}\right\} \tag{9.25}$$

The stiffness matrix of one element (the relationship between external forces, displacement of nodes and the shape of triangular element) is derived using the above equations as follows. Expressing Equation 9.25 in the form of a matrix,

$$\{u\} = [T]\{a\} \tag{9.26}$$

where,

$$\{u\} = \begin{Bmatrix} u_i \\ v_i \\ u_j \\ v_j \\ u_k \\ v_k \end{Bmatrix}, \quad [T] = \begin{bmatrix} 1 & x_i & y_i & 0 & 0 & 0 \\ 0 & 0 & 0 & 1 & x_i & y_i \\ 1 & x_j & y_j & 0 & 0 & 0 \\ 0 & 0 & 0 & 1 & x_j & y_j \\ 1 & x_k & y_k & 0 & 0 & 0 \\ 0 & 0 & 0 & 1 & x_k & y_k \end{bmatrix}, \quad \{a\} = \begin{Bmatrix} \alpha_1 \\ \alpha_2 \\ \alpha_3 \\ \alpha_4 \\ \alpha_5 \\ \alpha_6 \end{Bmatrix}$$

expressing $\{a\}$ by $\{u\}$, α_1, α_2 and so on are expressed by coordinates and displacements of nodes as follows.

$$\{a\} = [T]^{-1}\{u\} \tag{9.27}$$

where,

$$[T]^{-1} =$$

$$\frac{1}{2\Delta} \begin{bmatrix} x_j y_k - x_k y_j & 0 & x_k y_i - x_i y_k & 0 & x_i y_j - x_j y_i & 0 \\ y_j - y_k & 0 & y_k - y_i & 0 & y_i - y_j & 0 \\ x_k - x_j & 0 & x_i - x_k & 0 & x_j - x_i & 0 \\ 0 & x_j y_k - x_k y_j & 0 & x_k y_i - x_i y_k & 0 & x_i y_j - x_j y_i \\ 0 & y_j - y_k & 0 & y_k - y_i & 0 & y_i - y_j \\ 0 & x_k - x_j & 0 & x_i - x_k & 0 & x_j - x_i \end{bmatrix} \tag{9.28}$$

Δ is the area of triangular element, which is expressed as

$$2\Delta = \begin{vmatrix} 1 & x_i & y_i \\ 1 & x_j & y_j \\ 1 & x_k & y_k \end{vmatrix} \tag{9.29}$$

where nodes are numbered counterclockwise with i, j and k.

Strains ε_x, ε_y and γ_{xy} are expressed in terms of u and v as follows.

$$\varepsilon_x = \frac{\partial u}{\partial x}, \quad \varepsilon_y = \frac{\partial v}{\partial y}, \quad \gamma_{xy} = \frac{\partial u}{\partial y} + \frac{\partial v}{\partial x} \tag{9.30}$$

Thus, strains calculated by Equations 9.24 and 9.30 are constant, as follows.

$$\varepsilon_x = \alpha_2, \quad \varepsilon_y = \alpha_6, \quad \gamma_{xy} = \alpha_3 + \alpha_5 \tag{9.31}$$

Expressing Equation 9.31 in the form of a matrix, we have

$$\left\{ \begin{array}{c} \varepsilon_x \\ \varepsilon_y \\ \gamma_{xy} \end{array} \right\} = \begin{bmatrix} 0 & 1 & 0 & 0 & 0 & 0 \\ 0 & 0 & 0 & 0 & 0 & 1 \\ 0 & 0 & 1 & 0 & 1 & 0 \end{bmatrix} \left\{ \begin{array}{c} \alpha_1 \\ \alpha_2 \\ \alpha_3 \\ \alpha_4 \\ \alpha_5 \\ \alpha_6 \end{array} \right\} \tag{9.32}$$

$$(3 \times 6) \qquad\qquad (6 \times 1)$$

Namely,

$$\{\varepsilon\} = [B]\{\alpha\} \tag{9.33}$$

Using Equation (9.27) to Equation (9.33),

$$\{\varepsilon\} = [B][T]^{-1}\{u\} = [N]\{u\} \tag{9.34}$$

where

$$[N] = [B][T]^{-1}$$

Since the right hand side of Equation 9.34 does not include the terms x and y, the strains are constant inside a triangular plate element.

Now, the principle of virtual work is applied to calculate the stiffness matrix. When we apply virtual displacements to a body in equilibrium condition, we call the left hand side of Equation 8.27 the external work and call the right hand side of Equation 8.27 the internal work. The internal work W_{int} per unit volume is

$$w_{int} = \sigma_x \delta\varepsilon_x + \sigma_y \delta\varepsilon_y + \tau_{xy}\delta\gamma_{xy} = \{\sigma\}^T\{\delta\varepsilon\} \tag{9.35}$$

Introducing Equation 9.21 into Equation 9.35,

$$w_{int} = \{\varepsilon\}^T[D]\{\delta\varepsilon\}$$

Using Equation 9.34:

$$w_{int} = \{u\}^T[N]^T[D][N]\{\delta u\} \tag{9.36}$$

Therefore, the total internal work for one element is

$$W_{int} = \int^V w_{int}d\text{vol} = \{u\}^T\left[\int\int\int [N]^T[D][N]dx\cdot dy\cdot dz\right]\{\delta u\} \tag{9.37}$$

On the other hand, the external work W_{ext} is

$$W_{ext} = \{F\}^T\{\delta u\} \tag{9.38}$$

where $\{F\}$ is the column vector of the forces at nodal points. Namely,

$$\{F\} = \begin{Bmatrix} X_i \\ Y_i \\ X_j \\ Y_j \\ X_k \\ Y_k \end{Bmatrix} \tag{9.39}$$

Equating the internal work to the external work,

$$\{F\}^T\{\delta u\} = \{u\}^T\left[\int\int\int [N]^T[D][N]dx\cdot dy\cdot dz\right]\{\delta u\} \tag{9.40}$$

Since Equation 9.40 holds for arbitrary virtual displacements,

$$\{F\}^T = \{u\}^T\left[\int\int\int [N]^T[D][N]dx\cdot dy\cdot dz\right]$$

Namely,

$$\{F\} = \left[\iiint [N]^T [D][N] dx \cdot dy \cdot dz \right] \{u\} \tag{9.41}$$

[…] of Equation 9.41 is the stiffness matrix we want to calculate (see Equation 9.17). Denoting the stiffness matrix by $[k]$,

$$[k] = \iiint [N]^T [D][N] dx \cdot dy \cdot dz \tag{9.42}$$

For the case of plate with constant thickness t_0,

$$[k] = t_0 \iint [N]^T [D][N] dx \cdot dy \tag{9.43}$$

Since the integral function does not include the terms x and y, assuming the area of triangular element by Δ,

$$[k] = t_0 \Delta [N]^T [D][N] \tag{9.44}$$

$[k]$ is the matrix of 6×6 as shown by Equation 9.18 and the components of the matrix can be determined by Equations 9.28, 9.32 and 9.34 as follows.

Regarding $[N]$, we have

$$[N] = [B][T]^{-1} = \frac{1}{2\Delta} \begin{bmatrix} y_j - y_k & 0 & y_k - y_i & 0 & y_i - y_j & 0 \\ 0 & x_k - x_j & 0 & x_i - x_k & 0 & x_j - x_i \\ x_k - x_j & y_j - y_k & x_i - x_k & y_k - y_i & x_j - x_i & y_i - y_j \end{bmatrix} \tag{9.45}$$

And since $[D]$ is given by Equation 9.22, all the components of $[k]$ are summarized as Table 9.1.

Now, let us confirm the physical meaning of $[k]$ when the components $[k]$ are interpreted intuitively as equivalent to the meaning of the spring constant k for a one dimensional problem (Figure 9.10). For this purpose, the intuitive interpretation of k_{11} of one component of $[k]$ is examined as follows:

$$k_{11} = \frac{t_0 E}{4\Delta(1-\nu^2)} \left\{ (y_j - y_k)^2 + \frac{1-\nu}{2}(x_k - x_j)^2 \right\}$$

Considering

$$\Delta = \frac{1}{2}(x_k - x_j) \cdot (y_i - y_j), \quad (y_j - y_k) = 0$$

Table 9.1

$$c = \frac{t_0 E}{4\Delta(1-v^2)}$$

$$k_{11} = c\left\{ (y_j - y_k)^2 + \frac{1-v}{2}(x_k - x_j)^2 \right\}$$

$$k_{12} = c\left\{ v(y_j - y_k)\cdot(x_k - x_j) + \frac{1-v}{2}(x_k - x_j)\cdot(y_i - y_k) \right\}$$

$$k_{13} = c\left\{ (y_j - y_k)\cdot(y_k - y_i) + \frac{1-v}{2}(x_k - x_j)\cdot(x_i - x_k) \right\}$$

$$k_{14} = c\left\{ v(y_j - y_k)\cdot(x_i - x_k) + \frac{1-v}{2}(x_k - x_j)\cdot(y_k - y_i) \right\}$$

$$k_{15} = c\left\{ (y_j - y_k)\cdot(y_i - y_j) + \frac{1-v}{2}(x_k - x_j)\cdot(x_j - x_i) \right\}$$

$$k_{16} = c\left\{ v(y_j - y_k)\cdot(x_j - x_i) + \frac{1-v}{2}(x_k - x_j)\cdot(y_i - y_j) \right\}$$

$$k_{21} = k_{12}$$

$$k_{22} = c\left\{ (x_k - x_j)^2 + \frac{1-v}{2}(y_j - y_k)^2 \right\}$$

$$k_{23} = c\left\{ v(x_k - x_j)\cdot(y_k - y_i) + \frac{1-v}{2}(y_j - y_k)\cdot(x_i - x_k) \right\}$$

$$k_{24} = c\left\{ (x_k - x_j)\cdot(x_i - x_k) + \frac{1-v}{2}(y_j - y_k)\cdot(y_k - y_i) \right\}$$

$$k_{25} = c\left\{ v(x_k - x_j)\cdot(y_i - y_j) + \frac{1-v}{2}(y_j - y_k)\cdot(x_j - x_i) \right\}$$

$$k_{26} = c\left\{ (x_k - x_j)\cdot(x_j - x_i) + \frac{1-v}{2}(y_j - y_k)\cdot(y_i - y_j) \right\}$$

$$k_{31} = k_{13}, \quad k_{32} = k_{23}$$

$$k_{33} = c\left\{ (y_k - y_i)^2 + \frac{1-v}{2}(x_i - x_k)^2 \right\}$$

$$k_{34} = c\left\{ v(y_k - y_i)\cdot(x_i - x_k) + \frac{1-v}{2}(x_i - x_k)\cdot(y_k - y_i) \right\}$$

$$k_{35} = c\left\{ (y_k - y_i)\cdot(y_i - y_j) + \frac{1-v}{2}(x_i - x_k)\cdot(x_j - x_i) \right\}$$

$$k_{36} = c\left\{ v(y_k - y_i)\cdot(x_j - x_i) + \frac{1-v}{2}(x_i - x_k)\cdot(y_i - y_j) \right\}$$

$$k_{41} = k_{14}, \quad k_{42} = k_{24}, \quad k_{43} = k_{34}$$

$$k_{44} = c\left\{ (x_i - x_k)^2 + \frac{1-v}{2}(y_k - y_i)^2 \right\}$$

$$k_{45} = c\left\{ v(x_i - x_k)\cdot(y_i - y_j) + \frac{1-v}{2}(y_k - y_i)\cdot(x_j - x_i) \right\}$$

$$k_{46} = c\left\{ (x_i - x_k)\cdot(x_j - x_i) + \frac{1-v}{2}(y_k - y_i)\cdot(y_i - y_j) \right\}$$

$$k_{51} = k_{15}, \quad k_{52} = k_{25}, \quad k_{53} = k_{35}, \quad k_{54} = k_{45}$$

$$k_{55} = c\left\{ (y_i - y_j)^2 + \frac{1-v}{2}(x_j - x_i)^2 \right\}$$

$$k_{56} = c\left\{ v(y_i - y_j)\cdot(x_j - x_i) + \frac{1-v}{2}(x_j - x_i)\cdot(y_i - y_j) \right\}$$

$$k_{61} = k_{16}, \quad k_{62} = k_{26}, \quad k_{63} = k_{36}, \quad k_{64} = k_{46}, \quad k_{65} = k_{56}$$

$$k_{66} = c\left\{ (x_j - x_i)^2 + \frac{1-v}{2}(y_i - y_j)^2 \right\}$$

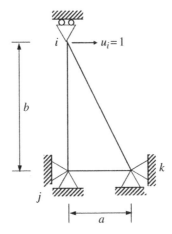

Figure 9.10

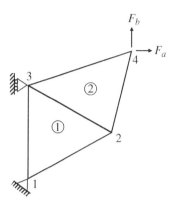

Figure 9.11

we have

$$k_{11} = \frac{t_0 E}{4(1+\nu)} \cdot \frac{(x_k - x_j)}{(y_i - y_j)} = \frac{t_0 E}{4(1+\nu)} \cdot \frac{a}{b}$$

This result matches the intuitive prediction that the force X_i, for producing displacement $u_i = 1$ at node i, should be necessarily larger for large a and smaller for large b. It is intuitively acceptable that X_i should be larger for large t_0 and E.

Readers are encouraged to confirm the physical meaning of other components of $[k]$ in the same way.

9.2.4 Stiffness Matrix of the Total Structure

In order to solve the problem for a total plane structure, individual stiffness matrices of triangular elements are composed into one total stiffness matrix. The method of assembly is essentially the same as the procedure explained for the case of composing two springs A and B. Figure 9.11 shows a plane structure which is divided into two triangular elements and four nodes. Numbering the nodes and elements is shown in Figure 9.11.

The boundary conditions of the structure of Figure 9.11 are defined as follows:

$$\left.\begin{array}{l} \text{The displacements in } x \text{ direction at nodes 1 and 3}: u = 0. \\ \qquad \text{The displacement in } y \text{ direction at node 1}: v = 0. \\[2ex] \text{The external force in } x \text{ direction at node 4}: F_a = 0. \\ \text{The external force in } y \text{ direction at node 4}: F_b = 0. \end{array}\right\} \qquad (9.46)$$

Paying attention to element ①, the relationship between the nodal force $\{f^1\}$ and the nodal displacement $\{\delta^1\}$ is expressed by

$$\{f^1\} = [k^1]\{\delta^1\} \tag{9.47}$$

The detail of Equation 9.47 is

$$
\begin{Bmatrix} X_1^1 \\ Y_1^1 \\ X_2^1 \\ Y_2^1 \\ X_3^1 \\ Y_3^1 \end{Bmatrix}
=
\begin{bmatrix}
k_{11}^1 & k_{12}^1 & k_{13}^1 & k_{14}^1 & k_{15}^1 & k_{16}^1 \\
k_{21}^1 & k_{22}^1 & \cdots & \cdots & \cdots & k_{26}^1 \\
\vdots & \vdots & \vdots & \vdots & \vdots & \vdots \\
\vdots & \vdots & \vdots & \vdots & \vdots & \vdots \\
\vdots & \vdots & \vdots & \vdots & \vdots & \vdots \\
k_{61}^1 & \cdots & \cdots & \cdots & \cdots & k_{66}^1
\end{bmatrix}
\begin{Bmatrix} u_1 \\ v_1 \\ u_2 \\ v_2 \\ u_3 \\ v_3 \end{Bmatrix}
\tag{9.48}
$$

where, regarding the expression of $X^m{}_n$, $Y^m{}_n$, ..., m is the element number and n is the node number. The values for $k^m{}_{11}$, $k^m{}_{12}$, ..., $k^m{}_{66}$ can be calculated by the equations already shown in Table 9.1.

Considering the boundary conditions of Equation 9.46, Equation 9.48 is written as follows:

$$
\left.
\begin{aligned}
X_1^1 &= k_{13}^1 u_2 + k_{14}^1 v_2 + k_{16}^1 v_3, \quad Y_1^1 = k_{23}^1 u_2 + k_{24}^1 v_2 + k_{26}^1 v_3 \\
X_2^1 &= k_{33}^1 u_2 + k_{34}^1 v_2 + k_{36}^1 v_3, \quad Y_2^1 = k_{43}^1 u_2 + k_{44}^1 v_2 + k_{46}^1 v_3 \\
X_3^1 &= k_{53}^1 u_2 + k_{54}^1 v_2 + k_{56}^1 v_3, \quad Y_3^1 = k_{63}^1 u_2 + k_{64}^1 v_2 + k_{66}^1 v_3
\end{aligned}
\right\}
\tag{9.49}
$$

As with element ①, regarding element ② we have

$$
\begin{Bmatrix} X_3^2 \\ Y_3^2 \\ X_2^2 \\ Y_2^2 \\ X_4^2 \\ Y_4^2 \end{Bmatrix}
=
\begin{bmatrix}
k_{11}^2 & k_{12}^2 & k_{13}^2 & k_{14}^2 & k_{15}^2 & k_{16}^2 \\
k_{21}^2 & k_{22}^2 & \cdots & \cdots & \cdots & k_{26}^2 \\
\vdots & \vdots & \vdots & \vdots & \vdots & \vdots \\
\vdots & \vdots & \vdots & \vdots & \vdots & \vdots \\
\vdots & \vdots & \vdots & \vdots & \vdots & \vdots \\
k_{61}^2 & \cdots & \cdots & \cdots & \cdots & k_{66}^2
\end{bmatrix}
\begin{Bmatrix} u_3 \\ v_3 \\ u_2 \\ v_2 \\ u_4 \\ v_4 \end{Bmatrix}
\tag{9.50}
$$

Considering the boundary conditions Equation (9.46),

$$\left. \begin{aligned}
X_3^2 &= k_{12}^2 v_3 + k_{13}^2 u_2 + k_{14}^2 v_2 + k_{15}^2 u_4 + k_{16}^2 v_4 \\
Y_3^2 &= k_{22}^2 v_3 + k_{23}^2 u_2 + k_{24}^2 v_2 + k_{25}^2 u_4 + k_{26}^2 v_4 \\
X_2^2 &= k_{32}^2 v_3 + k_{33}^2 u_2 + k_{34}^2 v_2 + k_{35}^2 u_4 + k_{36}^2 v_4 \\
Y_2^2 &= k_{42}^2 v_3 + k_{43}^2 u_2 + k_{44}^2 v_2 + k_{45}^2 u_4 + k_{46}^2 v_4 \\
X_4^2 &= k_{52}^2 v_3 + k_{53}^2 u_2 + k_{54}^2 v_2 + k_{55}^2 u_4 + k_{56}^2 v_4 \\
Y_4^2 &= k_{62}^2 v_3 + k_{63}^2 u_2 + k_{64}^2 v_2 + k_{65}^2 u_4 + k_{66}^2 v_4
\end{aligned} \right\} \tag{9.51}$$

Considering the boundary conditions and resultant force condition at nodes,

$$\left. \begin{aligned}
0 &= X_2^1 + X_2^2, \quad 0 = Y_2^1 + Y_2^2, \quad 0 = Y_3^1 + Y_3^2 \\
F_a &= X_4^2, \quad F_b = X_4^2
\end{aligned} \right\} \tag{9.52}$$

The expression for the total matrix by composing Equations 9.49 and 9.51 can be made as follows:

$$X_1^1 = k_{13}^1 u_2 + k_{14}^1 v_2 + k_{16}^1 v_3 \tag{9.53-1}$$

$$Y_1^1 = k_{23}^1 u_2 + k_{24}^1 v_2 + k_{26}^1 v_3 \tag{9.53-2}$$

$$X_2^1 + X_2^2 = \left(k_{33}^1 + k_{33}^2\right) u_2 + \left(k_{34}^1 + k_{34}^2\right) v_2 + \left(k_{36}^1 + k_{32}^2\right) v_3 + k_{35}^2 u_4 + k_{36}^2 v_4 \tag{9.53-3}$$

$$Y_2^1 + Y_2^2 = \left(k_{43}^1 + k_{43}^2\right) u_2 + \left(k_{44}^1 + k_{44}^2\right) v_2 + \left(k_{46}^1 + k_{42}^2\right) v_3 + k_{45}^2 u_4 + k_{46}^2 v_4 \tag{9.53-4}$$

$$X_3^1 + X_3^2 = \left(k_{53}^1 + k_{13}^2\right) u_2 + \left(k_{54}^1 + k_{14}^2\right) v_2 + \left(k_{56}^1 + k_{12}^2\right) v_3 + k_{15}^2 u_4 + k_{16}^2 v_4 \tag{9.53-5}$$

$$Y_3^1 + Y_3^2 = \left(k_{63}^1 + k_{23}^2\right) u_2 + \left(k_{64}^1 + k_{24}^2\right) v_2 + \left(k_{66}^1 + k_{22}^2\right) v_3 + k_{25}^2 u_4 + k_{26}^2 v_4 \tag{9.53-6}$$

$$X_4^2 = k_{53}^2 u_2 + k_{54}^2 v_2 + k_{52}^2 v_3 + k_{55}^2 u_4 + k_{56}^2 v_4 \tag{9.53-7}$$

$$X_4^2 = k_{63}^2 u_2 + k_{64}^2 v_2 + k_{62}^2 v_3 + k_{65}^2 u_4 + k_{66}^2 v_4 \tag{9.53-8}$$

Equation 9.53 is composed of eight sub-equations. The reason is that the number of nodes for the problem (Figure 9.11) is four.

Since five sub-equations, (9.53-3), (9.53-4), (9.53-6), (9.53-7) and (9.53-8) of Equation 9.53, correspond to the boundary conditions of Equation 9.52, the values of the left hand side of Equation 9.53 can be replaced by these boundary conditions. The unknowns in Equation 9.53 are u_2, v_2, v_3, u_4 and v_4, and the values can be obtained by solving the simultaneous equation with the five sub-equations.

Once the displacements are determined, the reaction forces of the structure of Figure 9.11 can be determined by inputting those values into Equation 9.53 (at (9.53-1), (9.53-2) and (9.53-5)).

Expressing this case by the matrix form, we have

$$
\begin{Bmatrix} 0 \\ 0 \\ 0 \\ F_a \\ F_b \end{Bmatrix} =
\begin{bmatrix}
\left(k^1_{33}+k^2_{33}\right) & \left(k^1_{34}+k^2_{34}\right) & \left(k^1_{36}+k^2_{32}\right) & k^2_{35} & k^2_{36} \\
\left(k^1_{43}+k^2_{43}\right) & \left(k^1_{44}+k^2_{44}\right) & \left(k^1_{46}+k^2_{42}\right) & k^2_{45} & k^2_{46} \\
\left(k^1_{63}+k^2_{23}\right) & \left(k^1_{64}+k^2_{24}\right) & \left(k^1_{66}+k^2_{22}\right) & k^2_{25} & k^2_{26} \\
k^2_{53} & k^2_{54} & k^2_{52} & k^2_{55} & k^2_{56} \\
k^2_{63} & k^2_{64} & k^2_{62} & k^2_{65} & k^2_{66}
\end{bmatrix}
\begin{Bmatrix} u_2 \\ v_2 \\ v_3 \\ u_4 \\ v_4 \end{Bmatrix}
\tag{9.54}
$$

Equation 9.54 is a linear simultaneous equation to determine the displacements u_2, v_2, v_3, u_4 and v_4 which are used to determine the reaction forces with Equation 9.53 (at 9.53-1, 9.53-2 and 9.53-5).

Equation 9.54 can be rewritten in the following simple form.

$$
\{F\}=[K]\{\delta\}
\tag{9.55}
$$

$\{F\}$: Column vector of known nodal forces
$[K]$: Stiffness matrix
$\{\delta\}$: Column vector of nodal displacements

Once the displacements are determined by Equation 9.54, the strains are calculated by Equation 9.34 in terms of $[N]$ given by Equation 9.45. The stresses are determined with the strains in terms of Equation 9.23 (Hooke's law).

By programming the procedure described above in the software, solutions for arbitrary shape and loading condition can be obtained automatically. In this calculation procedure, due to the relationship $k_{lm}=k_{ml}$ in one element, it should be noted that $[K]$ is also a symmetric matrix.

The solution for a simple problem like Figure 9.11 with only five unknowns can be solved by hand calculation. However, with an increasing number of elements, the number of degrees of freedom of the simultaneous equation becomes huge and the calculation must be performed using a computer. In plane problems, the number of degrees of freedom is twice the number of nodes. Thus, to solve a problem of 100 nodes, a system of 200 simultaneous equations must be solved.

9.2.5 Expression of Boundary Conditions and Basic Knowledge for Element Meshing

The boundary condition for FEM calculation should be as close as to those of real problems before meshing the structure in question. In general, there are two essential conditions for defining displacements at a node, as shown in Figure 9.12a, b. Figure 9.12a is the condition expressing the completely fixed node and Figure 9.12b is the condition expressing a condition fixed in one direction with free movement in another direction which is perpendicular to the fixed

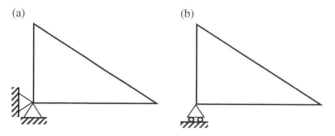

Figure 9.12

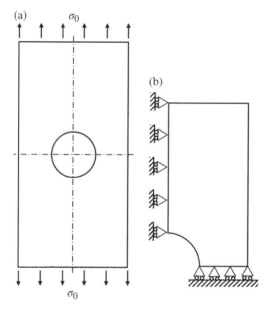

Figure 9.13

direction. The supporting condition of a node like Figure 9.12b is applied to a boundary where the shear stress is free (see the Example problem 3.1 in Chapter 3 of Part I).

As shown in the Example problem 3.1 in Chapter 3 of Part I, for the problem like Figure 9.13a there exists no shear stress τ_{xy} along the symmetry line. Knowing this nature, we can pick up one-quarter of the plate as Figure 9.13b and give supporting conditions like Figure 9.12 to the symmetry lines which limit the calculation into a smaller domain.

In the problem like Figure 9.13a, the stress distribution varies steeply near the hole due to stress concentration. Since we assumed the stress inside the element is constant, it is naturally necessary to put smaller elements near the hole to approximate the steep stress distribution. Figure 9.14 shows an example of such a mesh pattern. Although a solution can be obtained in FEM with any mesh pattern, it is necessary to carry out appropriate meshing to obtain an accurate result.

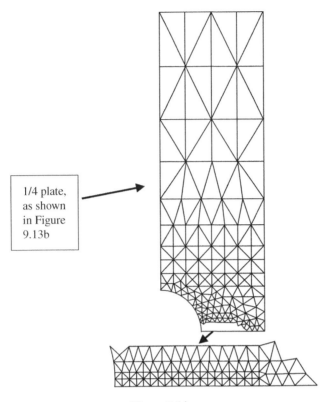

Figure 9.14

Due to the development of FEM theory, modern commercial FEM software is installed with more sophisticated elements and automatic meshing programs. These sophisticated elements are equipped with displacement functions which allow nonconstant stress and strain inside elements with various shapes other than triangular under the compatibility condition. The treatment of more complicated boundary conditions such as contact interfaces with friction are also installed. Automatic meshing programs are very useful for saving analytical time especially for the analysis of 3D problems. On the other hand, the modern progress of FEM software created another negative influence on structural analysis society, because these softwares are often used without understanding the basic knowledge of elasticity which is absolutely necessary to obtain accurate results and to avoid simple mistakes. It must be understood that studying the basic knowledge of theory of elasticity and FEM is crucially important for correct usage of modern FEM softwares.

Problems of Chapter 9

1. Regarding the components of the stiffness matrix, verify that the relationship $k_{lm} = k_{ml}$ holds.
2. Express the simultaneous equation to solve Figure 9.15 in the form of Equation 9.54.

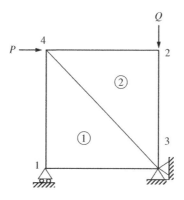

Figure 9.15

Assume that the components of the stiffness matrices k_{ij} are prepared as follows:

Element ①: Node composition 1, 2, 3	Element ②: Node composition 4, 3, 2

$$\begin{bmatrix} k_{11}^{①} & k_{12}^{①} & \cdots & \cdots & \cdots & k_{16}^{①} \\ k_{21}^{①} & \cdots & \cdots & \cdots & \cdots & \cdots \\ \vdots & \vdots & \vdots & \vdots & \vdots & \vdots \\ \vdots & \vdots & \vdots & \vdots & \vdots & \vdots \\ \vdots & \vdots & \vdots & \vdots & \vdots & \vdots \\ k_{61}^{①} & \cdots & \cdots & \cdots & \cdots & k_{66}^{①} \end{bmatrix} \qquad\qquad \begin{bmatrix} k_{11}^{②} & k_{12}^{②} & \cdots & \cdots & \cdots & k_{16}^{②} \\ k_{21}^{②} & \cdots & \cdots & \cdots & \cdots & \cdots \\ \vdots & \vdots & \vdots & \vdots & \vdots & \vdots \\ \vdots & \vdots & \vdots & \vdots & \vdots & \vdots \\ \vdots & \vdots & \vdots & \vdots & \vdots & \vdots \\ k_{61}^{②} & \cdots & \cdots & \cdots & \cdots & k_{66}^{②} \end{bmatrix}$$

3. Solve the problem of Figure 9.16 by FEM. Solve the simultaneous equation numerically with a calculator by the Gaus–Seidel method. Assume plate thickness $t = 1$, Poisson's ratio $\nu = 0.3$ and Young's modulus $E = 2 \times 10^4$ MPa. The unit of the nodal forces is N. The compositions of nodes for two elements are ① 1, 2, 3 and ② 3, 2, 4.

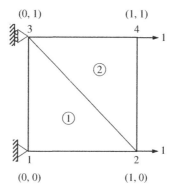

Figure 9.16

4. Compose the total stiffness matrix for the problem of Figure 9.17. Make the matrix $[18 \times 18]$ without considering the condition for the known displacements $u_1 = 0$, $v_1 = 0$, $u_2 = 0$ and $u_3 = 0$. The number of nodes for eight elements are denoted as follows:

① 1, 4, 5; ② 1, 5, 2; ③ 2, 5, 3; ④ 3, 5, 6 ⑤ 5, 4, 7; ⑥ 5, 7, 8; ⑦ 5, 8, 9; ⑧ 5, 9, 6

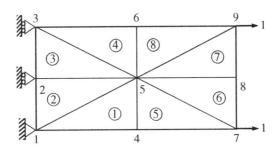

Figure 9.17

10

Bending of Plates

The concept of a thin plate or a thick plate in the theory of elasticity is relative. The absolute size of plates necessarily cannot be the measure for judging a thin or thick plate. Thin plate in elasticity is defined as a plate which has a smaller thickness relative to other dimensions of the plate. Furthermore, the theory in this chapter treats cases in which the deflection of plates is smaller than the thickness. The following assumptions are adopted which are similar to the beam bending problems of the strength of materials.

Assumptions

1. The plane perpendicular to the central plane of the plate remains perpendicular after deformation.
2. The central plane retains zero strain after deformation.
3. The normal stress in the direction perpendicular to the plane of the plate is neglected, because the values are relatively small compared with other stresses.

10.1   Simple Examples of Plate Bending

First we study a case in which a rectangular plate having length l in the x direction and a much longer length in the y direction is bent to a cylindrical shape, as shown in Figure 10.1. We call this case *cylindrical bending*. As shown in Figure 10.2, we define the plate thickness by h and take the central plane of the plate as the origin of the z axis, that is, $z=0$ is the neutral plane of the plate. The deflection w of the plate is defined in the z direction, that is the z direction is the positive direction.

Based on assumptions 1–3, the the relationship of the states before and after deformations are illustrated as Figure 10.3. Denoting the curvature radius of the central plane by ρ, ρ is approximately expressed with w as follows:

Theory of Elasticity and Stress Concentration, First Edition. Yukitaka Murakami.
© 2017 John Wiley & Sons, Ltd. Published 2017 by John Wiley & Sons, Ltd.

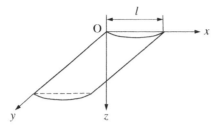

Figure 10.1

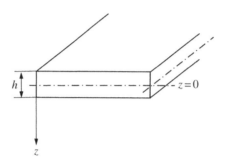

Figure 10.2

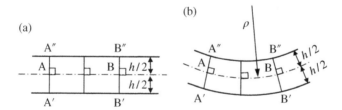

Figure 10.3

$$\frac{1}{\rho} = -\frac{d^2w}{dx^2} \tag{10.1}$$

Referring to Figure 10.4, strain in the x direction ε_x is

$$\varepsilon_x = \frac{A'B' - AB}{AB} = \frac{(\rho + z)d\theta - \rho d\theta}{\rho d\theta} = \frac{z}{\rho} \tag{10.2}$$

From Equations 10.1 and 10.2,

$$\varepsilon_x = -z\frac{d^2w}{dx^2} \tag{10.3}$$

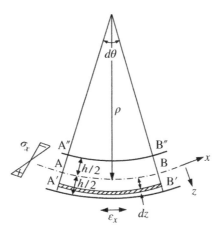

Figure 10.4

Considering assumption 3, $\sigma_z = 0$ and from Hooke's law,

$$\varepsilon_x = \frac{1}{E}\left(\sigma_x - \nu\sigma_y\right), \quad \varepsilon_y = \frac{1}{E}\left(\sigma_y - \nu\sigma_x\right)$$

Since $\varepsilon_y = 0$ for the case of cylindrical bending as in Figure 10.1, we have $\sigma_y = \nu\sigma_x$. Then,

$$\varepsilon_x = \frac{1-\nu^2}{E}\sigma_x$$

And

$$\sigma_x = \frac{E\varepsilon_x}{1-\nu^2} = -\frac{Ez}{1-\nu^2}\cdot\frac{d^2w}{dx^2} \tag{10.4}$$

$$\sigma_y = -\frac{\nu Ez}{1-\nu^2}\cdot\frac{d^2w}{dx^2} \tag{10.5}$$

Since the bending moment for plate bending is in general defined by the quantity for unit length, the bending moments M_x and M_y (see Figure 10.5) which are correlated with σ_x and σ_y respectively, are as follows.

$$M_x = \int_{-\frac{h}{2}}^{\frac{h}{2}} \sigma_x \cdot 1 \cdot z\,dz = -\int_{-\frac{h}{2}}^{\frac{h}{2}} \frac{Ez^2}{1-\nu^2}\cdot\frac{d^2w}{dx^2}\,dz$$

$$= -\frac{Eh^3}{12(1-\nu^2)}\cdot\frac{d^2w}{dx^2} = -D\frac{d^2w}{dx^2} \tag{10.6}$$

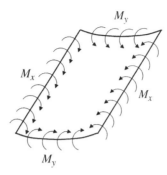

Figure 10.5

$$M_y = \int\limits_{-h/2}^{h/2} \sigma_y \cdot 1 \cdot z dz \tag{10.7}$$

where

$$D = \frac{Eh^3}{12(1-\nu^2)} \tag{10.8}$$

where D is called the *flexural rigidity of the plate*.

For the cylindrical bending, we have from the relationship of Equations 10.4 and 10.5:

$$M_y = \nu M_x \tag{10.9}$$

Equation 10.9 means that nonzero M_y is necessary to keep $\varepsilon_y = 0$. In contrast, it must be noted that, if $M_y = 0$, $\varepsilon_y \neq 0$ (which means the deformation of the plate is not cylindrical).

Writing Equation 10.6 as

$$D\frac{d^2w}{dx^2} = -M_x \tag{10.10}$$

we compare this equation with the differential equation for the beam bending of the strength of material, that is

$$EI\frac{d^2w}{dx^2} = -M, \quad I = \frac{h^3}{12}(\text{plate width} = 1) \tag{10.11}$$

It is noted that the deflection for the same value of bending moment, a plate has more resistance than a beam due to the difference between D and EI.

The plate bending for $M_y = 0$ in Figure 10.5 is called the *simple bending of plate*. In order to express the shape of the bent plate, the radii ρ of curvature of the neutral axis of the x and y axes are denoted by ρ_x and ρ_y, respectively. $\rho_y = \infty$ only for the cylindrical bending. ρ_y has a finite value for other cases.

Therefore, ρ_x and ρ_y are defined as follows:

$$\frac{1}{\rho_x} = -\frac{d^2 w}{dx^2}, \quad \frac{1}{\rho_y} = -\frac{d^2 w}{dy^2} \tag{10.12}$$

Since $\varepsilon_x = z/\rho_x$ and $\varepsilon_y = z/\rho_y$, and $M_y = 0$, that is $\sigma_y = 0$ for the simple bending, we have from Hooke's law:

$$\frac{z}{\rho_x} = \frac{\sigma_x}{E}, \quad \frac{z}{\rho_y} = -\nu\frac{\sigma_x}{E} \tag{10.13}$$

Then,

$$M_x = \int_{-h/2}^{h/2} \sigma_x \cdot 1 \cdot z\,dz = \int_{-h/2}^{h/2} \frac{E}{\rho_x} z^2\,dz = \frac{(1-\nu^2)D}{\rho_x} \tag{10.14}$$

Rewriting these equations,

$$\frac{1}{\rho_x} = \frac{M_x}{(1-\nu^2)D}, \quad \frac{1}{\rho_y} = -\frac{\nu M_x}{(1-\nu^2)D} = -\nu\frac{1}{\rho_x} \tag{10.15}$$

Thus, if the curvature of the x axis is positive, that of the y axis is negative.

10.2 General Problems of Plate Bending

Figure 10.6 shows the deformation of general bending problems for $M_x \neq 0$ and $M_y \neq 0$.

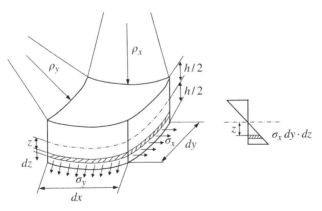

Figure 10.6

By Hooke's law and Figure 10.6, we have

$$\sigma_x = \frac{E_z}{1-\nu^2}\left(\varepsilon_x + \nu\varepsilon_y\right), \quad \sigma_y = \frac{E}{1-\nu^2}\left(\varepsilon_y + \nu\varepsilon_x\right)$$

$$\sigma_x = \frac{Ez}{1-\nu^2}\left(\frac{1}{\rho_x} + \nu\frac{1}{\rho_y}\right), \quad \sigma_y = \frac{Ez}{1-\nu^2}\left(\frac{1}{\rho_y} + \nu\frac{1}{\rho_x}\right) \tag{10.16}$$

Furthermore, by Equations 10.6 and 10.7,

$$\left.\begin{array}{l} M_x = D\left(\dfrac{1}{\rho_x} + \nu\dfrac{1}{\rho_y}\right) = -D\left(\dfrac{\partial^2 w}{\partial x^2} + \nu\dfrac{\partial^2 w}{\partial y^2}\right) \\[4mm] M_y = D\left(\dfrac{1}{\rho_y} + \nu\dfrac{1}{\rho_x}\right) = -D\left(\dfrac{\partial^2 w}{\partial y^2} + \nu\dfrac{\partial^2 w}{\partial x^2}\right) \end{array}\right\} \tag{10.17}$$

10.3 Transformation of Bending Moment and Torsional Moment

First of all, we define the sign of bending moment positive when a tensile normal stress acts at the section below the neutral plane ($z > 0$), as in Figure 10.7a. Namely, the sign of bending moment is the same as the sign of normal stress below the neutral plane. The bending moment is in general illustrated as Figure 10.7b or c. In Figure 10.7b, the bending moment acting as a screw driver rotated clockwise is defined positive. The arrows in Figure 10.7c are illustrated so that the image intuitively fits Figure 10.7a and b.

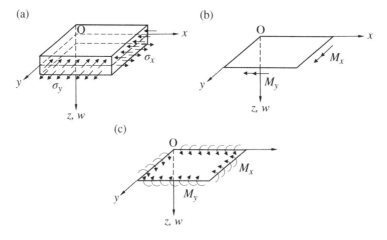

Figure 10.7

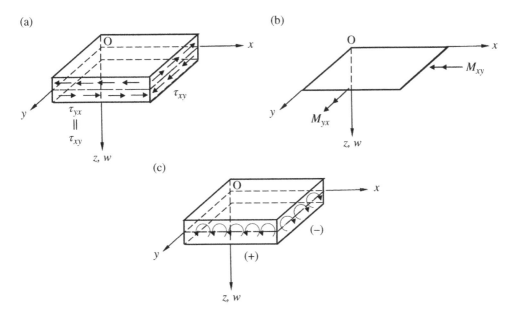

Figure 10.8

If twisting moment is applied to a plate, shear stress is produced as shown in Figure 10.8a. The sign of τ_{xy} or τ_{yx} is defined as consistent with the definition of plane problems. Thus, postive shear stress acting on the plane below the neutral plane ($z > 0$) is illustrated in Figure 10.8a. The sign of twisting moment is defined as follows. Figure 10.8b shows the action of twisting moment with arrows which are directed consistent with the clockwise rotation of a screw driver. The twisting moment is defined positive when the arrow is directed to the outward normal of the section and is negative when the direction is opposite, namely $M_{xy} = -M_{yx}$. Figure 10.8c shows twisting moment with the same type of arrows of Figure 10.7c for bending moment. The twisting moments M_{xy} and M_{yx} shown in Figure 10.8b are defined as the quantity for a unit length of the section.

According to the above definitions, the stresses at $z = h/2$ are calculated by

$$\sigma_x = \frac{M_x}{h^2/6}, \quad \sigma_y = \frac{M_y}{h^2/6}, \quad \tau_x = -\frac{M_{xy}}{h^2/6} = \frac{M_{yx}}{h^2/6}, \quad \left(z = \frac{h}{2}\right) \tag{10.18}$$

When M_x, M_y and M_{xy} or M_{yx} are known, by referring to Figure 10.9 the transformation of the bending moment and twisting moment at an arbitrary section can be done in the same way as a plane problem (see Equation 1.12) by the following equation, where it should be noted that M_{yx} has the same sign as τ_{xy}, $M_{n\xi}$ has the same sign as $\tau_{\xi\eta}$ and M_{xy} has the opposite sign to $M_{\xi\eta}$.

$$\left.\begin{array}{l} M_\xi = M_x\cos^2\theta + M_y\sin^2\theta - 2M_{xy}\cos\theta\cdot\sin\theta \\[6pt] M_\eta = M_x\sin^2\theta + M_y\cos^2\theta + 2M_{xy}\cos\theta\cdot\sin\theta \\[6pt] M_{\xi\eta} = \left(M_x - M_y\right)\cos\theta\cdot\sin\theta + M_{xy}\left(\cos^2\theta - \sin^2\theta\right) \end{array}\right\} \tag{10.19}$$

(a)   (b)

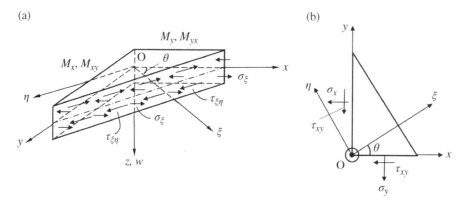

Figure 10.9

Problem 10.3.1   Illustrate the deformation shapes of a square thin plate in the following cases.

1. $M_x = M_0$, $M_y = 0$, $M_{xy} = 0$
2. $M_x = M_0$, $M_y = \nu M_0$, $M_{xy} = 0$
3. $M_x = M_0$, $M_y = M_0$, $M_{xy} = 0$
4. $M_x = M_0$, $M_y = -M_0$, $M_{xy} = 0$
5. $M_x = 0$, $M_y = 0$, $M_{xy} = M_0$

Problem 10.3.2   Verify the following equation:

$$M_{xy} = D(1-\nu)\frac{\partial^2 w}{\partial x \cdot \partial y} \tag{10.20}$$

Problem 10.3.3   Verify that the strain energy U to be stored in a plate with thickness h, area A under bending moment M_x, M_y and $M_{xy} = 0$ is given by the following equation:

$$U = \frac{1}{2}DA\left[\left(\frac{\partial^2 w}{\partial x^2}\right)^2 + \left(\frac{\partial^2 w}{\partial y^2}\right)^2 + 2\nu\frac{\partial^2 w}{\partial x^2}\cdot\frac{\partial^2 w}{\partial y^2}\right] \tag{10.21}$$

Verify that U for the case of $M_{xy} \neq 0$ is given by the following equation:

$$U = \frac{1}{2}DA\left[\left(\frac{\partial^2 w}{\partial x^2} + \frac{\partial^2 w}{\partial y^2}\right)^2 - 2(1-\nu)\left\{\frac{\partial^2 w}{\partial x^2}\cdot\frac{\partial^2 w}{\partial y^2} - \left(\frac{\partial^2 w}{\partial x \cdot \partial y}\right)^2\right\}\right] \tag{10.22}$$

Problem 10.3.4   Illustrate the deformation shape of a square plate for the case of Figure 10.10 in which concentrated forces are applied at the corners of the square plate.

Determine the major radius of curvature of the plate after deformation.
Which case of 1–5 in Problem 10.3.1 is similar to the case of Figure 10.10.

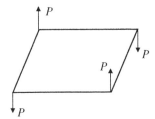

Figure 10.10

10.4 Differential Equations for a Plate Subjected to Surface Loads and their Applications

Figure 10.11a shows a plate subjected to uniform loading q on the surface. In this case, shear stresses τ_{xy} and τ_{yx} are produced in the direction of the thickness of the plate. The shear forces Q_x and Q_y are defined by integrating τ_{xy} and τ_{yx} per unit length of the plate. Since Figure 10.11a can be resolved to Figure 10.11b and Figure 10.11c, the equilibrium equations for each case are written as follows.

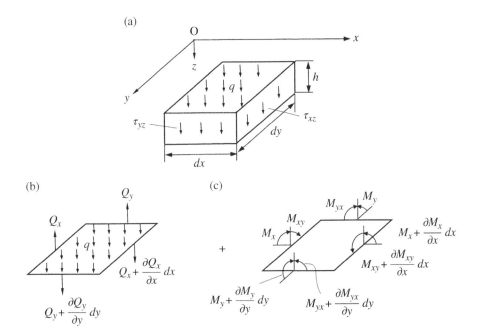

Figure 10.11

Considering the equilibrium condition of the plate element of $dxdy$ in the z direction,

$$\left[\left(Q_x+\frac{\partial Q_x}{\partial x}dx\right)-Q_x\right]dy+\left[\left(Q_y+\frac{\partial Q_y}{\partial y}dy\right)-Q_y\right]dx+qdx\cdot dy=0$$

Namely,

$$\frac{\partial Q_x}{\partial x}+\frac{\partial Q_y}{\partial y}+q=0 \tag{10.23}$$

Considering the equilibrium condition of rotation about the x axis of the plate element $dxdy$,

$$\left[\left(M_{xy}+\frac{\partial M_{xy}}{\partial x}dx\right)-M_{xy}\right]dy+\left[-\left(M_y+\frac{\partial M_y}{\partial y}dy\right)+M_y\right]dx$$

$$+\left(Q_y+\frac{\partial Q_y}{\partial y}dy\right)dx\cdot dy+\left[-Q_xdy+\left(Q_x+\frac{\partial Q_x}{\partial x}dx\right)dy\right]\frac{dy}{2}$$

$$+qdx\cdot dy\frac{dy}{2}=0$$

Neglecting the higher order terms in the above equation, we have

$$\frac{\partial M_{xy}}{\partial x}-\frac{\partial M_y}{\partial y}+Q_y=0 \tag{10.24}$$

Likewise, considering the equilibrium condition of rotation about the y axis,

$$\left[\left(M_x+\frac{\partial M_x}{\partial x}dx\right)-M_x\right]dy+\left[\left(M_{yx}+\frac{\partial M_{yx}}{\partial y}dy\right)-M_{yx}\right]dx$$

$$-\left(Q_x+\frac{\partial Q_x}{\partial y}dx\right)dy\cdot dx+\left[Q_ydx-\left(Q_y+\frac{\partial Q_y}{\partial y}dy\right)dx\right]\frac{dx}{2}$$

$$-qdx\cdot dy\frac{dx}{2}=0$$

Neglecting the higher order terms, we have

$$\frac{\partial M_x}{\partial x}+\frac{\partial M_{yx}}{\partial y}-Q_x=0 \tag{10.25}$$

Since regarding the condition of Figure 10.11a, forces in the x and y directions are 0 and the condition of rotation about the z axis is satisfied, Equations 10.23 to 10.25 completely describe the equilibrium conditions.

Considering $M_{xy} = -M_{yx}$, Q_x and Q_y can be removed from Equations 10.23 to 10.25 and then,

$$\frac{\partial^2 M_x}{\partial x^2} + \frac{\partial^2 M_y}{\partial y^2} - 2\frac{\partial^2 M_{xy}}{\partial x \cdot \partial y} = -q \tag{10.26}$$

Substituting Equations 10.17 and 10.20 into Equation 10.26,

$$\frac{\partial^4 w}{\partial x^4} + 2\frac{\partial^4 w}{\partial x^2 \cdot \partial y^2} + \frac{\partial^4 w}{\partial y^4} = \frac{q}{D} \text{ (Lagrange, 1811)} \tag{10.27}$$

The above equation is expressed in some cases for simplicity like:

$$\Delta\Delta w = \frac{q}{D}, \text{ or } \nabla^4 w = \frac{q}{D} \tag{10.28}$$

where $\Delta = \partial^2/\partial x^2 + \partial^2/\partial y^2$, $\nabla^2 = \partial^2/\partial x^2 + \partial^2/\partial y^2$.

Example problem 10.1
Determine the deflection of a simply supported long plate with width a under a uniformly distributed load q (Figure 10.12).

Solution
From Equation 10.27,

$$\frac{d^4 w}{dx^4} = \frac{q}{D}, \quad \frac{d^3 w}{dx^3} = \frac{q}{D}x + C_1,$$

$$\frac{d^2 w}{dx^2} = \frac{1}{2} \cdot \frac{q}{D}x^2 + C_1 x + C_2$$

Considering $M_x = 0$ at $x = 0$, $C_2 = 0$.

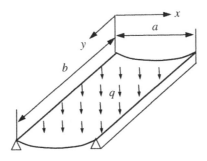

Figure 10.12

$$\frac{dw}{dx} = \frac{1}{6} \cdot \frac{q}{D} x^3 + \frac{1}{2} C_1 x^2 + C_3, \quad w = \frac{1}{24} \cdot \frac{q}{D} x^4 + \frac{1}{6} C_1 x^3 + C_3 x + C_4$$

Since $w = 0$ at $x = 0$, $C_4 = 0$.
Considering $M_x = 0$, that is $d^2w/dx^2 = 0$, at $x = a$.

$$\frac{1}{2} \cdot \frac{q}{D} a^2 + C_1 a = 0, \quad \text{namely } C_1 = -\frac{1}{2} \cdot \frac{q}{D} a$$

Since $w = 0$ at $x = a$,

$$\frac{1}{24} \cdot \frac{q}{D} a^4 - \frac{1}{12} \cdot \frac{q}{D} a^4 + C_3 a = 0, \quad \text{namely } C_3 = \frac{1}{24} \cdot \frac{q}{D} a^3$$

$$w = \frac{q}{24D} \left(x^4 - 2ax^3 + a^3 x \right)$$

w has a maximum value at $x = a/2$ and the value is

$$w_{max} = \frac{5}{384} \cdot \frac{q}{D} a^4$$

Example problem 10.2
Explain how to determine the deflection for a plate having finite a and b in Figure 10.12.

Solution
The essence of Levy's solution method [1] is briefly introduced in the following. The solution w for Equation 10.27 is separated into two parts as follows:

$$w = w_1 + w_2$$

Here, w_1 and w_2 are expressed as follows.

$$w_1 = \frac{q}{24D} \left(x^4 - 2ax^3 + a^3 x \right) \ldots \text{solution for Example problem 10.1}$$

$$w_2 = \sum_{m=1,3,5,\cdots}^{\infty} Y_m \sin \frac{m\pi x}{a}$$

Substituting w into Equation 10.27, we have

$$\sum_{m=1,3,5,\cdots}^{\infty} \left(Y_m^{(4)} - 2 \frac{m^2 \pi^2}{a^2} Y_m^{(2)} + \frac{m^4 \pi^4}{a^4} Y_m \right) \sin \frac{m\pi x}{a} = 0$$

Since the above equation must hold for arbitrary value of x,

$$Y_m^{(4)} - 2 \frac{m^2 \pi^2}{a^2} Y_m^{(2)} + \frac{m^4 \pi^4}{a^4} Y_m = 0$$

Solving this equation,

$$Y_m = \frac{qa^4}{D}\left(A_m\cosh\frac{m\pi y}{a} + B_m\frac{m\pi y}{a}\sinh\frac{m\pi y}{a} + C_m\sinh\frac{m\pi y}{a} + D_m\frac{m\pi y}{a}\cosh\frac{m\pi y}{a}\right)$$

The unknown constants $A_m \sim D_m$ can be determined by boundary conditions.

Example problem 10.3

Determine the deflection for the plate of Figure 10.12 with finite a and b under distributed loading $q = q_0\sin(\pi x/a) \cdot \sin(\pi y/b)$.

Solution

This problem can be solved easily compared with the case of $q_0 = $ constant. The basic differential equation can be written as follows:

$$\frac{\partial^4 w}{\partial x^4} + 2\frac{\partial^4 w}{\partial x^2 \cdot \partial y^2} + \frac{\partial^4 w}{\partial y^4} = \frac{q_0}{D}\sin\frac{\pi x}{a}\cdot\sin\frac{xy}{b} \tag{a}$$

The boundary conditions are:

$$\left.\begin{array}{l} w = 0,\ M_x = 0\ \ (\partial^2 w/\partial x^2 = 0)\ at\ x = 0,\ x = a \\ w = 0,\ M_y = 0\ \ (\partial^2 w/\partial y^2 = 0)\ at\ y = 0,\ y = b \end{array}\right\} \tag{b}$$

If we assume the deflection as

$$w = C\sin\frac{\pi x}{a}\cdot\sin\frac{\pi y}{b} \tag{c}$$

The equation (c) satisfies the conditions of (b). Substituting (c) into (a),

$$\pi^4\left(\frac{1}{a^2} + \frac{1}{b^2}\right)^2 C = \frac{q_0}{D}, \quad \text{namely } C = \frac{q_0}{\pi^4(1/a^2 + 1/b^2)^2 D} \tag{d}$$

Finally, we have

$$w = \frac{q_0}{\pi^4(1/a^2 + 1/b^2)^2 D}\sin\frac{\pi x}{a}\cdot\sin\frac{\pi y}{b} \tag{e}$$

10.5   Boundary Conditions in Plate Bending Problems

Although some boundary conditions in plate bending problems are similar to those of the strength of materials, there are other boundary conditions related to τ_{xy}, τ_{xz} and τ_{yz}, that is M_{xy}, M_{yx}, Q_x and Q_y which should be strictly treated as 3D problems. However, the strict

3D treatment of plate bending problems makes the solution very complicated; and an alternative convenient approximate and useful practical solution called *classical theory* has been established in which the boundary conditions are approximately satisfied.

The boundary conditions of plate bending problems are classified into the following three cases:

1. *Built in edge*
    The condition for the fixed edge of a plate along $x = a$ is expressed as follows:

$$(w)_{x=a} = 0, \quad \left(\frac{\partial w}{\partial x}\right)_{x=a} = 0 \qquad (10.29)$$

2. *Simply supported edge*
    The condition for simply supported edge along $x = a$ is

$$(w)_{x=a} = 0, \quad (M_x)_{x=a} = \left(\frac{\partial^2 w}{\partial x^2} + \nu \frac{\partial^2 w}{\partial y^2}\right)_{x=a} = 0 \qquad (10.30)$$

Considering $\left(\partial^2 w / \partial y^2\right)_{x=a} = 0$, the second condition can be written as $\left(\partial^2 w / \partial x^2\right)_{x=a} = 0$.

3. *Free edge*
    The condition for free edge in plate bending problems has a different characteristic compared with those of the elementary theory of strength of materials and the plane problem. This nature of the boundary condition is a unique point for plate bending problems.

It is natural to define the free edge condition so that the following three quantities are zero:

$$(M_x)_{x=a} = 0, \quad (M_{xy})_{x=a} = 0, \quad (Q_x)_{x=a} = 0$$

These three conditions are too many for solving the classical theory of plate bending. The reason is that only the curvature of the neutral plane of a plate receives attention and the local deformation of a plate is not considered in the theory. Even if the local deformation cannot strictly be treated, it is useful to obtain a solution within practically sufficient accuracy by combining the three conditions into two conditions from the viewpoint of the contribution to overall deformation. In this regard, we focus on the relationship between M_{xy} and Q_x.

As shown in Figure 10.13, from the viewpoint of overall deformation of a plate M_{xy} is approximately equivalent to the loading of Figure 10.13b. The effect of the difference between two statically equivalent loads produces a different stress and strain state at the edge of the plate but the difference is small at the internal position at the distance of the order of plate thickness h from the edge (Saint Venant's principle). Therefore, as shown in Figure 10.14a the distribution of τ_{xy} at a section is equivalent to a pair of the distribution of shear forces. Since the part of the distribution of shear forces is cancelled due to the opposite sign, summarizing the effect at both the sections $x = $ constant and $y = $ constant, as shown in Figure 10.14c, the final remaining effect is expressed with an equivalent concentrated force $P = 2M_{xy}$ acting at the corner of the plate. Therefore, if M_{xy} is not constant and varies, as in Figure 10.15, the remaining shear force after cancellation due to the opposite signs is $-\partial M_{xy} / \partial y$ (Kirchhoff, 1850).

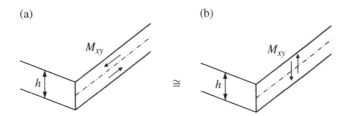

Figure 10.13

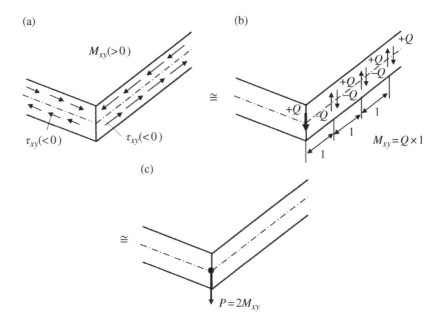

Figure 10.14

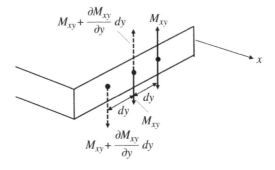

Figure 10.15

Finally, the free edge condition at $x = a$ required by the combination of shear force and twisting moment can be written as follows:

$$V_x = \left(Q_x - \frac{\partial M_{xy}}{\partial y}\right)_{x=a} = 0 \text{(the quantity per unit length)}$$

From Equation 10.25:

$$Q_x = \frac{\partial M_x}{\partial x} + \frac{\partial M_{yx}}{\partial y}$$

Substituting Equations 10.17 and 10.20 into the above equation,

$$V_x = -D\left[\frac{\partial^3 w}{\partial x^3} + (2-\nu)\frac{\partial^3 w}{\partial x \cdot \partial y^2}\right]_{x=a} = 0$$

Therefore, the free edge condition for plate bending is summarized by the following two equations.

$$\left(\frac{\partial^2 w}{\partial x^2} + \nu\frac{\partial^2 w}{\partial y^2}\right)_{x=a} = 0, \quad \left[\frac{\partial^3 w}{\partial x^3} + (2-\nu)\frac{\partial^3 w}{\partial x \cdot \partial y^2}\right]_{x=a} = 0 \tag{10.31}$$

The theory of plate bending based on above mentioned approximation is called *the classical theory* or *Kirchhoff theory*.

10.6   Polar Coordinate Expression of the Quantities of Plate Bending

As well as plane problems, there are various plate bending problems to be treated with the polar coordinate. The related useful quantities in the polar coordinate are summarized in the following:

$$\nabla^4 w = \left(\frac{\partial^2}{\partial r^2} + \frac{1}{r}\cdot\frac{\partial}{\partial r} + \frac{1}{r^2}\cdot\frac{\partial^2}{\partial \theta^2}\right)\left(\frac{\partial^2 w}{\partial r^2} + \frac{1}{r}\cdot\frac{\partial w}{\partial r} + \frac{1}{r^2}\cdot\frac{\partial^2 w}{\partial \theta^2}\right) = \frac{q}{D} \tag{10.32}$$

$$M_r = -D\left(\frac{\partial^2 w}{\partial x^2} + \nu\frac{\partial^2 w}{\partial y^2}\right)_{\theta=0} = -D\left[\frac{\partial^2 w}{\partial r^2} + \nu\left(\frac{1}{r}\cdot\frac{\partial w}{\partial r} + \frac{1}{r^2}\cdot\frac{\partial^2 w}{\partial \theta^2}\right)\right] \tag{10.33}$$

$$M_\theta = -D\left(\frac{\partial^2 w}{\partial y^2} + \nu\frac{\partial^2 w}{\partial x^2}\right)_{\theta=0} = -D\left[\left(\frac{1}{r}\cdot\frac{\partial w}{\partial r} + \frac{1}{r^2}\cdot\frac{\partial^2 w}{\partial \theta^2}\right) + \nu\frac{\partial^2 w}{\partial r^2}\right] \tag{10.34}$$

$$M_{r\theta} = D(1-\nu)\cdot\left(\frac{\partial^2 w}{\partial x \cdot \partial y}\right)_{\theta=0} = D(1-\nu)\cdot\left[\frac{1}{r}\cdot\frac{\partial^2 w}{\partial r \cdot \partial \theta} - \frac{1}{r^2}\cdot\frac{\partial w}{\partial \theta}\right] \tag{10.35}$$

$$Q_x = \frac{\partial M_x}{\partial x} + \frac{\partial M_{yx}}{\partial y} = -D\frac{\partial}{\partial x}\left(\frac{\partial^2 w}{\partial x^2} + \nu\frac{\partial^2 w}{\partial y^2}\right) - D(1-\nu)\frac{\partial^3 w}{\partial x \cdot \partial y^2}$$

$$= -D\frac{\partial}{\partial x}\left(\frac{\partial^2 w}{\partial x^2} + \frac{\partial^2 w}{\partial y^2}\right) = -D\frac{\partial}{\partial x}\left(\nabla^2 w\right) \tag{10.36}$$

$$Q_y = -\frac{\partial M_{xy}}{\partial x} + \frac{\partial M_y}{\partial y} = -D\frac{\partial}{\partial y}\left(\frac{\partial^2 w}{\partial x^2} + \frac{\partial^2 w}{\partial y^2}\right) = -D\frac{\partial}{\partial y}\left(\nabla^2 w\right) \tag{10.37}$$

$$Q_r = -D\frac{\partial}{\partial r}\left(\nabla^2 w\right) \tag{10.38}$$

$$Q_\theta = -D\frac{\partial}{r\partial\theta}\left(\nabla^2 w\right) \tag{10.39}$$

Expressing the boundary conditions at $r = a$ in the polar coordinate,

1. Fixed end: $(w)_{r=a} = 0,\quad \left(\dfrac{\partial w}{\partial r}\right)_{r=a} = 0$ \hfill (10.40)

2. Simply supported end: $(w)_{r=a} = 0,\ (M_r)_{r=a} = 0$ \hfill (10.41)

3. Free edge: $(M_r)_{r=a} = 0,\ V = \left(Q_r - \dfrac{\partial M_{r\theta}}{r\partial\theta}\right)_{r=a} = 0$ \hfill (10.42)

10.7 Stress Concentration in Plate Bending Problems

10.7.1 Stress Concentration at a Circular Hole in Bending a Wide Plate

This problem (Figure 10.16) can be solved with the polar coordinate equations in the previous section in the same way as the plane problem (Section 6.5 of Part I).

The moment concentration factor K_t at $\theta = \pm\pi/2$ at the edge of the hole where the bending moment M_θ has its maximum value is expressed by

$$K_t = \frac{5+3\nu}{3+\nu} \tag{10.43}$$

M_θ at $\theta = 0$ and π at the edge of hole is shown in Figure 10.16 (see Chapter 6 in Part II for a more detailed explanation).

10.7.2 Stress Concentration at an Elliptical Hole in Bending a Wide Plate

As well as the plane problem, it is not easy to solve the stress concentration problem for plate containing an elliptical hole [2]. The moment concentration factor K_t at point A and B of Figure 10.17 is obtained as

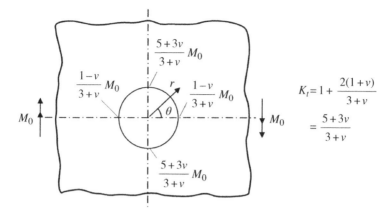

Figure 10.16

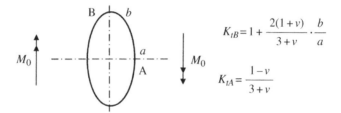

Figure 10.17

$$K_{tB} = 1 + \frac{2(1+\nu)}{3+\nu} \cdot \frac{b}{a} \tag{10.44}$$

$$K_{tA} = \frac{1-\nu}{3+\nu} \tag{10.45}$$

K_t given by Equation 10.44 for the case $a = b$ naturally coincides with Equation 10.43. The bending moment M_θ at point A is the same regardless of a and b, as is the case of a plane problem. As shown in several examples in plane problems (Chapter 6 of Part I), the results of Equations 10.43 to 10.45 can be applied to various practical problems of plate bending. The concept of the equivalent ellipse is naturally applicable also to bending problems (see Chapter 6 in Part II for a more detailed explanation).

10.8   Bending of a Circular Plate

This section studies the axisymmetric stress and deflection of a circular plate loaded on its surface. The assumptions described in the beginning of Chapter 10 of Part I are assumed to be applicable also in the problem of a circular plate.

In axisymmetric deformation problems there is no displacement in the θ direction and the displacements and stresses are a function of only the r coordinate. Therefore, $\tau_{r\theta} = 0$ and $\tau_{\theta z} = 0$. Excluding the terms related to θ in Equation 10.32,

$$\left(\frac{\partial^2}{\partial r^2} + \frac{1}{r} \cdot \frac{\partial}{\partial r}\right) \cdot \left(\frac{\partial^2 w}{\partial r^2} + \frac{1}{r} \cdot \frac{\partial w}{\partial r}\right) = \frac{q}{D} \tag{10.46}$$

Expanding this equation,

$$\frac{d^4 w}{dr^4} + \frac{2}{r} \cdot \frac{d^3 w}{dr^3} - \frac{1}{r^2} \cdot \frac{d^2 w}{dr^2} + \frac{1}{r^3} \cdot \frac{dw}{dr} = \frac{q}{D} \tag{10.47}$$

The solution for this differential equation can be obtained by the same procedure used for Equation 6.22. It follows that

$$w = w_0 + C_1 + C_2 \log r + C_3 r^2 + C_4 r^2 \log r \tag{10.48}$$

where w_0 is the special solution expressed as

$$w_0 = \frac{q r^4}{64D} \tag{10.49}$$

$C_1 \sim C_4$ are constants to be determined by the boundary conditions.
Excluding the terms of θ in Equations 10.33, 10.34 and 10.38,

$$M_r = -D\left(\frac{d^2 w}{dr^2} + \nu \frac{1}{r} \cdot \frac{dw}{dr}\right), \quad M_\theta = -D\left(\frac{1}{r} \cdot \frac{dw}{dr} + \nu \frac{d^2 w}{dr^2}\right) \tag{10.50}$$

$$Q_r = -D\frac{d}{dr}\left(\frac{d^2 w}{dr^2} + \frac{1}{r} \cdot \frac{dw}{dr}\right) \tag{10.51}$$

Example problem 10.4
Determine the deflection of a circular plate with fixed circumference under a uniform distributed load q (Figure 10.18).

Solution
From the general solution Equation 10.48,

$$w = \frac{q r^4}{64D} + C_1 + C_2 \log r + C_3 r^2 + C_4 r^2 \log r$$

Considering $w = $ finite at $r = 0$ (the center of the plate), $C_2 = 0$. Thus,

$$w' = \frac{q r^3}{16D} + 2C_3 r + C_4 (2r \log r + r)$$

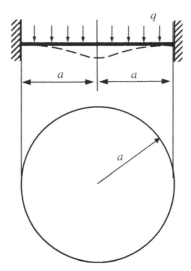

Figure 10.18

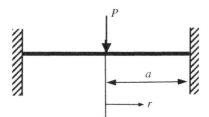

Figure 10.19

$$w'' = \frac{3qr^2}{16D} + 2C_3 + C_4(2\log r + 3)$$

Considering $M_r =$ finite at $r = 0$, $C_4 = 0$ (or, from $Q_r \propto r$, $C_4 = 0$).
Since $w' = 0$ at $r = a$, $C_3 = -qa^2/32D$ and also since $w = 0$, $C_1 = qa^4/64D$. Finally, we have

$$w = \frac{q}{64D}\left(a^2 - r^2\right)^2 \tag{10.52}$$

From this solution, M_r (or σ_r) and M_θ (or σ_θ) can be calculated with Equation 10.50.

Example problem 10.5
Determine the deflection of a circular plate with fixed circumference under a concentrated load P at the center of a plate (Figure 10.19).

Solution
The general solution for this problem is written as

$$w = C_1 + C_2 \log r + C_3 r^2 + C_4 r^2 \log r$$

The conditions for determining $C_1 \sim C_4$ are as follows.

1. $w' \to 0$ as $r \to 0$.
2. $w = 0$ at $r = a$.
3. $w' = 0$ at $r = a$.
4. $2\pi r Q_r + P = 0$ irrespective of r (see Figure 10.20).
   From these conditions, we have

$$C_1 = \frac{Pa^2}{16\pi D}, \quad C_2 = 0, \quad C_3 = -\frac{P}{16\pi D}(2\log a + 1), \quad C_4 = \frac{P}{8\pi D}$$

Finally,

$$w = \frac{Pr^2}{8\pi D}\log\frac{r}{a} + \frac{P}{16\pi D}(a^2 - r^2) \tag{10.53}$$

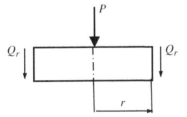

Figure 10.20

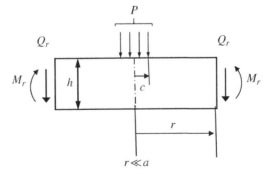

Figure 10.21

However, we need to treat the neighbourhood of the center of plate $(r=0)$ as a singular area, because the real situation near the center is like that shown in Figure 10.21 and the situation is beyond the assumptions of classical theory. The exact solution for the neighbourhood of the center must be solved as a 3D problem.

Example problem 10.6

Determine the deflection of a simply supported circular plate under a uniform distributed load q (Figure 10.22).

Solution

The general solution for this problem is written as

$$w = \frac{qr^4}{64D} + C_1 + C_2\log r + C_3 r^2 + C_4 r^2 \log r$$

The conditions for determining $C_1 \sim C_4$ are as follows.

1. $w' = 0$ and $M_r = M_\theta =$ finite at $r = 0$.
2. $w = 0$ and $M_r = 0$ at $r = a$.

From the first condition of 1, $C_2 = 0$ and $w' = 0$ is also satisfied. Other constants are determined as follows.

$$C_1 = \frac{(5+\nu)}{64D(1+\nu)}qa^4, \quad C_3 = -\frac{(3+\nu)}{32D(1+\nu)}qa^2, \quad C_4 = 0$$

Then,

$$w = \frac{q(a^2 - r^2)}{64D}\left[\frac{5+\nu}{1+\nu}a^2 - r^2\right] \tag{10.54}$$

Example problem 10.7

Determine the deflection of a simply supported circular plate containing a circular hole at the center when force is loaded at the edge of the hole (Figure 10.23).

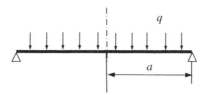

q

a

Figure 10.22

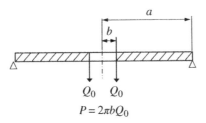

Figure 10.23

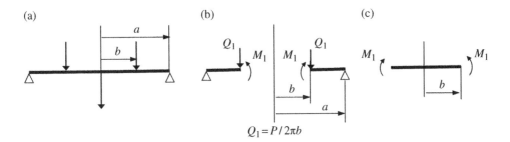

$Q_1 = P/2\pi b$

Figure 10.24

Solution
The conditions for determining unknown constants are as follows.

1. $w=0$ and $M_r=0$ at $r=a$.
2. $M_r=0$ at $r=b$.
3. $Q_r=-Q_0 b/r$ $(r>b)$.

From these conditions, we have

$$w = \frac{Pr^2}{8\pi D}\left(\log\frac{r}{a}-1\right) - \frac{Ar^2}{4} - B\log\frac{r}{a} + C \tag{10.55}$$

$$A = \frac{P}{4\pi D}\left(\frac{1-v}{1+v} - \frac{2b^2}{a^2-b^2}\log\frac{b}{a}\right), \quad B = -\frac{(1+v)}{(1-v)}\cdot\frac{P}{4\pi D}\cdot\frac{a^2 b^2}{a^2-b^2}\log\frac{b}{a}$$

$$C = \frac{Pa^2}{8\pi D}\left(1 + \frac{1}{2}\cdot\frac{1-v}{1+v} - \frac{b^2}{a^2-b^2}\log\frac{b}{a}\right)$$

Example problem 10.8
Determine the deflection of a simply supported circular plate in which concentrated force is loaded at the circumference of $r=b$ (Figure 10.24a).

Solution

The problem of Figure 10.24a can be separated into two problems: Figure 10.24b and c. Denoting Figure 10.24b as Problem A and Figure 10.24c as Problem B, the solutions (i.e. deflections and gradients) for Problems A and B must be continuous and the bending moment M_r must be equal at $r = b$.

Although the general solution is

$$w = C_1 + C_2 \log r + C_3 r^2 + C_4 r^2 \log r$$

the values of unknown constants for Problem A and B are different.

Therefore, the variables are treated separately with subscripts A and B, and the problems are solved separately in the following solutions.

Problem A

$$w_A = C_{1A} + C_{2A} \log r + C_{3A} r^2 + C_{4A} r^2 \log r$$

The boundary conditions for Problem A are

1. $Q_r = -P/2\pi b$, $M_r = M_1$, $(\partial w_A / \partial r)_{r=b} = (\partial w_B / \partial r)_{r=b}$ at $r = b$.
2. $Q_r = -P/2\pi a$, $M_r = 0$, $w_A = 0$ at $r = a$.

$$w_A{}' = \frac{C_{2A}}{r} + 2C_{3A} r + 2C_{4A} r \log r + C_{4A} r$$

$$w_A{}'' = -\frac{C_{2A}}{r^2} + 2C_{3A} + 2C_{4A} \log r + 3C_{4A}$$

$$w_A{}''' = \frac{2C_{2A}}{r^3} + \frac{2C_{4A}}{r}$$

$$Q_r = -D\left(w''' + \frac{1}{r} w'' - \frac{1}{r^2} w' \right) = -D\frac{4C_{4A}}{r}$$

Then, $C_{4A} = P/8\pi D$ (a)

From $(M_r)_{r=a} = 0$,

$$-\frac{1-v}{a^2} C_{2A} + 2(1+v)C_{3A} + \{2(1+v)\log a + 3 + v\}C_{4A} = 0$$ (b)

From $(w_A)_{r=a} = 0$,

$$C_{1A} + C_{2A} \log a + C_{3A} a^2 + C_{4A} a^2 \log a = 0$$ (c)

Problem B

$$w_B = C_{1B} + C_{2B} \log r + C_{3B} r^2 + C_{4B} r^2 \log r$$

$$\text{From } (w_B)_{r=0} = \text{finite, } C_{2B} = 0 \tag{d}$$

$$\text{From } (M_r)_{r=0} = (M_\theta)_{r=0} = \text{finite, } C_{4B} = 0 \tag{e}$$

$$(M_r)_{r=b} = -D(1+\nu) \cdot 2C_{3B} = M_1 \tag{f}$$

Now, from $(w_A)_{r=b} = (w_B)_{r=b}$,

$$C_{1A} + C_{2A} \log b + C_{3A} b^2 + C_{4A} b^2 \log b = C_{1B} + C_{3B} b^2 \tag{g}$$

From $(M_{rA})_{r=b} = (M_{rB})_{r=b}$,

$$-D\left[-\frac{1-\nu}{b^2} C_{2A} + 2(1+\nu)C_{3A} + \{2(1+\nu)\log b + 3 + \nu\}C_{4A} \right]$$

$$= -D(1+\nu) \cdot 2C_{3B} \tag{h}$$

From $(w'_A)_{r=b} = (w'_B)_{r=b}$,

$$\frac{C_{2A}}{b} + 2C_{3A}b + C_{4A}b(2\log b + 1) = 2C_{3B}b \tag{i}$$

Thus, by determining the unknown constants using the above conditions, we have the solutions.

For $r \geq b$,

$$w_A = \frac{P}{8\pi D}\left[(a^2 - r^2) \cdot \left(1 + \frac{1}{2} \cdot \frac{1-\nu}{1+\nu} \cdot \frac{a^2 - b^2}{a^2}\right) + (b^2 + r^2)\log\frac{r}{a} \right] \tag{10.56}$$

For $r \leq b$,

$$w_B = \frac{P}{8\pi D}\left[(b^2 + r^2)\log\frac{b}{a} + (a^2 - b^2)\frac{(3+\nu)a^2 - (1-\nu)r^2}{2(1+\nu)a^2} \right] \tag{10.57}$$

Problems of Chapter 10

1. A constant bending moment M_n is applied along the outer periphery of a plate of arbitrary shape. Verify that the bending moment in the plate is M_n and the twisting moment is 0 everywhere and in every direction in the plate.

   Verify that the deformed form of the plate under this loading condition becomes a part of a sphere.

2. Calculate the maximum tensile stress and the location of the maximum tensile stress in Example problem 10.4 (Figure 10.18), where the radius of the circular plate is a, thickness is h, Young's modulus is E and Poisson's ratio is ν.

3. Determine the deflection when a simply supported circular plate with radius a is subjected to a uniform distributed load q inside radius c at the central part of the plate.

4. Determine the deflection when a circular plate with radius a is fixed at the outer periphery and is subjected to a concentrated load q along the ring $r = b$, where the total amount of the load is P, that is $P = 2\pi b q$.

5. Determine the deflection u_0 at the center of a simply supported circular plate with radius a when the circular plate is subjected to a concentrated load P at an arbitrary point at $r = b$. Pay attention that the load is applied eccentrically to the circular plate.

References

[1] S. P. Timoshenko and S. Woinowsky-Krieger (1959) *Theory of Plates and Shells*, 2nd edn, McGraw–Hill, New York, p. 113.

[2] Goodier, J.N. (1936) *Phil. Mag.*, **68**:7–22; and Washizu (1952), *Trans. Jpn Soc. Mech. Eng.*, **41**:18–68.

11

Deformation and Stress in Cylindrical Shells

11.1 Basic Equations

A cylindrical thin pressure vessel (as shown in Figure 11.1) and a water or natural liquid gas reservoir tank (as shown in Figure 11.2) are commonly in practical use. The practical problems of a thin pressure vessel like Figure 11.1 can be solved by a combination of the theory of circular plate in Chapter 10 of Part I with the theory of cylindrical shell.

Although thin conical vessels and thin axisymmetric vessels having varying diameters are used in practical cases, such a theory is not explained in this chapter. If the reader completely understands the theory of this chapter, the reader will be able to study easily the extended theory by themself.

In order to solve the basic differential equation for cylindrical shells, first attention must be paid to the equilibrium condition as well as the problems of elasticity studied in the previous chapters. The next step is to pay attention to the relationship between forces and deformations or stresses and strains, and to reduce the relationship to the differential equation with regard to displacement.

This chapter treats only problems axisymmetric both in geometry and forces.

11.1.1 Equilibrium Condition

Figure 11.3 shows the equilibrium condition for an infinitesimal element ABCD in a thin cylindrical shell with radius a and under internal pressure p. Figure 11.4a and b separately illustrate the forces and bending moments acting on ABCD.

Due to the symmetry of the problem, the circumferential displacement is 0 and the forces, bending moments and radial displacement u_r do not vary with circumferential direction θ and twisting moments do not act.

Theory of Elasticity and Stress Concentration, First Edition. Yukitaka Murakami.
© 2017 John Wiley & Sons, Ltd. Published 2017 by John Wiley & Sons, Ltd.

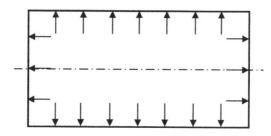

Figure 11.1

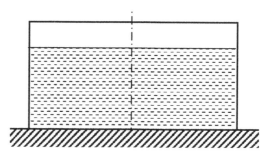

Figure 11.2

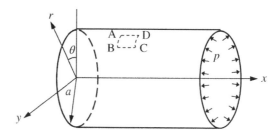

Figure 11.3

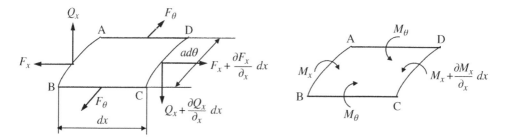

Figure 11.4

Therefore, the equilibrium conditions for ABCD are satisfied only under the action of (1) force in the x direction, (2) force in radial direction and (3) moment of rotation along the y axis. The equilibrium conditions of (1), (2) and (3) are expressed as follows.

$$\left.\begin{aligned}
&\left(F_x + \frac{dF_x}{dx}dx\right)ad\theta - F_x ad\theta = 0 \\
&\left(Q_x + \frac{dQ_x}{dx}dx\right)ad\theta - Q_x ad\theta + F_\theta dx \cdot d\theta - pad\theta \cdot dx = 0 \\
&\left(M_x + \frac{dM_x}{dx}dx\right)ad\theta - M_x ad\theta - \left(Q_x + \frac{dQ_x}{dx}dx\right)ad\theta \cdot dx \\
&\quad + \frac{1}{2}pad\theta(dx)^2 = 0
\end{aligned}\right\} \tag{11.1}$$

Neglecting higher order terms,

$$\left.\begin{aligned}
&\frac{dF_x}{dx}ad\theta \cdot dx = 0 \\
&\left(\frac{dQ_x}{dx} + \frac{F_\theta}{a} - p\right)ad\theta \cdot dx = 0 \\
&\left(\frac{dM_x}{dx} - Q_x\right)ad\theta \cdot dx = 0
\end{aligned}\right\} \tag{11.2}$$

From the first equation of Equation 11.2:

$$F_x = \text{constant} \tag{11.3}$$

The value of the constant in Equation 11.3 depends on the problem. However, this problem is simple, because it is the case that the cylinder is subjected to only simple axial tension or compression and the solution can be obtained easily.

In general problems, such a solution is added by simple superposition. In the following discussion, the analysis is proceeded assuming $F_x = 0$.

In conclusion, the following equations are derived as the basis for the equilibrium conditions:

$$\frac{dQ_x}{dx} + \frac{1}{a}F_\theta = p, \quad \frac{dM_x}{dx} - Q_x = 0 \tag{11.4}$$

11.1.2   Forces and Deformations

Two equations were obtained by the equilibrium condition as Equation 11.4. However, Equation 11.4 contains three unknowns (F_θ, Q_x, M_x). This nature of the problem is the same as the statically indeterminate problems in the elementary theory of strength of materials.

Equation 10.17 shows the relationship between bending moment and curvature and displacement for a thin flat plate under no twisting moment. It must be noted that the cylindrical shells to be analyzed here have initial curvature a in the θ direction as their radius of shell.

Although in the problem of Figure 11.3 the radius of cylinder varies due to the action of internal pressure p, this deformation is due to circumferential elongation produced by σ_θ and is not produced by bending moment M_θ. Consequently, when we apply Equation 10.17 to the infinitesimal element ABCD in Figure 11.3, we need to interpret the problem as the cylinder is bent to the shape of a cylinder of $\rho_x = \infty$ and $\rho_y = a$ by a certain initial moment and then the new deformation is produced by the internal pressure p of Figure 11.3.

Therefore, putting $w = -u_r$ (u_r: displacement in the radial direction), we have

$$\left.\begin{aligned}
M_x &= -D\left[\frac{\partial^2 w}{\partial x^2} + \nu\left(\frac{\partial w}{r\partial r} + \frac{1}{r^2}\cdot\frac{\partial^2 w}{\partial\theta^2}\right)\right] \\[2ex]
M_\theta &= -D\left[\left(\frac{\partial w}{r\partial r} + \frac{1}{r^2}\cdot\frac{\partial^2 w}{\partial\theta^2}\right) + \nu\frac{\partial^2 w}{\partial x^2}\right]
\end{aligned}\right\}
\tag{11.5}$$

Considering $\partial u_r/\partial\theta = -\partial w/\partial\theta = 0$ and $\partial u_r/\partial r = -\partial w/\partial r = 0$, Equation 11.5 can be reduced to

$$M_x = -D\frac{d^2 w}{dx^2}, \ M_\theta = -\nu D\frac{d^2 w}{dx^2} = \nu M_x
\tag{11.6}$$

Next, consider the relationship between the forces F_x, F_θ and deformation. Considering the relationship between deformations, Hooke's law, stress and strain, we have the following equations, where u is the displacement in the x direction.

$$\varepsilon_x = \frac{du}{dx}, \ \varepsilon_\theta = \frac{u_r}{a} = -\frac{w}{a}
\tag{11.7}$$

$$\sigma_x = \frac{E}{1-\nu^2}(\varepsilon_x + \nu\varepsilon_\theta), \ \ \sigma_y = \frac{E}{1-\nu^2}(\varepsilon_\theta + \nu\varepsilon_x)
\tag{11.8}$$

Then,

$$\left.\begin{aligned}
F_x &= \sigma_x\cdot h\cdot 1 = \frac{Eh}{1-\nu^2}(\varepsilon_x + \nu\varepsilon_\theta) = \frac{Eh}{1-\nu^2}\left(\frac{du}{dx} - \nu\frac{w}{a}\right) \\[2ex]
F_\theta &= \sigma_\theta\cdot h\cdot 1 = \frac{Eh}{1-\nu^2}(\varepsilon_\theta + \nu\varepsilon_x) = \frac{Eh}{1-\nu^2}\left(-\frac{w}{a} + \nu\frac{du}{dx}\right)
\end{aligned}\right\}
\tag{11.9}$$

Since only the problem of $F_x = 0$ is treated as expressed in terms of Equation 11.3 and the related explanation, we have $du/dx = \nu w/a$, and then

$$F_\theta = -\frac{Eh}{a}w
\tag{11.10}$$

11.1.3   Basic Differential Equation

From the equilibrium condition and Equations 11.4, 11.6 and 11.10, which were obtained from the relationship between forces and deformations, the following differential equation about w is derived as

$$\frac{d^2}{dx^2}\left(D\frac{d^2 w}{dx^2}\right) + \frac{Eh}{a^2} w = -p \tag{11.11}$$

For the case of $h = $ constant,

$$\frac{d^4 w}{dx^4} + \frac{Eh}{Da^2} w = -\frac{p}{D} \tag{11.12}$$

In order to simplify the expression of the solution of the differential equation (Equation 11.12), the following new parameter β is introduced.

$$\frac{d^4 w}{dx^4} + 4\beta^4 w = -\frac{p}{D} \tag{11.13}$$

where

$$\beta^4 = \frac{Eh}{4Da^2} = \frac{3(1-\nu^2)}{a^2 h^2}, \; or \; \beta = \frac{[3(1-\nu^2)]^{1/4}}{\sqrt{ah}} \tag{11.14}$$

Thus, the solution of Equation 11.13 is expressed as follows:

$$w = e^{\beta x}(C_1 \cos \beta x + C_2 \sin \beta x) + e^{-\beta x}(C_3 \cos \beta x + C_4 \sin \beta x) - \frac{p}{4\beta^4 D} \tag{11.15}$$

or

$$w = \cosh \beta x(C_1' \cos \beta x + C_2' \sin \beta x) + \sinh \beta x(C_3' \cos \beta x + C_4' \sin \beta x) - \frac{p}{4\beta^4 D} \tag{11.16}$$

$C_1 \sim C_4$ and $C_1' \sim C_4'$ are determined from boundary conditions.
 The following quantities will be used frequently for the solution of various problems.

$$\frac{dw}{dx} = \beta e^{\beta x}(C_1 \cos \beta x + C_2 \sin \beta x)$$

$$+ \beta e^{\beta x}(-C_1 \sin \beta x + C_2 \cos \beta x) - \beta e^{-\beta x}(C_3 \cos \beta x + C_4 \sin \beta x)$$

$$+ \beta e^{-\beta x}(-C_3 \sin \beta x + C_4 \cos \beta x)$$

$$\frac{d^2 w}{dx^2} = 2\beta^2 e^{\beta x}(-C_1 \sin \beta x + C_2 \cos \beta x) - 2\beta^2 e^{-\beta x}(-C_3 \sin \beta x + C_4 \cos \beta x)$$

$$\frac{d^3w}{dx^3} = 2\beta^3 e^{\beta x}(-C_1 \sin \beta x + C_2 \cos \beta x) - 2\beta^3 e^{\beta x}(C_1 \cos \beta x + C_2 \sin \beta x)$$

$$+ 2\beta^3 e^{-\beta x}(-C_3 \sin \beta x + C_4 \cos \beta x) + 2\beta^3 e^{-\beta x}(C_3 \cos \beta x + C_4 \sin \beta x)$$

11.2   Various Problems of Cylindrical Shells

Example problem 11.1

Determine the deflection and bending moment for the problem of Figure 11.5, where P is the force per unit length along the circumferential direction.

Solution

The constant $C_1 \sim C_4$ of the general solution of Equation 11.15 are determined from the boundary conditions.

As $x \rightarrow \infty$, $w \rightarrow 0$. And then, $C_1 = 0$ and $C_2 = 0$.

Therefore

$$w = e^{-\beta x}(C_3 \cos \beta x + C_4 \sin \beta x) \tag{a}$$

From $M_x = 0$ at $x = 0$,

$$-D\left(\frac{d^2w}{dx^2}\right)_{x=0} = 0 \tag{b}$$

Considering $Q_x = -P$ at $x = 0$ and Equation 11.4,

$$\left(\frac{dM_x}{dx}\right)_{x=0} = -D\left(\frac{d^3w}{dx^3}\right)_{x=0} = p \tag{c}$$

From (a), (b) and (c),

$$C_3 = -\frac{P}{2\beta^3 D}, \quad C_4 = 0 \tag{d}$$

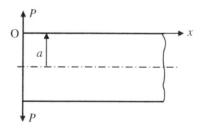

Figure 11.5

Finally, we have

$$w = -\frac{P}{2\beta^3 D}e^{-\beta x}\cos\beta x \qquad\qquad\text{(e)}$$

$$M_x = \frac{P}{\beta}e^{-\beta x}\sin\beta x \qquad\qquad\text{(f)}$$

Example problem 11.2

Determine the deflection and the bending moment for the problem of Figure 11.6.

Solution

As $x \to \infty$, $w \to 0$. And then, $C_1 = 0$ and $C_2 = 0$.

Therefore

$$w = e^{-\beta x}(C_3\cos\beta x + C_4\sin\beta x) \qquad\qquad\text{(a)}$$

From $M_x = M_0$ at $x = 0$,

$$-D\left(\frac{d^2 w}{dx^2}\right)_{x=0} = M_0 \qquad\qquad\text{(b)}$$

Considering $Q_x = 0$ at $x = 0$,

$$-D\left(\frac{d^3 w}{dx^3}\right)_{x=0} = 0 \qquad\qquad\text{(c)}$$

From (a), (b) and (c),

$$C_3 = -\frac{M_0}{2\beta^2 D}, \quad C_4 = \frac{M_0}{2\beta^2 D} \qquad\qquad\text{(d)}$$

Finally, we have

$$w = \frac{M_0}{2\beta^2 D}e^{-\beta x}(\sin\beta x - \cos\beta x) \qquad\qquad\text{(e)}$$

$$M_x = M_0 e^{-\beta x}(\cos\beta x + \sin\beta x) \qquad\qquad\text{(f)}$$

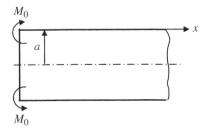

Figure 11.6

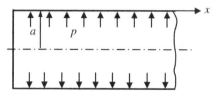

Figure 11.7

Example problem 11.3
Determine the deflection w for the problem of Figure 11.7.

Solution
From Example problems 11.1 and 11.2, $C_1 = C_2 = C_3 = C_4 = 0$. Thus,

$$w = -\frac{p}{4\beta^4 D} = -\frac{pa^2}{Eh}$$

The same solution can be obtained by another way as follows.
From the equilibrium condition,

$$\sigma_\theta = \frac{pa}{h}$$

Therefore, $\varepsilon_\theta = \sigma_\theta/E = pa/Eh$ and

$$w = -a\varepsilon_\theta = -\frac{pa^2}{Eh} \quad \text{(displacement in the outer radial direction)}$$

Problems of Chapter 11

1. Determine the deflection w and the bending moment M_x for the problem of Figure 11.8.

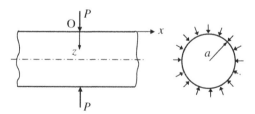

Figure 11.8

2. Determine the deflection w and the bending moment M_x near the fixed end of the cylindrical shell in Figure 11.9 for the case $a \ll l$.

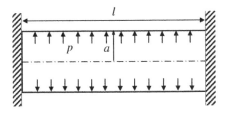

Figure 11.9

3. Determine the maximum stress which occurs at the connection part of the cylindrical shell and the circular plate in Figure 11.10 for the case $a \ll l$. The thickness is h for both the cylindrical shell and the circular plate.

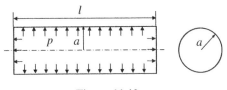

Figure 11.10

(This problem can be used as an alternative and conservative estimation for the cylindrical shell connected by a curved end plate.)

4. Derive the differential equation for the beam placed on an elastic foundation and subjected to a distributed load p as in Figure 11.11. Show that the equation has the same form as Equation 11.12. Assume the elastic constant of the foundation as k.

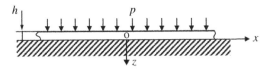

Figure 11.11

5. Figure 11.12 shows a thin pipe which has n equally spaced slits with length l at the free end where a rigid circular plate with radius $(a + \delta)$ is fit into the inner wall with radius a. Determine the maximum bending stress at length l from the free end. Assume $\delta \ll a$ and n is sufficiently large so that the section divided by slits can be approximated to a rectangular shape. The friction between the rigid circular plate and the pipe can be ignored. The elastic constants of the pipe are Young's modulus E and Poisson's ratio ν.

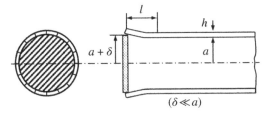

Figure 11.12

12

Thermal Stress

Thermal stress arises due to non-uniform temperature distribution in machines and structures. Non-uniform temperature distribution produces non-uniform thermal expansion or contraction from location to location in structures. Although the part with a temperature higher than its surroundings tends to expand, the surrounding part tends to prevent the expansion. If free expansion and contraction are allowed, no thermal stress is produced. Thermal stress is produced either by the surrounding structures preventing these free thermal deformations or by inversely influencing the surrounding structures by thermal expansion or contraction. Therefore, the amount of thermal stress is proportional to the variation of deformation from the free deformation corresponding to the zero stress state of a structure.

12.1 Thermal Stress in a Rectangular Plate – Simple Examples of Thermal Stress

Let us give a temperature change in a rectangular plate with width h, length l, thickness b and coefficient of thermal expansion α. In order to understand the nature of thermal stress, two cases are explained, one with a free end (Figure 12.1a) and the other with a fixed end (Figure 12.1b).

12.1.1 Constant Temperature Rise $\Delta T(x,y) = T_0$ (Constant)

In this case, the thermal strains in Figure 12.1a are $\varepsilon_{xT} = \varepsilon_{yT} = \alpha T_0$ and $\gamma_{xyT} = 0$, the stresses are all zero, that is $\sigma_x = \sigma_y = \tau_{xy} = 0$.

Theory of Elasticity and Stress Concentration, First Edition. Yukitaka Murakami.
© 2017 John Wiley & Sons, Ltd. Published 2017 by John Wiley & Sons, Ltd.

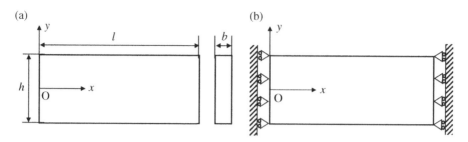

Figure 12.1

In the case of Figure 12.1b, since the longitudinal elongation $\delta = \alpha T_0 l$ is constrained, stress arises. The state of Figure 12.1b is equivalent to a plate which is elongated from l to $l(1 + \alpha T_0)$ and afterward the plate is compressed by a compressive load P to the initial length l. Therefore, we have the following equation:

$$-\delta = \frac{Pl(1 + \alpha T_0)}{Ebh(1 + \alpha T_0)^2} \cong \frac{Pl}{Ebh} \tag{12.1}$$

Thus, P is expressed as

$$P = -\alpha T_0 Ebh \tag{12.2}$$

Finally, the thermal stress in the plate of Figure 12.1b is

$$\sigma_x = -\alpha T_0 E, \quad \sigma_y = 0, \quad \tau_{xy} = 0 \tag{12.3}$$

12.1.2   Linear Temperature Distribution $\Delta T(x,y) = 2T_0 y/h$

Since the elongation and contraction are function of y, the stress is not clear even in Figure 12.1a.

In the case of Figure 12.1b, since the thermal strain $\varepsilon_{xT} = 2\alpha T_0 y/h$ is constrained, the stress σ_x is calculated as

$$\sigma_x = -\frac{2\alpha T_0 E y}{h} \tag{12.4}$$

Although in this state the resultant force in the x direction is 0 at the fixed end $x = 0$ or $x = l$, the bending moment expressed by the following equation is produced.

$$M = \frac{bh^2}{6} \alpha T_0 E \tag{12.5}$$

where the sign of M is defined positive for bending the plate convex downward.

The state of Figure 12.1a can be made by superimposing a moment having opposite sign of Equation 12.5 to the state of Figure 12.1b. By this superposition, the fixed ends of Figure 12.1b become free ends and the internal stress expressed by Equation 12.4 is cancelled. Then, consequently it is understood that thermal stress does not arise in Figure 12.1a.

Thus, thermal stress does not arise in a plate having a linear temperature distribution and the plate freely expands or contracts linearly with temperature distribution and is eventually bent to a circular shape. It must be noted that a plate with linear temperature distribution as $\Delta T(x,y) = 2T_0 y/h$ does not produce thermal stress in the plate of Figure 12.1a, because this problem is often misunderstood due to a simple reason for nonconstant temperature distribution.

12.1.3   Quadratic Temperature Distribution $\Delta T(x,y) = 2T_0(y/h)^2$

As in Section 12.1.2, it is not easy to understand the thermal stress in the plate of Figure 12.1a only from the temperature distribution.

Likewise Section 12.2.1, we start from the case of Figure 12.1b. Since the thermal strain in the x direction $\varepsilon_{xT} = \alpha T_0(2y/h)^2$ is constrained, σ_x becomes as follows.

$$\sigma_x = -\alpha T_0 E(2y/h)^2 \tag{12.6}$$

Although in this state the bending moment is 0 at the fixed end $x=0$ or $x=l$ due to the quadratic stress distribution of Equation 12.6, the resultant force P in the x direction expressed by the following equation is produced.

$$P = \int_{-h/2}^{h/2} \sigma_x b\, dy = -\frac{bh}{3}\alpha T_0 E \tag{12.7}$$

In order to find the stress in the plate of Figure 12.1a, a tensile force equal to the value of the compressive force acting on the fixed end of the plate of Figure 12.1b is superimposed to the state of Figure 12.1b. Applying this tensile stress, the following stress is produced in the plate.

$$\sigma'_x = -\frac{P}{bh} = \frac{1}{3}\alpha T_0 E \tag{12.8}$$

Superposing Equation 12.6 on Equation 12.8, the stress in the plate of Figure 12.1a is

$$\sigma_x = \frac{1}{3}\alpha T_0 E - \alpha T_0 E\left(\frac{2y}{h}\right)^2 \tag{12.9}$$

If the tensile force superimposed in the above derivation is replaced by a uniformly distributed tensile stress, the ends of the plate at $x=0$ and $x=l$ do not satisfy the condition of a free end and the stress expressed by Equation 12.9 remains at both ends of the plate. However, since the resultant force obtained by integrating Equation 12.9 along the end of the plate becomes 0, if we realise a free end by superposing a stress with the opposite sign to Equation 12.9 at the ends, the

influence is limited near the end of the plate due to the Saint Venant's principle and is not transferred to the main part of the plate.

Therefore, we can understand that the stress distribution in the plate of Figure 12.1a is given
by Equation 12.9 except for the vicinity of the fee ends.

In the above problems, the mechanism of generation of thermal stress can be understood
from the viewpoint of Hooke's law as follows. Assuming temperature variation by T,

$\alpha T =$ strain produced by free expansion or contraction,

Final strain = strain by free expansion (or contraction) + strain produced by stress.

The strain produced by stress obeys Hooke's law. Therefore, the above relationship can be
expressed as follows.

Orthogonal coordinate system

$$\left.\begin{aligned}
\varepsilon_x &= \alpha T + \frac{1}{E}\left[\sigma_x - \nu\left(\sigma_y + \sigma_z\right)\right] \\
\varepsilon_y &= \alpha T + \frac{1}{E}\left[\sigma_y - \nu\left(\sigma_z + \sigma_x\right)\right] \\
\varepsilon_z &= \alpha T + \frac{1}{E}\left[\sigma_z - \nu\left(\sigma_x + \sigma_y\right)\right]
\end{aligned}\right\}$$

(12.10)

$$\gamma_{xy} = \tau_{xy}/G, \gamma_{yz} = \tau_{yz}/G, \gamma_{zx} = \tau_{zx}/G$$

(12.11)

Cylindrical coordinate system

$$\left.\begin{aligned}
\varepsilon_r &= \alpha T + \frac{1}{E}\left[\sigma_r - \nu\left(\sigma_\theta + \sigma_z\right)\right] \\
\varepsilon_\theta &= \alpha T + \frac{1}{E}\left[\sigma_\theta - \nu\left(\sigma_z + \sigma_r\right)\right] \\
\varepsilon_z &= \alpha T + \frac{1}{E}\left[\sigma_z - \nu\left(\sigma_r + \sigma_\theta\right)\right]
\end{aligned}\right\}$$

(12.12)

$$\gamma_{r\theta} = \tau_{r\theta}/G, \gamma_{\theta z} = \tau_{\theta z}/G, \gamma_{zr} = \tau_{zr}/G$$

(12.13)

The above relationships mean that if the strains are equal to the values of free expansion (or
contraction), there is no constraint and the thermal stresses corresponding to the second terms of
Equation 12.10 or Equation 12.12 are not generated. It must be noted that the strains produced
by free expansion and contraction do not contribute to shear strains.

Problem 12.1.1   Determine the thermal stress for the case of $\Delta T(x, y) = T_0(2y/h)^n$ in
Figure   12.1.   Using   the   solution,   determine   the   thermal   stress   for   the   case
of $\Delta T(x, y) = T_0\{1 - (2y/h)^n\}$.

Problem 12.1.2   Verify that in order to solve thermal stress problems by the finite element
method (FEM), it is necessary to apply the forces at nodes expressed by the following equation
(*d*vol means integral over a volume.).

$$\{F\} = -\int [N]^T [D]\{\varepsilon_0\} d\text{vol}$$

where $\{F\}$ is the nodal force for an element, $\{\varepsilon_0\}$ is the thermal strain without constraint and for example in plane stress problems it is expressed as

$$\{\varepsilon_0\} = \alpha T \begin{Bmatrix} \alpha T \\ \alpha T \\ 0 \end{Bmatrix}$$

12.2 Thermal Stress in a Circular Plate

Thermal stress in a circular plate with temperature distribution expressed by a function only of radial distance r (Figure 12.2) is investigated.

This is a plane stress problem with temperature distribution as

$$T = T(r) \tag{12.14}$$

As already described, if the thermal stresses are generated by the constraint of thermal deformation, the stresses σ_r, σ_θ and $\tau_{r\theta}$ may be expected to arise. However, from the symmetric nature of the problem, $\tau_{r\theta} = 0$ and the equilibrium equation is written as follows (see Equation 4.6):

$$\frac{d\sigma_r}{dr} + \frac{\sigma_r - \sigma_\theta}{r} = 0 \tag{12.15}$$

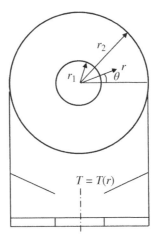

Figure 12.2

Putting $\sigma_z = 0$ in Equation 12.12 and expressing σ_r and σ_θ with αT and ε_r and ε_θ, we have

$$\left.\begin{array}{l} \sigma_r = \dfrac{E}{1-\nu^2}\left[\varepsilon_r + \nu\varepsilon_\theta - (1+\nu)\alpha T\right] \\[4mm] \sigma_\theta = \dfrac{E}{1-\nu^2}\left[\varepsilon_\theta + \nu\varepsilon_r - (1+\nu)\alpha T\right] \end{array}\right\} \tag{12.16}$$

Substituting Equation 12.16 into Equation 12.15,

$$r\frac{d}{dr}(\varepsilon_r + \nu\varepsilon_\theta) + (1-\nu)\cdot(\varepsilon_r - \varepsilon_\theta) - (1+\nu)\alpha r\frac{dT}{dr} = 0 \tag{12.17}$$

Expressing ε_r and ε_θ with deformation u in the radial direction,

$$\varepsilon_r = \frac{du}{dr}, \qquad\qquad \varepsilon_\theta = \frac{u}{r} \tag{12.18}$$

Substituting these equations into Equation 12.17,

$$\frac{d^2u}{dr^2} + \frac{1}{r}\cdot\frac{du}{dr} - \frac{u}{r^2} = (1+\nu)\alpha\frac{dT}{dr} \tag{12.19}$$

Putting $r = e^t$ and determining the general solution u_1 of Equation 12.19,

$$u_1 = C_1 e^t + C_2 e^{-t} = C_1 r + \frac{C_2}{r} \tag{12.20}$$

In order to determine the special solution u_2, we put $u = e^t f$ and then

$$\frac{d^2f}{dt^2} + 2\frac{df}{dt} = (1+\nu)\alpha\frac{dT}{dt} \tag{12.21}$$

$$\frac{df}{dt} + 2f = (1+\nu)\alpha T \tag{12.22}$$

Denoting D as $D = d/dt$,

$$f = \frac{(1+\nu)\alpha}{D+2}T = (1+\nu)\alpha e^{-2t}\int_{t_1}^{t} e^{2t}T dt$$

And then,

$$u_2 = (1+\nu)\alpha e^{-t}\int_{r_1}^{r} e^t T dr = (1+\nu)\alpha\frac{1}{r}\int_{r_1}^{r} rT dr \tag{12.23}$$

Finally, the solution of Equation 12.19 is

$$u = u_1 + u_2 = C_1 r + \frac{C_2}{r} + (1+\nu)\alpha \frac{1}{r} \int_{r_1}^{r} Tr\,dr \tag{12.24}$$

This can be used as the solution for various problems having various boundary conditions in terms of C_1 and C_2.

The stresses σ_r and σ_θ are determined as follows by calculating ε_r and ε_θ from u of Equation 12.24 and substituting them into Equation 12.16.

$$\left.\begin{array}{l} \sigma_r = \dfrac{E}{1-\nu^2}\left[C_1(1+\nu) - C_2(1-\nu)\dfrac{1}{r^2}\right] - \alpha E\dfrac{1}{r^2}\displaystyle\int_{r_1}^{r} Tr\,dr \\[4mm] \sigma_\theta = \dfrac{E}{1-\nu^2}\left[C_1(1+\nu) + C_2(1-\nu)\dfrac{1}{r^2}\right] - \alpha ET + \alpha E\dfrac{1}{r^2}\displaystyle\int_{r_1}^{r} Tr\,dr \end{array}\right\} \tag{12.25}$$

Example problem 12.1   Simply supported circular plate without hole.

From the condition of the finite displacement at $r=0$, $C_2=0$. Also from $\sigma_r=0$ at $r=r_2$,

$$C_1 = (1-\nu)\frac{\alpha}{r_2^2}\int_0^{r_2} Tr\,dr$$

Then

$$\left.\begin{array}{l} \sigma_r = \alpha E\left(\dfrac{1}{r_2^2}\displaystyle\int_0^{r_2} Tr\,dr - \dfrac{1}{r^2}\displaystyle\int_0^{r} Tr\,dr\right) \\[4mm] \sigma_\theta = \alpha E\left(\dfrac{1}{r_2^2}\displaystyle\int_0^{r_2} Tr\,dr - T + \dfrac{1}{r^2}\displaystyle\int_0^{r} Tr\,dr\right) \end{array}\right\} \tag{12.26}$$

where the following integral is to be considered.

$$\lim_{r\to 0}\frac{1}{r}\int_0^{r} Tr\,dr = 0, \quad \lim_{r\to 0}\frac{1}{r^2}\int_0^{r} Tr\,dr = \frac{1}{2}T(0)$$

Problem 12.2.1   The temperature in a circular plate with free outer edge without hole is highest at the center and linearly drops from the center to the outer periphery. Determine the circumferential tensile stress σ_θ produced at the outer edge under the condition of the temperature difference $100\,^\circ\text{C}$, $\alpha = 1.0 \times 10^{-5}\,1/^\circ\text{C}$, and $E = 196\,\text{GPa}$. (Note: the value of σ_θ is the same irrespective of the size of the plate.)

Problem 12.2.2 Determine the thermal stress when the temperature distribution is given by $T = T_0\{1 - (r/r_o)^n\}$ in the problem similar to the previous problem. In addition, if $T_0 = $ constant, judge in which case $n < 1$ or $n > 1$ the stress σ_θ at the outer periphery has larger value.

Problem 12.2.3 Derive the solution for the thermal stress in the form of Equation 12.26 for a free edge condition at inner and outer circumferential edge, at $r = r_1$ and $r = r_2$.

12.3 Thermal Stress in a Cylinder

Thermal stress is investigated for a cylinder which has a thickness larger than the diameter of the circular plate in Figure 12.2. The axial displacement w is assumed to be completely constrained. In this condition, $\varepsilon_z = 0$ and the problem is equivalent to a plane strain problem. Various practical thermal stress problems can be estimated by these two solutions, that is those for a circular plate explained in the previous section and the solution for a cylinder to be explained in this section.

Assuming $\varepsilon_z = 0$, and then from Equation 12.12,

$$\sigma_z = \nu(\sigma_r + \sigma_\theta) - \alpha TE \tag{12.27}$$

Substituting Equation 12.27 into Equation 12.12,

$$\left.\begin{aligned}
\varepsilon_r &= (1+\nu)\alpha T + \frac{1-\nu^2}{E}\left(\sigma_r - \frac{\nu}{1-\nu}\sigma_\theta\right) \\
\varepsilon_\theta &= (1+\nu)\alpha T + \frac{1-\nu^2}{E}\left(\sigma_\theta - \frac{\nu}{1-\nu}\sigma_r\right)
\end{aligned}\right\} \tag{12.28}$$

Expressing ε_r and ε_θ with displacement u, the equilibrium equation (Equation 12.15) is written as follows in the same way as the previous section.

$$\frac{d^2u}{dr^2} + \frac{1}{r}\cdot\frac{du}{dr} - \frac{1}{r^2}u = \frac{1+\nu}{1-\nu}\alpha\frac{dT}{dr} \tag{12.29}$$

Since the form of this differential equation is the same as Equation 12.19 except for the coefficient on the right-hand side of the equation, the solution can be written as follows:

$$u = C_1 r + \frac{C_2}{r} + \frac{1+\nu}{1-\nu}\alpha\frac{1}{r}\int_{r_1}^{r} Trdr \tag{12.30}$$

$$\left.\begin{aligned}
\sigma_r &= \frac{E}{(1+\nu)\cdot(1-2\nu)}C_1 - \frac{E}{(1+\nu)}\cdot\frac{C_2}{r^2} - \frac{\alpha E}{1-\nu}\cdot\frac{1}{r^2}\int_{r_1}^{r} Trdr \\
\sigma_\theta &= \frac{E}{(1+\nu)\cdot(1-2\nu)}C_1 + \frac{E}{(1+\nu)}\cdot\frac{C_2}{r^2} - \frac{\alpha ET}{1-\nu} + \frac{\alpha E}{1-\nu}\cdot\frac{1}{r^2}\int_{r_1}^{r} Trdr
\end{aligned}\right\} \tag{12.31}$$

The unknown constants C_1 and C_2 are determined from boundary conditions in the same way as in the previous section.

Example problem 12.2   A solid cylinder with free circumferential surface.
From $u =$ finite at $r=0$, $C_2=0$. From $\sigma_r=0$ at $r=r_2$,

$$C_1 = \frac{(1+\nu)\cdot(1-2\nu)}{1-\nu}\alpha\frac{1}{r_2{}^2}\int_0^{r_2} Tr dr$$

Then,

$$u = \frac{(1+\nu)\cdot(1-2\nu)}{1-\nu}\alpha\frac{r}{r_2{}^2}\int_0^{r_2} Tr dr + \frac{1+\nu}{1-\nu}\alpha\frac{1}{r}\int_0^{r} Tr dr \tag{12.32}$$

$$\sigma_r = \frac{\alpha E}{1-\nu}\cdot\frac{1}{r_2{}^2}\int_0^{r_2} Tr dr - \frac{\alpha E}{1-\nu}\cdot\frac{1}{r^2}\int_0^{r} Tr dr \tag{12.33}$$

$$\sigma_\theta = \frac{\alpha E}{1-\nu}\cdot\frac{1}{r_2{}^2}\int_0^{r_2} Tr dr - \frac{\alpha ET}{1-\nu} + \frac{\alpha E}{1-\nu}\cdot\frac{1}{r^2}\int_0^{r} Tr dr \tag{12.34}$$

σ_z is calculated by substituting Equations 12.33 and 12.34 into Equation 12.27.

Problems of Chapter 12

1. Determine the upper bound of the tensile thermal stress which is generated at the wall of a pressure vessel with temperature $300\,^\circ\mathrm{C}$ when cooling water with temperature $20\,^\circ\mathrm{C}$ is instantaneously supplied on a small circular spot of the wall of the pressure vessel. Use these data: coefficient of thermal expansion $\alpha = 1.0 \times 10^{-5}\,1/^\circ\mathrm{C}$, Young's modulus $E = 206$ GPa and Poisson's ratio $\nu = 0.3$.
2. A metal B is coated with thin layer on a plastic sphere A with diameter d. This sphere is put into a high temperature liquid. The liquid pressure is p. The temperature rise of the sphere is increased uniformly by ΔT. The coefficients of thermal expansion for A and B are α_A and α_B, Young's modulus E_A and E_B and Poisson's ratio ν_A and ν_B, respectively. Determine the circumferential stress σ_θ and radial stress σ_r in the material B.
3. Determine the thermal stress around the central part of the rectangular plate with free end as Figure 12.1a for the following temperature distribution in the plate.

$$T = T_1\sin(\pi y/h) + T_2\cos(\pi y/h)$$

Assume the coefficient of thermal expansion as α and Young's modulus as E.

13

Contact Stress

The deformation of two bodies in contact is influenced by each other. Stress produced by the force acting at the contact surface is called *contact stress*. Contact stress plays an important role in practical problems, such as the contact between train wheel and rail, gears and bearings.

The basic solution for contact stress analysis is the stress produced by a point force applied on the edge of a semi-infinite plate for two dimensional (2D) problems (Figure 13.1) and the stress produced by a point force applied on the surface of a semi-infinite body for three dimensional (3D) problems (Figure 13.2). Since in general, the area in contact between two bodies is sufficiently small compared to the surface of the bodies, the problems can be analyzed by superposing the basic solutions for Figures 13.1 or 13.2 within practically sufficient accuracy. The theory of contact stress based on this concept is known as the *Hertz contact theory*.

Once the pressure distribution at the contact surface is determined, the stresses at other places can be calculated by integrating the basic solution of Figures 13.1 or 13.2 over the contact surface. If frictional force acts on the contact surface, the solution for a point force acting tangentially to the contact surface must be used as the basic solution. However, in this chapter only the problems with forces acting normal to the contact surface are treated.

If the solution method for problems with normal pressure is understood, the same method can be applied to friction problems. Therefore, the question in this chapter is how to determine pressure distribution at the contact surface. One method is to assume pressure distribution with a practically reasonable form. More exact method is to use the basic solutions for two bodies in contact and at the same time to take into account the deformation of the contact surfaces of two bodies.

Theory of Elasticity and Stress Concentration, First Edition. Yukitaka Murakami.
© 2017 John Wiley & Sons, Ltd. Published 2017 by John Wiley & Sons, Ltd.

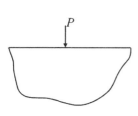

Figure 13.1

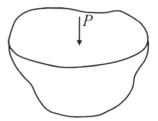

Figure 13.2

13.1   2D Contact Stress

13.1.1   Stress and Deformation for Uniform Pressure Distribution

The stress and deformation for uniform pressure distribution on the free edge of a semi-infinite plate of Figure 13.3 is the basis for the analysis of various contact problems. The solution is derived as follows.

Based on the solution (Equation 6.88) for a point force acting on a free edge of a semi-infinite plate, the stress produced by the distributed loading $q(\xi) = q_0(\xi_1 \le \xi \le \xi_2)$ is derived as follows:

$$
\left.
\begin{aligned}
\sigma_x &= -\frac{2q_0}{\pi}\int_{\xi_1}^{\xi_2} \frac{(x-\xi)^2 y}{\left\{(x-\xi)^2 + y^2\right\}^2}\,d\xi \\[2mm]
\sigma_y &= -\frac{2q_0}{\pi}\int_{\xi_1}^{\xi_2} \frac{y^3}{\left\{(x-\xi)^2 + y^2\right\}^2}\,d\xi \\[2mm]
\tau_{xy} &= -\frac{2q_0}{\pi}\int_{\xi_1}^{\xi_2} \frac{(x-\xi)y^2}{\left\{(x-\xi)^2 + y^2\right\}^2}\,d\xi
\end{aligned}
\right\}
\tag{13.1}
$$

Integrating these equations by putting $t = x-\xi$,

$$
\left.
\begin{aligned}
\sigma_x &= \frac{q_0}{\pi}\left[-\frac{ty}{(t^2+y^2)} + \arctan\frac{t}{y}\right]_{t_1}^{t_2} \\[2mm]
\sigma_y &= \frac{q_0}{\pi}\left[\frac{ty}{(t^2+y^2)} + \arctan\frac{t}{y}\right]_{t_1}^{t_2} \\[2mm]
\tau_{xy} &= -\frac{q_0}{\pi}\left[\frac{y^2}{(t^2+y^2)}\right]_{t_1}^{t_2}
\end{aligned}
\right\}
\tag{13.2}
$$

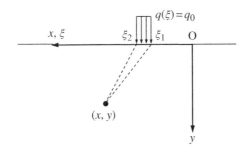

Figure 13.3

where $t_1 = x - \xi_1$ and $t_2 = x - \xi_2$.

Then, the stress along the edge of the infinite plate is expressed as:

1. $\sigma_x = 0$ and $\sigma_y = 0$ at $y = 0$, $x < \xi_1$ and $x > \xi_2$.
2. $\sigma_x = \sigma_y = -q_0$ at $y = 0$, $\xi_1 < x < \xi_2$.

The result of (2) must be noted as the characteristic nature of a contact stress field. Considering the following equation:

$$\frac{\partial u}{\partial x} = \varepsilon_x = \frac{1}{E}\left(\sigma_x - \nu\sigma_y\right) \tag{13.3}$$

And substituting σ_x and σ_y of Equation 13.2 into Equation 13.3, and by integrating, we have the following equation

$$u = \frac{q_0}{\pi E}\left[-y\log\left(t^2 + y^2\right) + (1-\nu)t \text{ arc } \tan\frac{t}{y}\right]_{t_1}^{t_2} + f(y) \tag{13.4}$$

where $f(y)$ is a function only of y. Likewise,

$$\frac{\partial v}{\partial y} = \varepsilon_y = \frac{1}{E}\left(\sigma_y - \nu\sigma_x\right) \tag{13.5}$$

From this equation, we have

$$v = \frac{q_0}{E}\left[t\log\left(t^2 + y^2\right) - (1-\nu)t \, \log|t| + (1-\nu)y \, \arctan\frac{t}{y}\right]_{t_1}^{t_2} + g(x) \tag{13.6}$$

where $g(x)$ is a function only of x.

Here, with $\gamma_{xy} = \partial u / \partial y + \partial v / \partial x$, the following equation is derived with Equations 13.4 and 13.6.

$$\gamma_{xy} = \frac{q_0}{\pi E}\left[-2(1+\nu)\frac{y^2}{(t^2 + y^2)}\right]_{t_1}^{t_2} + \frac{q_0}{\pi E}[-(1-\nu)\log|t|]_{t_1}^{t_2} + \frac{\partial g(x)}{\partial x} + \frac{\partial f(y)}{\partial y} \tag{13.7}$$

On the other hand, since $\gamma_{xy} = \tau_{xy}/G = 2(1+\nu)\tau_{xy}/E$, from the third line of Equation 13.2,

$$\gamma_{xy} = \frac{q_0}{\pi E}\left[-2(1+\nu)\frac{y^2}{(t^2+y^2)}\right]_{t_1}^{t_2} \tag{13.8}$$

Equating Equation 13.7 to Equation 13.8,

$$\frac{\partial f(y)}{\partial y} = 0 \tag{13.9}$$

$$\frac{\partial g(x)}{\partial x} = -\frac{q_0}{\pi E}[-(1-\nu)\log|t|]_{t_1}^{t_2} \tag{13.10}$$

Then,

$$g(x) = \frac{q_0}{\pi E}[(1-\nu)\{t \log|t|-t\}]_{t_1}^{t_2} + C_1 \tag{13.11}$$

where C_1 is a constant. Finally, υ is derived as follows:

$$\upsilon = \frac{q_0}{\pi E}\left[t \log(t^2+y^2)-(1-\nu)t+(1-\nu)y \arctan\frac{t}{y}\right]_{t_1}^{t_2} + C_1 \tag{13.12}$$

Here, putting $y=0$, $d\xi = \xi_2 - \xi_1$ and $q_0 = q(\xi)$, and excluding the term for displacement as rigid body, the displacement at the surface of a semi-infinite plate due to distributed loading $q(\xi)$ acting on $d\xi$ is expressed as follows:

$$d\upsilon = -\frac{2}{\pi E}q(\xi)\log|x-\xi|d\xi \tag{13.13}$$

Therefore, if $q(\xi)$ is distributed from $\xi = a$ to $\xi = b$,

$$\upsilon = -\frac{2}{\pi E}\int_a^b q(\xi)\log|x-\xi|d\xi \tag{13.14}$$

The displacement υ at the center of the infinitesimal width 2ε of pressure distribution as shown in Figure 13.4 is calculated as follows.

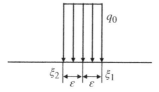

Figure 13.4

Putting $t_2 = -\varepsilon$, $t_1 = \varepsilon$ and $y = 0$, and $\varepsilon \to 0$,

$$v = -\frac{4q_0}{\pi E}\lim_{\varepsilon \to 0}\varepsilon \cdot \log\varepsilon = 0 \qquad (13.15)$$

This result means that, when we calculate the displacement v just below the load $q(\xi)$ distributed along a certain distance of ξ, the displacement at the central point of the infinitesimal region $d\xi$ within the distributed loading can be ignored.

13.1.2 Contact Stress by a Rigid Punch

By pushing down a rigid punch on the edge of a semi-infinite elastic plate as shown in Figure 13.5, a constant displacement is produced at the contact surface. The pressure distribution in this problem is determined as follows.

Assuming the distributed load at $\xi = -a \sim a$ as $q(\xi)$, the vertical displacement is expressed from Equation 13.14 as

$$v = -\frac{2}{\pi E}\int_{-a}^{a} q(\xi)\log|x-\xi|d\xi \qquad (13.16)$$

Therefore, the solution is to determine the load $q(\xi)$ which produces $v = v_0$ (constant) irrespective of the value of x for $|x| < a$. For the condition $v = v_0$, Equation 13.16 is a well known integral equation and the solution [1] is given by

$$q(\xi) = \frac{P}{\pi\sqrt{a^2 - \xi^2}} \qquad (13.17)$$

where P is the total load applied to the rigid punch.

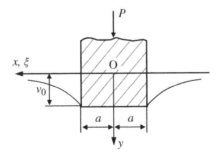

Figure 13.5

13.1.3 Stress Field by Elliptical Contact Stress Distribution

The pressure distribution at the contact surface between two circular plates or two circular cylinders, as in Figure 13.6, is often assumed in the form of a semi-ellipse or semi-circle, as in Figure 13.7.

Since the dimension of the contact surface is relatively small compared to the dimension of the cylinder, the contact stress can be approximately calculated as the stress in a semi-infinite plate subjected to the same pressure distribution. Therefore, including q_0 of Equation 13.1 in the integral as $q(\xi)$, the stress field of Figure 13.7 is derived as follows [2].

$$
\left.
\begin{aligned}
\sigma_x &= -\frac{q_0}{\pi} y \left[\frac{b^2 + 2x^2 + 2y^2}{b} F - \frac{2\pi}{b} + 3xG \right] \\[2mm]
\sigma_y &= -\frac{q_0}{\pi} y [bF + xG] \\[2mm]
\tau_{xy} &= \frac{q_0}{\pi} y^2 G
\end{aligned}
\right\}
\tag{13.18}
$$

where

$$
\left.
\begin{aligned}
F &= \frac{\pi}{K_1} \cdot \frac{1 + \sqrt{K_2/K_1}}{\sqrt{K_2/K_1}\sqrt{2\sqrt{K_2/K_1} + [(K_1 + K_2 - 4b^2)/K_1]}} \\[3mm]
G &= \frac{\pi}{K_1} \cdot \frac{1 - \sqrt{K_2/K_1}}{\sqrt{K_2/K_1}\sqrt{2\sqrt{K_2/K_1} + [(K_1 + K_2 - 4b^2)/K_1]}} \\[3mm]
K_1 &= (b-x)^2 + y^2, \quad K_2 = (b+x)^2 + y^2
\end{aligned}
\right\}
\tag{13.19}
$$

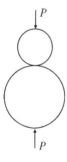

Figure 13.6

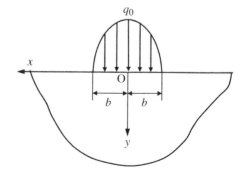

Figure 13.7

Regarding σ_z, we may regard $\sigma_z = 0$ for a circular plate due to the plane stress condition, and $\sigma_z \cong \nu(\sigma_x + \sigma_y)$ for circular cylinder which is close to the plane strain condition.

Since phenomena such as contact fatigue are related not only to the stress given by Equation 13.18 but also to shear stresses in other directions, the stress field near the contact surface must be investigated in detail using the stress transformation equation explained in Chapter 1. Recent advances in the fracture mechanics approach to contact fatigue problems associated with cracks under contact stress can be studied based on the basic theory in Chapter 6 and Part II of this book.

13.2   3D Contact Stress

The basic solution for the analysis of 3D contact problems is a point force applied on the surface of a semi-infinite body, as shown in Figure 13.8. Although the analytical procedure is the same as for 2D problems, the calculation becomes naturally more complicated.

The solution for the problem of Figure 13.8 was obtained by J. Boussinesq (1885). The solution is expressed as follows with the cylindrical coordinates of Figure 13.8.

$$
\left.
\begin{aligned}
\sigma_\gamma &= \frac{P}{2\pi}\left[(1-2\nu)\cdot\left\{\frac{1}{r^2} - \frac{z}{r^2}\left(r^2+z^2\right)^{-1/2}\right\} - 3r^2 z\left(r^2+z^2\right)^{-5/2}\right] \\[2mm]
\sigma_\theta &= \frac{P}{2\pi}(1-2\nu)\cdot\left[-\frac{1}{r^2} + \frac{z}{r^2}\left(r^2+z^2\right)^{-1/2} + z\left(r^2+z^2\right)^{-3/2}\right] \\[2mm]
\sigma_z &= -\frac{3P}{2\pi}z^3\left(r^2+z^2\right)^{-\frac{5}{2}} \\[2mm]
\tau_{\gamma z} &= -\frac{3P}{2\pi}rz^2\left(r^2+z^2\right)^{-\frac{5}{2}}
\end{aligned}
\right\}
\tag{13.20}
$$

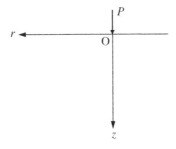

Figure 13.8

Displacement w in the z direction can be obtained by the same procedure as in a 2D case by integrating the strains as follows:

$$w = \frac{P}{2\pi E}\left[(1+v)z^2\left(r^2+z^2\right)^{-3/2} + 2(1-v)\cdot\left(r^2+z^2\right)^{-1/2}\right]$$ (13.21)

Considering the relationship between displacement u in the r direction and ε_θ as $\varepsilon_\theta = u/r$ (see Equation 2.25), and from Hooke's law,

$$u = \frac{(1-2v)\cdot(1+v)P}{2\pi Er}\left[z\left(r^2+z^2\right)^{-1/2} - 1 + \frac{1}{1-2v}r^2z\left(r^2+z^2\right)^{-3/2}\right]$$ (13.22)

At $z=0$ (surface), we have

$$w = \frac{P(1-v^2)}{\pi Er}$$ (13.23)

$$u = -\frac{(1-2v)\cdot(1+v)P}{2\pi Er}$$ (13.24)

13.2.1 Stress and Displacement Due to a Constant Distributed Loading

Based on the previously obtained solutions, the stresses and displacements produced by the load distributed over a certain region can be calculated by the same procedure as in the previous section [3].

As an example problem, the stress and displacement due to constant load q_0 distributed over a circular area with radius a as shown in Figure 13.9 are calculated in the following.

In order to calculate the stress at point A (r, θ) outside the area of load distribution, performing the integration by putting the origin of the coordinate at the center of the circle may be a first choice and natural. However, the method is not successful. It is interesting that performing the integral by using the polar coordinate system (s, φ) with the origin at a point A is an elegant and easier solution for this problem.

The equation for this calculation is given based on Equation 13.23 as follows:

$$dw = \frac{(1-v^2)q_0}{\pi E}\cdot\frac{sd\varphi\cdot ds}{s} = \frac{(1-v^2)q_0}{\pi E}d\varphi\cdot ds$$ (13.25)

Taking the total distributed load into consideration,

$$w = \frac{4(1-v^2)q_0}{\pi E}\int_{s_1}^{s_2}\int_{\varphi_1}^{\varphi_2}d\varphi\cdot ds$$ (13.26)

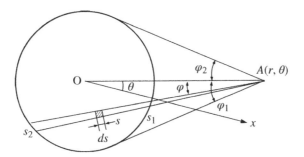

Figure 13.9

Neglecting the details of the integration, the result is expressed as follows:

$$w = \frac{4(1-\nu^2)q_0 r}{\pi E}\left[E\left(\frac{\pi}{2},k\right)-(1-k^2)F\left(\frac{\pi}{2},k\right)\right]$$ (13.27)

where $k^2 = a^2/r^2$, $F(\pi/2, k)$ and $E(\pi/2, k)$ are the complete elliptic integrals of the first and second kinds, respectively, which can be calculated by numerical tables or by computer subroutine libraries. $F(\pi/2, k)$ and $E(\pi/2, k)$ are expressed as

$$E\left(\frac{\pi}{2},k\right) = \int_0^{\pi/2}\sqrt{1-k^2\sin^2\theta}\,d\theta$$ (13.28)

$$F\left(\frac{\pi}{2},k\right) = \int_0^{\pi/2}\frac{d\theta}{\sqrt{1-k^2\,\sin^2\theta}}$$ (13.29)

The displacement inside the area of load distribution is determined by a similar method as

$$w = \frac{4(1-\nu^2)q_0 a}{\pi E}E\left(\frac{\pi}{2},k\right)$$ (13.30)

Also the stress distribution is given by

$$\left.\begin{array}{l}\sigma_z = q_0\left[-1+\dfrac{z^3}{(a^2+z^2)^{3/2}}\right] \\[4mm] \sigma_r = \dfrac{q_0}{2}\left[-(1+2\nu)+\dfrac{2(1+\nu)z}{\sqrt{a^2+z^2}}-\left(\dfrac{z}{\sqrt{a^2+z^2}}\right)^3\right]\end{array}\right\}$$ (13.31)

$$\sigma_\theta = \sigma_r$$

13.2.2   Contact Stress Due to a Rigid Circular Punch

The pressure distribution produced at the contact surface by giving a constant displacement to a rigid circular punch [4] is derived by a similar procedure to that in the 2D problem as follows

$$q(r) = \frac{P}{2\pi a \sqrt{a^2 - r^2}} \tag{13.32}$$

$$c = \frac{P(1-\nu^2)}{2aE} \tag{13.33}$$

where P is the total applied force and c is the vertical displacement of the punch.

13.2.3   Contact Between Two Spheres

In the contact problem between two elastic bodies having curved surfaces (Figures 13.10 and 13.11), the area of load distribution and intensity of load must be determined based on the conditions that the relative distance between two bodies and surface displacements must be compatibly adjusted to each other. In the solution, although the contact surface projected to the z direction is a circle with unknown radius a, the load distribution q (r) viewed on the r–z section is assumed to be an ellipse. Thus, replacing the area near the contact surface by a semi-infinite body, the basic equations to be applied are expressed as

$$w_1 + w_2 = c - (z_1 + z_2) \tag{13.34}$$

$$w_1 = \frac{(1-\nu_1^2)}{E_1} \iint q(r)ds \cdot d\varphi \tag{13.35}$$

$$w_2 = \frac{(1-\nu_2^2)}{E_2} \iint q(r)ds \cdot d\varphi \tag{13.36}$$

where c is the displacement by which certain points far from the contact surface in two spheres (e.g. the center of the spheres) approach each other. Subscripts 1 and 2 are the quantities related to spheres 1 and 2, respectively.

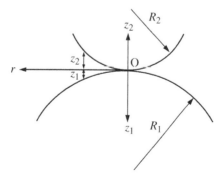

Figure 13.10

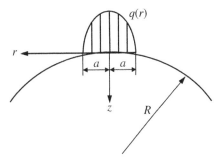

Figure 13.11

The result calculated by Hertz [5] for two spheres with radius R_1 and R_2 are given by the following equations.

Radius of the circular contact surface:

$$a = \left\{ \frac{3}{4} \cdot \frac{P[(1-\nu_1{}^2)/E_1 + (1-\nu_2{}^2)/E_2]R_1 R_2}{R_1 + R_2} \right\}^{1/3} \tag{13.37}$$

Maximum contact pressure:

$$q_0 = \frac{3}{2} \cdot \left(\frac{P}{\pi a^2} \right) \tag{13.38}$$

13.2.4   *Contact Between Two Cylinders*

Although the details of calculation are neglected[1], the contact pressure for two cylinders with radius R_1 and R_2 having a parallel axis becomes like Figure 13.7 and the stress distribution is expressed in the form of Equation 13.18. The width of contact ($2b$) has the following relationship with the load P' per unit axial length of the cylinders and other variables.

One half of the contact width:

$$b = \sqrt{\frac{4P'[(1-\nu_1{}^2)/E_1 + (1-\nu_2{}^2)/E_2]R_1 R_2}{R_1 + R_2}} \tag{13.39}$$

Maximum contact pressure:

$$q_0 = \frac{2P'}{\pi b} \tag{13.40}$$

---

[1] This is essentially a 2D problem; the same solution can be obtained as the extreme case of a 3D problem.

Problems of Chapter 13

1. Using Equation 13.17, determine the stress intensity factor for the problem of Figure 13.12 which shows deep double cracks facing each other.
2. Using Equation 13.32, determine the stress intensity factor for the problem of Figure 13.13 which shows deep 3D circumferential cracks.
3. Explain the reason for $\tau_{xy} = 0$ along the y axis under the contact loading of Figure 13.7. For this reason σ_x and σ_y are the principal stresses and accordingly the maximum shear stress acts at a plane inclined by 45° to the x and y axes. Denoting the maximum shear stress by $\tau_{45°}$, determine the location on the y axis and the value of $\tau_{45°}$.
4. Verify the stress field produced by a frictional force which acts from left to right under the contact load of Figure 13.7 is given by the following equations [2]. Use the notation f as the coefficient of friction.

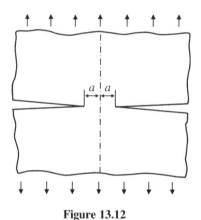

Figure 13.12

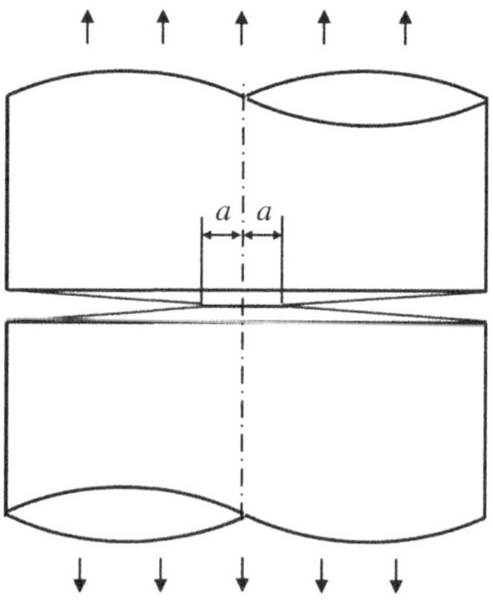

Figure 13.13

$$\sigma_x = -\frac{fq_0}{\pi}\left[\left(2x^2-2b^2-3y^2\right)G-\frac{2\pi x}{b}-\frac{2x\left(b^2-x^2-y^2\right)}{b}F\right]$$

$$\sigma_y = -\frac{fq_0}{\pi}y^2G$$

$$\tau_{xy} = \frac{fq_0}{\pi}y\left[\frac{\left(b^2+2x^2+y^2\right)}{b}F-\frac{2\pi}{b}+3xG\right]$$

(13.41)

where F and G are the same as those used in Equation 13.19.

References

[1] M. Sadowsky (1928) Z. Angew. Math. Mech., **8**:107.

[2] J. O. Smith and G. C. Liu (1953) J. Appl. Mech., **20**:157.

[3] S. P. Timoshenko and J. N. Goodier (1982) Theory of Elasticity, 3rd edn, McGraw–Hill International, New York, p. 403.

[4] S. P. Timoshenko and J. N. Goodier (1982) Theory of Elasticity, 3rd edn, McGraw–Hill International, New York, p. 408 (reporting work by J. Boussinesq, 1885).

[5] S. P. Timoshenko and J. N. Goodier (1982) Theory of Elasticity, 3rd edn, McGraw–Hill International, New York, p. 412.

Answers and Hints for Part I Problems

Chapter 1

1. $\theta \cong 1/20$.
2. Normal stress is $-p_0$ in all directions. Shear stress is 0 everywhere. Solve using the same method as Example problem 1.1.
3. $\sigma_\eta = -\sigma_0$. Solution 1: Use stress transformation equations. Solution 2: Use the stress invariant.
4. Solve by the same method as Example problem 1.1 and use the 3D stress transformation equations.
5. Solve by the same method as Example problem 1.2.
6. Calculate the principal stresses σ_1 and σ_2 ($\sigma_1 > \sigma_2$) in terms of p and T. The balloon cannot sustain compressive load. Then, when σ_2 becomes negative (compression), the shape of the balloon becomes unstable and buckling occurs by producing a spiral wrinkle having an angle θ_c perpendicular to the direction of σ_2, namely the direction of σ_1. Therefore, T_c is the value of T when σ_2 becomes 0. $T_c = \sqrt{2\pi}/8 \cdot d^3 p$. $\theta_c = 54.74$ degrees. It is very difficult to perform the perfect experiment of twisting the balloon, because gripping the balloon changes the stresses in the balloon and eventually influences the principal stresses. However, the direction of the initial wrinkle $\theta_c \cong 54.74$ degrees can be detected even in the imperfect experiment.

Theory of Elasticity and Stress Concentration, First Edition. Yukitaka Murakami.
© 2017 John Wiley & Sons, Ltd. Published 2017 by John Wiley & Sons, Ltd.

Chapter 2

1. Normal strains in three independent directions are necessary.
2. $(\varepsilon_x + \varepsilon_y)$. Solution 1: Calculate the area S as $S = 1/2 a(1 + \varepsilon_x)(1 + \varepsilon_y)\sin(\pi/2 - \gamma_{xy})$.
   Solution 2: Use the strain invariant.
4. $\varepsilon_{45°}$ is expressed with ε_x, ε_y and γ_{xy} as follows:

$$\varepsilon_{45°} = \varepsilon_x l_1^2 + \varepsilon_y m_1^2 + \gamma_{xy} l_1 m_1$$

$$= \frac{1}{2}\left(\varepsilon_x + \varepsilon_y + \gamma_{xy}\right)$$

Therefore,

$$\gamma_{xy} = 2\varepsilon_{45°} - \left(\varepsilon_x + \varepsilon_y\right)$$

From this equation and Equations (a), (b) and (c),

$$c_0 + c_1 x + c_2 y + c_3 x^2 + c_4 y^2 + c_5 xy$$
$$= 2\varepsilon_{45°} - \left\{(a_0 + b_0) + (a_1 + b_1)x + (a_2 + b_2)y + (a_3 + b_3)x^2 + (a_4 + b_4)y^2 + (a_5 + b_5)xy\right\},$$

where there are six unknowns. Since $\varepsilon_{45°}$ is measured at five points, there are five known conditions. In addition to these conditions, it is reasonable to satisfy the compatibility equation $\partial^2 \varepsilon_x/\partial y^2 + \partial^2 \varepsilon_y/\partial x^2 = \partial^2 \gamma_{xy}/\partial x \partial y$. Thus, the following equation is derived:

$$c_5 = 2(a_4 + b_3)$$

Thus, from these six conditions, $c_0 \sim c_5$ can be determined.

Chapter 3

2. $\sigma - 79.2$ MPa, $T - 2244$ N $\cdot$ m
3. $\tau_{xy} = 0$ at $x = \pm W$ and $y = \pm L$ (free surface); $\tau_{r\theta} = 0$ at the circumferential edge of the circular hole (free surface); $\tau_{xy} = 0$ at $y = 0$ (symmetric plane).
   Note that $\tau_{xy} \neq 0$ at the interface between the rigid disk and the plate (see Problem 5 of Chapter 5).
4. $\sigma_{zB} = \dfrac{E_B}{E_A} \cdot \dfrac{1 - \nu_A \nu_B}{1 - \nu_B^2}\sigma, \quad \sigma_{rB} = 0, \quad \sigma_{\theta B} = \dfrac{E_B}{E_A} \cdot \dfrac{\nu_B - \nu_A}{1 - \nu_B^2}\sigma$
5. (1) $\sigma_a = dp/(4h).\sigma_t = dp/(2h)$. (2) $\delta = (1 - 2\nu)dpl/(4Eh)$.

Chapter 4

2.

$$G\left[2\frac{\partial}{\partial x}\left(\varepsilon_x + \frac{\nu}{1-2\nu}e\right) + \frac{\partial\gamma_{xy}}{\partial y} + \frac{\partial\gamma_{zx}}{\partial z}\right] + X = 0$$

$$G\left[\frac{\partial\gamma_{xy}}{\partial x} + 2\frac{\partial}{\partial y}\left(\varepsilon_y + \frac{\nu}{1-2\nu}e\right) + \frac{\partial\gamma_{yz}}{\partial z}\right] + Y = 0$$

$$G\left[\frac{\partial\gamma_{zx}}{\partial x} + \frac{\partial\gamma_{yz}}{\partial y} + 2\frac{\partial}{\partial z}\left(\varepsilon_z + \frac{\nu}{1-2\nu}e\right)\right] + Z = 0$$

where $e = \varepsilon_x + \varepsilon_y + \varepsilon_z$.

$$\frac{\nu}{1-2\nu}G\frac{\partial e}{\partial x} + G\nabla^2 u + X = 0$$

$$\frac{\nu}{1-2\nu}G\frac{\partial e}{\partial y} + G\nabla^2 v + Y = 0$$

$$\frac{\nu}{1-2\nu}G\frac{\partial e}{\partial z} + G\nabla^2 w + Z = 0$$

3. The equilibrium conditions to be satisfied are:

$$\frac{\partial\sigma_x}{\partial x} + \frac{\partial\tau_{xy}}{\partial y} + \frac{\partial\tau_{zx}}{\partial z} = 0, \quad \frac{\partial\tau_{xy}}{\partial x} = 0, \quad \frac{\partial\tau_{zx}}{\partial x} = 0$$

Therefore, $\tau_{xy} = f(y,z)$, $\tau_{zx} = g(y,z)$, $\tau_{zx} = 0$ at $z = 0$, $\tau_{xy} = 0$ at $y = h$ and $\tau_{xy}/\tau_{zx} = h/a$ on the boundaries AB and AC.

Taking the following equation into consideration:

$$\sigma_x = \frac{6P}{ah^3}(2h-3y)\cdot(l-x)$$

we have

$$\tau_{xy} = \frac{6P}{ah^3}(h-y)y, \quad \tau_{zx} = \frac{6P}{ah^3}(h-y)z$$

Chapter 5

1. Pay attention to the stress distribution at the section A–A (refer to Figure P5.1). In the case of the rectangular plate the area higher than the nominal stress (dotted line) and the area below

the nominal stress must be equal. In the case of the cylinder, the area higher than the nominal stress near the central axis of the cylinder looks larger than the area at the outer part of the cylinder. This is because the stress distribution in the cylinder must satisfy the following equation.

$$\int_0^{r_1} (\sigma_z - \sigma_n) \cdot 2\pi r dr = \int_{r_1}^{W} (\sigma_n - \sigma_z) \cdot 2\pi r dr$$

(r_1 is the coordinate at $\sigma_z = \sigma_n$)

Namely, the infinitesimal area in the integral equation has relationship with r in terms of $2\pi r dr$.

2. Due to Saint Venant's principle, the difference in stress between two cases occurs at $x = (l-h) \sim l$. Therefore, the difference in deflection between two cases occurs for the case of $l \leq h$.

3. Concentrated force $= (\sigma_1 + \sigma_2)bs/2$, Concentrated moment $= bs^2(\sigma_2 - \sigma_1)/12$.

4. Define σ_n by $\sigma_n = P/W$ (refer to Figure P5.1).

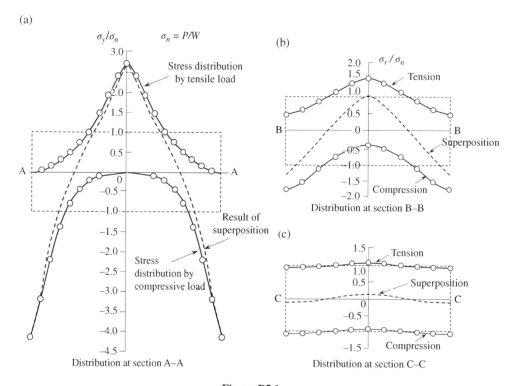

Figure P5.1

5.

$\sigma_{\xi A} = \sigma_{\xi B}$	Correct
$\sigma_{\eta A} = \sigma_{\eta B}$	Wrong
$\tau_{\xi\eta A} = \tau_{\xi\eta B}$	Correct
$\varepsilon_{\xi A} = \varepsilon_{\xi B}$	Wrong
$\varepsilon_{\eta A} = \varepsilon_{\eta B}$	Correct
$\gamma_{\xi\eta A} = \gamma_{\xi\eta B}$	Wrong

Chapter 6

Problem 6.4.1 $u = -\dfrac{2ab^2}{E(b^2 - a^2)} p_0$ at $r = a$. $u = -\dfrac{1}{E} \left[\dfrac{b^2 + a^2}{b^2 - a^2} - \nu \right] b p_0$ at $r = b$.

Problem 6.4.2 $\sigma_r = 0$, $\sigma_\theta = 2\sigma_0$.

Problem 6.5.1 Considering

$$\sigma_x = \sigma_0 \left(1 + \frac{a^2}{2r^2} + \frac{3a^4}{2r^4} \right)$$

along the y axis, σ_x at point A $(y = 2a)$ is:

$$\sigma_x = \sigma_0 + \frac{7}{32}\sigma_0$$

Thus, the difference from σ_0 is only $(7/32)\,\sigma_0$.

Problem 6.5.2 The stress field at the center of the circular plate is:

$$\sigma_x = \frac{(3 + \nu)\rho\omega^2 b^2}{8}, \quad \sigma_y = \sigma_x$$

When a small circular hole exists in the stress field, the stress at the edge of the circular hole is calculated as

$$\sigma_\theta = 3\sigma_x - \sigma_y \cong \frac{(3 + \nu)\rho\omega^2 b^2}{4}$$

This solution agrees with the exact solution for a hollow circular plate containing an extremely small circular hole.

Problem 6.5.3 It can be regarded approximately that the stress field around the circular hole A is not disturbed by the presence of the circular hole B. When the circular hole B does not exist, the stress field at the hole B influenced by the circular hole A is calculated as $\sigma_x^* = 4\sigma_0/3$, $\sigma_y^* = \sigma_0/3$. Assuming the circular hole B exists in such a stress field, the maximum stress at the edge of the hole is estimated as: $\sigma_{max} \cong 3\sigma_x^* - \sigma_y^* = 11\sigma_0/3$.

Problem 6.6.1 There is no shear stress τ_{xy} along the central line of Figure 6.13 (imagine a cross put along the center line.). Therefore, the difference between the center line of Figure 6.13 and the free surface edge of the semi-infinite plate of Figure 6.14 is that σ_x exists along the center line of Figure 6.13. If the stress σ_x is cancelled along the center line, the stress state becomes the same as Figure 6.14. However, since σ_x acts in the x direction and not in the tensile load direction (y direction), the cancellation of σ_x does not much increase the stress at notch root in the y direction. For example, $K_t = 3.0$ for a circular hole in an infinite plate and $K_t = 3.065$ for a semi-circular notch in a semi-infinite plate. The stress intensity factor K_I for an edge crack in a semi-infinite plate is 12% higher than that for a crack in an infinite plate.

Problem 6.6.2 $\sigma_{max} = \sigma_0 + (4a/b)\sigma_0$

Problem 6.7.1 Estimating the average stress σ_θ as $\sigma_\theta \cong ap_i/(b-a)$ and using the solution of Howland (Figure 6.16), the maximum stress can be calculated as $\sigma_{\theta max} \cong 98$ MPa.

Problem 6.8.1 $K_I = 3.36\sigma_0\sqrt{\pi a}$

Problem 6.9.2

$$\sigma_x = -\frac{2M}{\pi} \cdot \frac{2xy(x^2-y^2)}{(x^2+y^2)^3}, \quad \sigma_y = -\frac{2M}{\pi} \cdot \frac{4xy^3}{(x^2+y^2)^3}, \quad \tau_{xy} = -\frac{2M}{\pi} \cdot \frac{(3x^2-y^2)y^2}{(x^2+y^2)^3}$$

Problems of Chapter 6

1. $\nu\sigma_0(a^2+b^2)/2b^2$
2. **Solution 1** We denote the solid cylinder by A and the hollow cylinder by B. The shrink fit pressure is denoted by p. The stresses inside the solid cylinder are $\sigma_{rA} = \sigma_{\theta A} = -p$ and the stresses at the inner wall of the hollow cylinder are $\sigma_{rB} = -p$, $\sigma_{\theta B} = (b^2+a^2)p/(b^2-a^2)$.
   Considering $\varepsilon_\theta = u/a = (\sigma_\theta - \nu\sigma_r)/E$ and $\Delta = 2a[|\varepsilon_{\theta A}| + |\varepsilon_{\theta B}|]$ at $r=a$,

$$p = \frac{b^2-a^2}{4ab^2}E\Delta$$

Therefore,

$$\sigma_{rB} = -\frac{b^2-a^2}{4ab^2}E\Delta, \quad \sigma_{\theta B} = \frac{b^2+a^2}{4ab^2}E\Delta$$

Solution 2 If we add tensile stress $\sigma_{rB} = p$ at $r=b$ after shrink fitting, the stress σ_r at the boundary $r=a$ becomes $\sigma_r = 0$. In this condition, the solid cylinder can be freely pulled out and the inner diameter of B becomes larger by Δ than the initial state. This situation is written by equations as follows:

$$\varepsilon_{\theta B} = \frac{\Delta}{2a} = \left(\frac{\sigma_{\theta B}}{E}\right)_{r=a}$$

and

$$\sigma_{\theta B} = \frac{2b^2 p}{(b^2 - a^2)}$$

Then,

$$p = \frac{(b^2 - a^2)E\Delta}{4ab^2}$$

Since in this solution method we pay attention only to the deformation of the outer hollow cylinder without calculating the deformation of the inner solid cylinder, the calculation is simple and we can avoid calculation mistakes.

3. (a) $3.8\sigma_0$, (b) $3.8\sigma_0$, (c) $\left(1 + 2\sqrt{(a_1 + a_2)/a_1}\right)\sigma_0$.
4. A. $12P/\pi d$; B. $-20P/\pi d$.
5. The problem is replaced by the one in which a point force P and a concentrated moment $M = Ph$ is applied at the origin O.
   The stress field by P:

$$\sigma_x{}^P = -\frac{2P}{\pi} \cdot \frac{x^3}{r^4}, \quad \sigma_y{}^P = -\frac{2P}{\pi} \cdot \frac{xy^2}{r^4}, \quad \tau_{xy}{}^P = -\frac{2P}{\pi} \cdot \frac{x^2 y}{r^4} \tag{1}$$

The stress field by M:

$$\sigma_x{}^M = -\frac{2M}{\pi} \cdot \frac{2xy(x^2 - y^2)}{(x^2 + y^2)^3}, \quad \sigma_y{}^M = -\frac{2M}{\pi} \cdot \frac{4xy^3}{(x^2 + y^2)^3}$$

$$\tau_{xy}{}^M = -\frac{2M}{\pi} \cdot \frac{(3x^2 - y^2)y^2}{(x^2 + y^2)^3} \tag{2}$$

$$\sigma_x = \sigma_x{}^P + \sigma_x{}^M, \quad \sigma_y = \sigma_y{}^P + \sigma_y{}^M, \quad \tau_{xy} = \tau_{xy}{}^P + \tau_{xy}{}^M$$

Calculating the principal stresses σ_1 and σ_2 from these stresses,

$$\sigma_{max} = 3\sigma_1 - \sigma_2$$

6. $K_{\mathrm{I}} = \left(\sigma_0 - \frac{\sqrt{3}}{2}\tau_0\right)\sqrt{\pi a}, \quad K_{\mathrm{II}} = \frac{1}{2}\tau_0\sqrt{\pi a}$
7. $\theta_0 = -0.928\,\mathrm{rad} = -53.17°$
8. Expressing the stresses in pipe A:

$$\sigma_{rA} = C_{1A} + \frac{C_{2A}}{r^2}, \quad \sigma_{\theta A} = C_{1A} - \frac{C_{2A}}{r^2}$$

Likewise in pipe B:

$$\sigma_{rB} = C_{1B} + \frac{C_{2B}}{r^2}, \quad \sigma_{\theta B} = C_{1B} - \frac{C_{2B}}{r^2}$$

From the boundary conditions:

$$C_{2A} = -a^2(p + C_{1A})$$

$$C_{1B} = \frac{1}{(b^2-c^2)}\left[(b^2-a^2)C_{1A}-a^2p\right], \quad C_{2B} = -\frac{c^2}{(b^2-c^2)}\left[(b^2-a^2)C_{1A}-a^2p\right]$$

$$C_{1A} = -\frac{\left[\dfrac{1+\nu_A}{E_A}a^2 + \dfrac{1}{E_B}\{(1-\nu_B)b^2+(1+\nu_B)c^2\}\dfrac{a^2}{b^2-c^2}\right]p}{\left[\dfrac{1}{E_A}\{(1-\nu_A)b^2+(1+\nu_A)a^2\}-\dfrac{1}{E_B}\{(1-\nu_B)b^2+(1+\nu_B)c^2\}\dfrac{b^2-a^2}{b^2-c^2}\right]}$$

The stress at an arbitrary value of r can be calculated from these results.

Chapter 7

1. $Tl(a+b)/(2Gha^2b^2)$
2. $3Tl\log(a_2/a_1)/[t^3 G(a_2-a_1)]$. $\quad G = E/2(1+\nu)$
3. $\dfrac{4Ta}{\pi Ghd^3\left\{1+(4/3)(h/d)^2\right\}}$
4. $\dfrac{6T}{\pi GDh^3[1+3\pi D^2/\{8(1+\pi/2)h^2\}]}$

Chapter 8

Problem 8.1.4

1. $U_1 = \dfrac{1-\nu}{E}\sigma_0^2 \pi b^2, \quad U_2 = \dfrac{1}{E}\sigma_0^2 \pi b^2\left(1-\dfrac{a^2}{b^2}\right)\cdot\left[\left(1+\dfrac{a^2}{b^2}\right)-\nu\left(1-\dfrac{a^2}{b^2}\right)\right]$
2. $U_1 = U_2$

Problem 8.3.1 From the principle of the virtual work,

$$\int_{S_\sigma} \sigma_0 \delta u dS_\sigma = \int_V \sigma_x \delta \varepsilon_x dV$$

and considering $\delta\varepsilon_x = \partial(\delta u)/\partial x = 2\alpha x$, the above equation is

$$\sigma_0 \alpha l^2 A = \int_0^l \sigma_x(2\alpha x)Adx = \sigma_x \alpha l^2 A.$$

Then, $\sigma_x = \sigma_0$.

Problem 8.4.2 Assuming $w = a_1 + a_2 x + a_3 x^2 + a_4 x^3$, solve by the same way as the Example (Figure 8.8).: $w = (Pl/2EI)x^2 - (P/6EI)x^3$

Problem 8.5.1 $U = \dfrac{P^{4/3}l}{4\sqrt[3]{AE}}, \quad U_c = \dfrac{3P^{4/3}l}{4\sqrt[3]{AE}}$

Problem 8.6.2 By the reciprocal theorem $R\delta_B = P\delta_A + Q\delta_c$,

$$\delta_A = \frac{1}{P}(R\delta_B - Q\delta_c)$$

Alternative solution Assuming $AB = b$ and $BC = c$, calculate the displacement by the strength of materials and drive the above relationship.

Problems of Chapter 8

1. $\delta_0 = \dfrac{16Pl^3}{\pi^4 EI} = \dfrac{Pl^3}{6.088EI}$. The difference between this solution and $\dfrac{Pl^3}{6EI}$ (the conventional solution of the strength of materials) is approximately 1.5%.

Chapter 9

2.

$$\begin{bmatrix} k_{11}^{\textcircled{1}} & 0 & 0 & k_{15}^{\textcircled{1}} & k_{16}^{\textcircled{1}} \\ 0 & k_{55}^{\textcircled{2}} & k_{56}^{\textcircled{2}} & k_{51}^{\textcircled{2}} & k_{52}^{\textcircled{2}} \\ 0 & k_{65}^{\textcircled{2}} & k_{66}^{\textcircled{2}} & k_{61}^{\textcircled{2}} & k_{62}^{\textcircled{2}} \\ k_{51}^{\textcircled{1}} & k_{15}^{\textcircled{2}} & k_{16}^{\textcircled{2}} & k_{55}^{\textcircled{1}} + k_{11}^{\textcircled{2}} & k_{56}^{\textcircled{1}} + k_{12}^{\textcircled{2}} \\ k_{61}^{\textcircled{1}} & k_{26}^{\textcircled{2}} & k_{26}^{\textcircled{2}} & k_{65}^{\textcircled{1}} + k_{21}^{\textcircled{2}} & k_{66}^{\textcircled{1}} + k_{22}^{\textcircled{2}} \end{bmatrix} \begin{Bmatrix} u_1 \\ u_2 \\ v_2 \\ u_4 \\ v_4 \end{Bmatrix} = \begin{Bmatrix} 0 \\ 0 \\ -Q \\ P \\ 0 \end{Bmatrix}$$

3. Since the boundary conditions are $u_1 = 0$, $v_1 = 0$, $u_3 = 0$, $X_2 = 1$, $X_4 = 1$, the simultaneous equation is expressed as:

$$\begin{bmatrix} (k_{33}^1 + k_{33}^2) & (k_{34}^1 + k_{34}^2) & (k_{36}^1 + k_{32}^2) & k_{35}^2 & k_{36}^2 \\ (k_{43}^1 + k_{43}^2) & (k_{44}^1 + k_{44}^2) & (k_{46}^1 + k_{42}^2) & k_{45}^2 & k_{46}^2 \\ (k_{63}^1 + k_{23}^2) & (k_{64}^1 + k_{24}^2) & (k_{66}^1 + k_{22}^2) & k_{25}^2 & k_{26}^2 \\ k_{53}^2 & k_{54}^2 & k_{52}^2 & k_{55}^2 & k_{56}^2 \\ k_{63}^2 & k_{64}^2 & k_{62}^2 & k_{65}^2 & k_{66}^2 \end{bmatrix} \begin{Bmatrix} u_2 \\ v_2 \\ v_3 \\ u_4 \\ v_4 \end{Bmatrix} = \begin{Bmatrix} 1 \\ 0 \\ 0 \\ 1 \\ 0 \end{Bmatrix}$$

Determine the components of the stiffness matrix $[K]$.

Regarding the element $\textcircled{1}$, i, j and k in Table 9.1 correspond to the node numbers as $i \rightarrow 1$, $j \rightarrow 2$ and $k \rightarrow 3$. Thus,

$$c = \frac{tE}{4\Delta(1-\nu^2)} = \frac{1 \times 2 \times 10^4}{4 \times 1/2 \times (1-0.3^2)} = 10\,989.01 \cong 10\,990$$

$$k_{33}^1 = c\left\{ (y_3-y_1)^2 + \frac{1-\nu}{2}(x_1-x_3)^2 \right\} = c\left\{ (1-0)^2 + \frac{1-0.3}{2}(0-0)^2 \right\} = c$$

$$k_{34}^1 = c\left\{ \nu(y_3-y_1)\cdot(x_1-x_3) + \frac{1-\nu}{2}(x_1-x_3)\cdot(y_3-y_1) \right\} = 0$$

$$k_{36}^1 = c\left\{ \nu(y_3-y_1)\cdot(x_2-x_1) + \frac{1-\nu}{2}(x_1-x_3)\cdot(y_1-y_2) \right\}$$

$$= c\left\{ 0.3(1-0)\cdot(1-0) + \frac{1-\nu}{2}(0-0)\cdot(0-0) \right\} = 0.3c$$

$$k_{43}^1 = k_{34}^1 = 0$$

$$k_{44}^1 = c\left\{ (x_1-x_3)^2 + \frac{1-\nu}{2}(y_3-y_1)^2 \right\} = c\left\{ (0-0)^2 + \frac{1-0.3}{2}(1-0)^2 \right\} = 0.35c$$

$$k_{46}^1 = c\left\{ (x_1-x_3)\cdot(x_2-x_1) + \frac{1-\nu}{2}(y_3-y_1)\cdot(y_1-y_2) \right\}$$

$$= c\left\{ (0-0)\cdot(1-0) + \frac{1-0.3}{2}(1-0)\cdot(0-0) \right\} = 0$$

$$k_{63}^1 = k_{36}^1 = 0.3c, \quad k_{64}^1 = k_{46}^1 = 0$$

$$k_{66}^1 = c\left\{ (x_2-x_1)^2 + \frac{1-\nu}{2}(y_1-y_2)^2 \right\} = c\left\{ (1-0)^2 + \frac{1-0.3}{2}(0-0)^2 \right\} = c$$

Regarding element ②, i, j and k in Table 9.1 correspond to the node numbers as $i \to 3$, $j \to 2$ and $k \to 4$. c has the same value as element ①. Thus,

$$k_{22}^2 = c\left\{ (x_4-x_2)^2 + \frac{1-\nu}{2}(y_2-y_4)^2 \right\} = c\left\{ (1-1)^2 + \frac{1-0.3}{2}(0-1)^2 \right\} = 0.35c$$

$$k_{23}^2 = c\left\{ \nu(x_4-x_2)\cdot(y_4-y_3) + \frac{1-\nu}{2}(y_2-y_4)\cdot(x_3-x_4) \right\}$$

$$= c\left\{ 0.3(1-1)\cdot(1-1) + \frac{1-0.3}{2}(0-1)\cdot(0-1) \right\} = 0.35c$$

$$k_{24}^2 = c\left\{ (x_4-x_2)\cdot(x_3-x_4) + \frac{1-\nu}{2}(y_2-y_4)\cdot(y_4-y_3) \right\}$$

$$= c\left\{ (1-1)\cdot(0-1) + \frac{1-0.3}{2}(0-1)\cdot(1-1) \right\} = 0$$

$$k_{26}^2 = c\left\{ (x_4 - x_2)\cdot(x_2 - x_3) + \frac{1-\nu}{2}(y_2 - y_4)\cdot(y_3 - y_2) \right\}$$

$$= c\left\{ (1-1)\cdot(1-0) + \frac{1-0.3}{2}(0-1)\cdot(1-0) \right\} = -0.35c$$

$$k_{32}^2 = k_{23}^2 = 0.35c$$

$$k_{33}^2 = c\left\{ (y_4 - y_3)^2 + \frac{1-\nu}{2}(x_3 - x_4)^2 \right\} = c\left\{ (1-1)^2 + \frac{1-0.3}{2}(0-1)^2 \right\} = 0.35c$$

$$k_{34}^2 = c\left\{ \nu(y_4 - y_3)\cdot(x_3 - x_4) + \frac{1-\nu}{2}(x_3 - x_4)\cdot(y_4 - y_3) \right\}$$

$$= c\left\{ 0.3(1-1)\cdot(0-1) + \frac{1-0.3}{2}(0-1)\cdot(1-1) \right\} = 0$$

$$k_{35}^2 = c\left\{ (y_4 - y_3)\cdot(y_3 - y_2) + \frac{1-\nu}{2}(x_3 - x_4)\cdot(x_2 - x_3) \right\}$$

$$= c\left\{ (1-1)\cdot(1-0) + \frac{1-0.3}{2}(0-1)\cdot(1-0) \right\} = -0.35c$$

$$k_{36}^2 = c\left\{ \nu(y_4 - y_3)\cdot(x_2 - x_3) + \frac{1-\nu}{2}(x_3 - x_4)\cdot(y_3 - y_2) \right\}$$

$$= c\left\{ 0.3(1-1)\cdot(1-0) + \frac{1-0.3}{2}(0-1)\cdot(1-0) \right\} = -0.35c$$

$$k_{42}^2 = k_{24}^2 = 0, \quad k_{43}^2 = k_{34}^2 = 0$$

$$k_{44}^2 = c\left\{ (x_3 - x_4)^2 + \frac{1-\nu}{2}(y_4 - y_3)^2 \right\} = c\left\{ (0-1)^2 + \frac{1-0.3}{2}(1-1)^2 \right\} = c$$

$$k_{45}^2 = c\left\{ \nu(x_3 - x_4)\cdot(y_3 - y_2) + \frac{1-\nu}{2}(y_4 - y_3)\cdot(x_2 - x_3) \right\}$$

$$= c\left\{ 0.3(0-1)\cdot(1-0) + \frac{1-0.3}{2}(1-1)\cdot(1-0) \right\} = -0.3c$$

$$k_{46}^2 = c\left\{ (x_3 - x_4)\cdot(x_2 - x_3) + \frac{1-\nu}{2}(y_4 - y_3)\cdot(y_3 - y_2) \right\}$$

$$= c\left\{ (0-1)\cdot(1-0) + \frac{1-0.3}{2}(1-1)\cdot(1-0) \right\} = -c$$

$$k_{52}^2 = k_{25}^2 = -0.35c, \quad k_{53}^2 = k_{35}^2 = -0.35c, \quad k_{54}^2 = k_{45}^2 = -0.3c$$

$$k_{55}^2 = c\left\{ (y_3 - y_2)^2 + \frac{1-\nu}{2}(x_2 - x_3)^2 \right\} = c\left\{ (1-0)^2 + \frac{1-0.3}{2}(1-0)^2 \right\} = 1.35c$$

$$k_{56}^2 = c\left\{ \nu(y_3 - y_2)\cdot(x_2 - x_3) + \frac{1-\nu}{2}(x_2 - x_3)\cdot(y_3 - y_2)\right\}$$

$$= c\left\{ 0.3(1-0)\cdot(1-0) + \frac{1-0.3}{2}(1-0)\cdot(1-0)\right\} = 0.65c$$

$$k_{62}^2 = k_{26}^2 = -0.35c, \quad k_{63}^2 = k_{36}^2 = -0.35c, \quad k_{64}^2 = k_{46}^2 = -c, \quad k_{65}^2 = k_{56}^2 = 0.65c$$

$$k_{66}^2 = c\left\{ (x_2 - x_3)^2 + \frac{1-\nu}{2}(y_3 - y_2)^2\right\} = c\left\{ (1-0)^2 + \frac{1-0.3}{2}(1-0)^2\right\} = 1.35c$$

$[K]$ is expressed by the above values of k as follows:

$$[K] = c \begin{bmatrix} 1.35 & 0 & 0.65 & -0.35 & -0.35 \\ 0 & 1.35 & 0 & -0.3 & -1 \\ 0.65 & 0 & 1.35 & -0.35 & -0.35 \\ -0.35 & -0.3 & -0.35 & 1.35 & 0.65 \\ -0.35 & -1 & -0.35 & 0.65 & 1.35 \end{bmatrix}$$

Therefore, the simultaneous equation is:

$$[K]\begin{Bmatrix} u_2 \\ v_2 \\ v_3 \\ u_4 \\ v_4 \end{Bmatrix} = \begin{Bmatrix} 1 \\ 0 \\ 0 \\ 1 \\ 0 \end{Bmatrix}$$

Now, putting $cu_2 = u_2', cv_2 = v_2', cv_3 = v_3', cu_4 = u_4', cv_4 = v_4',$

$$\left.\begin{array}{l} 1.35u_2' \qquad\qquad\qquad +0.65v_3' \;\; -0.35u_4' \;\; -0.35v_4' \;=\; 1 \\ \qquad\qquad 1.35v_2' \qquad\qquad\qquad\quad -0.3u_4' \qquad -v_4' \;=\; 0 \\ +0.65u_2' \qquad\qquad\quad +1.35v_3' \;\; -0.35u_4' \;\; -0.35v_4' \;=\; 0 \\ -0.35u_2' \;\; -0.3v_2' \;\; -0.35v_3' \;\; +1.35u_4' \;\; +0.65v_4' \;=\; 1 \\ -0.35u_2' \qquad -v_2' \quad\; -0.35v_3' \;\; +0.65u_4' \;\; +1.35v_4' \;=\; 0 \end{array}\right\}$$

Here, we modify the above equations in the following form so that we can apply the Gauss–Seidel method for solving the simultaneous equation:

$$u_2' = \frac{1}{1.35}(1-0.65v_3'+0.35u_4'+0.35v_4')$$

$$v_2' = \frac{1}{1.35}(0.3u_4'+v_4')$$

$$v_3' = \frac{1}{1.35}(-0.65u_2'+0.35u_4'+0.35v_4')$$

$$u_4' = \frac{1}{1.35}(1+0.35u_2'+0.3v_2'+0.35v_3'-0.65v_4')$$

$$v_4' = \frac{1}{1.35}(0.35u_2'+v_2'+0.35v_3'-0.65u_4')$$

The Gauss–Seidel method can be achieved as follows. First, we determine u_2' by assuming $v_3' = u_4' = v_4' = 0$ in the first of the above equations. Next, we determine v_2' by putting $u_4' = v_4' = 0$ in the second equation. In the third equation, in order to determine v_3', we put all unknowns to 0 except for u_2' which was already determined in the previous step. In the following steps, we determine up to v_4'. Thus, we finish the first loop calculations.

In the second loop calculations, we input the values of v_3', u_4' and v_4' (which were determined in the first loop calculation) into the first equation and determine the new value of u_2'. In the following calculations, we use these new values. We repeat this procedure until the solutions converge. The condition of convergence of this method is that the k values at the diagonal locations in the simultaneous equation must be larger than other k values.

Table P9.1 shows the convergence of the solution by every step up to 20 loop calculations.

Table P9.1 Convergence and number of loop calculations, n

n	u_2'	v_2'	v_3'	u_4'	v_4'
0	0	0	0	0	0
1	7.407×10^{-1}	0	-3.567×10^{-1}	8.403×10^{-1}	-3.050×10^{-1}
2	1.051	-3.920×10^{-2}	-3.674×10^{-1}	1.056	-3.603×10^{-1}
3	1.098	-3.216×10^{-2}	-3.483×10^{-1}	1.101	-3.598×10^{-1}
4	1.101	-2.172×10^{-2}	-3.377×10^{-1}	1.107	-3.512×10^{-1}
5	1.099	-1.419×10^{-2}	-3.333×10^{-1}	1.105	-3.441×10^{-1}
6	1.099	-9.283×10^{-3}	-3.316×10^{-1}	1.103	-3.392×10^{-1}
7	1.098	-6.107×10^{-3}	-3.308×10^{-1}	1.102	-3.360×10^{-1}
8	1.099	-4.033×10^{-3}	-3.304×10^{-1}	1.101	-3.338×10^{-1}
9	1.099	-2.670×10^{-3}	-3.302×10^{-1}	1.100	-3.324×10^{-1}
10	1.099	-1.768×10^{-3}	-3.300×10^{-1}	1.100	-3.315×10^{-1}
11	1.099	-1.171×10^{-3}	-3.299×10^{-1}	1.099	-3.309×10^{-1}
12	1.099	-7.762×10^{-4}	-3.298×10^{-1}	1.099	-3.305×10^{-1}
13	1.099	-5.144×10^{-4}	-3.298×10^{-1}	1.099	-3.302×10^{-1}
14	1.099	-3.408×10^{-4}	-3.297×10^{-1}	1.099	-3.300×10^{-1}
15	1.099	-2.259×10^{-4}	-3.297×10^{-1}	1.099	-3.299×10^{-1}
16	1.099	-1.497×10^{-4}	-3.297×10^{-1}	1.099	-3.298×10^{-1}
17	1.099	-9.918×10^{-5}	-3.297×10^{-1}	1.099	-3.298×10^{-1}
18	1.099	-6.573×10^{-5}	-3.297×10^{-1}	1.099	-3.297×10^{-1}
19	1.099	-4.355×10^{-5}	-3.297×10^{-1}	1.099	-3.297×10^{-1}
20	1.099	-2.886×10^{-5}	-3.297×10^{-1}	1.099	-3.297×10^{-1}

From $c = 10990$,

$$u_2 = u_2'/c = 1.099/10990 = 1 \times 10^{-4}$$

$$v_2 = v_2'/c = -2.886 \times 10^{-5}/10990 = -2.626 \times 10^{-9} \cong 0$$

$$v_3 = v_3'/c = -3.297 \times 10^{-1}/10990 = -3 \times 10^{-5}$$

$$u_4 = u_4'/c = 1.099/10990 = 1 \times 10^{-4}$$

$$v_4 = v_4'/c = -3.297 \times 10^{-1}/10990 = -3 \times 10^{-5}$$

The stresses and strains for the individual elements are determined as follows:

$$\{\varepsilon\} = [B][T]^{-1}\{u\}$$

The concrete expression for the above equation is:

$$
\begin{Bmatrix} \varepsilon_x \\ \varepsilon_y \\ \gamma_{xy} \end{Bmatrix} = \frac{1}{2\Delta}
\begin{bmatrix}
y_j - y_k & 0 & y_k - y_i & 0 & y_i - y_j & 0 \\
0 & x_k - x_j & 0 & x_i - x_k & 0 & x_j - x_i \\
x_k - x_j & y_j - y_k & x_i - x_k & y_k - y_i & x_j - x_i & y_i - y_j
\end{bmatrix}
\begin{Bmatrix} u_i \\ v_i \\ u_j \\ v_j \\ u_k \\ v_k \end{Bmatrix},
$$

where $i = 1$, $j = 2$ and $k = 3$.

Substituting the numerical values, regarding element ①, we have:

$$
\begin{Bmatrix} \varepsilon_x \\ \varepsilon_y \\ \gamma_{xy} \end{Bmatrix} = \frac{1}{2 \times \frac{1}{2}}
\begin{bmatrix}
(0-1) & 0 & (1-0) & 0 & (0-0) & 0 \\
0 & (0-1) & 0 & (0-0) & 0 & (1-0) \\
(0-1) & (0-1) & (0-0) & (1-0) & (1-0) & (0-0)
\end{bmatrix}
$$

$$
\begin{Bmatrix} 0 \\ 0 \\ 1 \times 10^{-4} \\ 0 \\ 0 \\ -3 \times 10^{-5} \end{Bmatrix} = \begin{Bmatrix} 1 \times 10^{-4} \\ -3 \times 10^{-5} \\ 0 \end{Bmatrix}
$$

Table P9.2

	u_1	v_1	u_2	v_2	u_3	v_3	u_4	v_4	u_5	v_5	u_6	v_6	u_7	v_7	u_8	v_8	u_9	v_9
X_1	①k_{11} ②k_{11}	①k_{12} ②k_{12}	②k_{15}	②k_{16}			①k_{13}	①k_{14}	①k_{15} ②k_{13}	①k_{16} ②k_{14}								
Y_1	①k_{21} ②k_{21}	①k_{22} ②k_{22}	②k_{25}	②k_{26}			①k_{23}	①k_{24}	①k_{25} ②k_{23}	①k_{26} ②k_{24}								
X_2	②k_{51}	②k_{52}	②k_{55} ③k_{11}	②k_{56} ③k_{12}	③k_{15}	③k_{16}			②k_{53} ③k_{13}	②k_{54} ③k_{14}								
Y_2	②k_{61}	②k_{62}	②k_{65} ③k_{21}	②k_{66} ③k_{22}	③k_{25}	③k_{26}			②k_{63} ③k_{23}	②k_{64} ③k_{24}								
X_3			③k_{51}	③k_{52}	③k_{55} ④k_{11}	③k_{56} ④k_{12}			③k_{53} ④k_{13}	③k_{54} ④k_{14}	④k_{15}	④k_{16}						
Y_3			③k_{61}	③k_{62}	③k_{65} ④k_{21}	③k_{66} ④k_{22}			③k_{63} ④k_{23}	③k_{64} ④k_{24}	④k_{25}	④k_{26}						
X_4	①k_{31}	①k_{32}					①k_{33} ⑤k_{33}	①k_{34} ⑤k_{34}	①k_{35} ⑤k_{31}	①k_{36} ⑤k_{32}			⑤k_{35}	⑤k_{36}				
Y_4	①k_{41}	①k_{42}					①k_{43} ⑤k_{43}	①k_{44} ⑤k_{44}	①k_{45} ⑤k_{41}	①k_{46} ⑤k_{42}			⑤k_{45}	⑤k_{46}				
X_5	①k_{51} ②k_{31}	①k_{52} ②k_{32}	②k_{35} ③k_{31}	②k_{36} ③k_{32}	③k_{35} ④k_{31}	③k_{36} ④k_{32}	①k_{53} ⑤k_{13}	①k_{54} ⑤k_{14}	①k_{55} ②k_{33} ③k_{33} ④k_{33} ⑤k_{11} ⑥k_{11} ⑦k_{11} ⑧k_{11}	①k_{56} ②k_{34} ③k_{34} ④k_{34} ⑤k_{12} ⑥k_{12} ⑦k_{12} ⑧k_{12}	④k_{35} ⑦k_{13}	④k_{36} ⑦k_{14}	⑤k_{15} ⑥k_{13}	⑤k_{16} ⑥k_{14}	⑥k_{15} ⑧k_{15}	⑥k_{16} ⑧k_{16}	⑦k_{15} ⑧k_{13}	⑦k_{16} ⑧k_{14}
Y_5	①k_{61} ②k_{41}	①k_{62} ②k_{42}	②k_{45} ③k_{41}	②k_{46} ③k_{42}	③k_{45} ④k_{41}	③k_{46} ④k_{42}	①k_{63} ⑤k_{23}	①k_{64} ⑤k_{24}	①k_{65} ②k_{43} ③k_{43} ④k_{43} ⑤k_{21} ⑥k_{21} ⑦k_{21} ⑧k_{21}	①k_{66} ②k_{44} ③k_{44} ④k_{44} ⑤k_{22} ⑥k_{22} ⑦k_{22} ⑧k_{22}	④k_{45} ⑦k_{23}	④k_{46} ⑦k_{24}	⑤k_{25} ⑥k_{23}	⑤k_{26} ⑥k_{24}	⑥k_{25} ⑧k_{25}	⑥k_{26} ⑧k_{26}	⑦k_{25} ⑧k_{23}	⑦k_{26} ⑧k_{24}
X_6					④k_{51}	④k_{52}			④k_{53} ⑦k_{31}	④k_{54} ⑦k_{32}	④k_{55} ⑦k_{33}	④k_{56} ⑦k_{34}					⑦k_{35}	⑦k_{36}
Y_6					④k_{61}	④k_{62}			④k_{63} ⑦k_{41}	④k_{64} ⑦k_{42}	④k_{65} ⑦k_{43}	④k_{66} ⑦k_{44}					⑦k_{45}	⑦k_{46}
X_7							⑤k_{53}	⑤k_{54}	⑤k_{51} ⑥k_{31}	⑤k_{52} ⑥k_{32}			⑤k_{55} ⑥k_{33}	⑤k_{56} ⑥k_{34}	⑥k_{35}	⑥k_{36}		
Y_7							⑤k_{63}	⑤k_{64}	⑤k_{61} ⑥k_{41}	⑤k_{62} ⑥k_{42}			⑤k_{65} ⑥k_{43}	⑤k_{66} ⑥k_{44}	⑥k_{45}	⑥k_{46}		
X_8									⑥k_{51} ⑧k_{51}	⑥k_{52} ⑧k_{52}			⑥k_{53}	⑥k_{54}	⑥k_{55} ⑧k_{55}	⑥k_{56} ⑧k_{56}	⑧k_{53}	⑧k_{54}
Y_8									⑥k_{61} ⑧k_{61}	⑥k_{62} ⑧k_{62}			⑥k_{63}	⑥k_{64}	⑥k_{65} ⑧k_{65}	⑥k_{66} ⑧k_{66}	⑧k_{63}	⑧k_{64}
X_9									⑦k_{51} ⑧k_{31}	⑦k_{52} ⑧k_{32}	⑦k_{53}	⑦k_{54}			⑧k_{35}	⑧k_{36}	⑦k_{55} ⑧k_{33}	⑦k_{56} ⑧k_{34}
Y_9									⑦k_{61} ⑧k_{41}	⑦k_{62} ⑧k_{42}	⑦k_{63}	⑦k_{64}			⑧k_{45}	⑧k_{46}	⑦k_{65} ⑧k_{43}	⑦k_{66} ⑧k_{44}

$$
\left\{
\begin{array}{c}
\sigma_x \\
\sigma_y \\
\tau_{xy}
\end{array}
\right\}
=
\frac{E}{1-\nu^2}
\begin{bmatrix}
1 & \nu & 0 \\
\nu & 1 & 0 \\
0 & 0 & (1-\nu)/2
\end{bmatrix}
\left\{
\begin{array}{c}
\varepsilon_x \\
\varepsilon_y \\
\gamma_{xy}
\end{array}
\right\}
=
\frac{2\times10^4}{1-0.3^2}
\begin{bmatrix}
1 & 0.3 & 0 \\
0.3 & 1 & 0 \\
0 & 0 & 0.35
\end{bmatrix}
$$

$$
\left\{
\begin{array}{c}
1\times10^{-4} \\
-3\times10^{-5} \\
0
\end{array}
\right\}
=
\left\{
\begin{array}{c}
2 \\
0 \\
0
\end{array}
\right\}
$$

Regarding element ②, $i=3, j=2$ and $k=4$,

$$
\left\{
\begin{array}{c}
\varepsilon_x \\
\varepsilon_y \\
\gamma_{xy}
\end{array}
\right\}
=
\frac{1}{2\Delta}
\begin{bmatrix}
-1 & 0 & 0 & 0 & 1 & 0 \\
0 & 0 & 0 & -1 & 0 & 1 \\
0 & -1 & -1 & 0 & 1 & 1
\end{bmatrix}
\left\{
\begin{array}{c}
0 \\
-3\times10^{-5} \\
1\times10^{-4} \\
0 \\
1\times10^{-4} \\
-3\times10^{-5}
\end{array}
\right\}
=
\left\{
\begin{array}{c}
1\times10^{-4} \\
-3\times10^{-5} \\
0
\end{array}
\right\}
$$

$$
\left\{
\begin{array}{c}
\sigma_x \\
\sigma_y \\
\tau_{xy}
\end{array}
\right\}
=
\frac{E}{1-\nu^2}
\begin{bmatrix}
1 & \nu & 0 \\
\nu & 1 & 0 \\
0 & 0 & (1-\nu)/2
\end{bmatrix}
\left\{
\begin{array}{c}
\varepsilon_x \\
\varepsilon_y \\
\gamma_{xy}
\end{array}
\right\}
=
\left\{
\begin{array}{c}
2 \\
0 \\
0
\end{array}
\right\}
$$

4. See Table P9.2, where the sign $+$ is neglected. It follows that $k_{11}^①k_{11}^②$ means $k_{11}^① + k_{11}^②$.

Chapter 10

Problem 10.3.2

Paying attention to the section of $y=$ constant, the neutral axis is inclined by dw/dx to the x axis by which the point at $z=z$ moves by $u=-z(\partial w/\partial x)$ in the x direction.

Likewise, paying attention to the section $x=$ constant, $v=-z(\partial w/\partial y)$. As $\gamma_{xy}=(\partial u/\partial y)+(\partial v/\partial x)$, $\tau_{xy}=\tau_{yz}=-2Gz(\partial^2 w/\partial x\cdot\partial y)$. Then,

$$
M_{xy}=-M_{yz}=-\int_{-h/2}^{h/2}\tau_{xy}z dz=2G\frac{\partial^2 w}{\partial x\cdot\partial y}\int_{-h/2}^{h/2}z^2 dz-D(1-\nu)\frac{\partial^2 w}{\partial x\cdot\partial y}
$$

Problem 10.3.3

Strain energy is equivalent to the work done by bending moment and twisting moments. The work by M_x per dy is

$$
\frac{1}{2}M_x dy\theta_y=\frac{1}{2}M_x dy\frac{dx}{\rho_x}=-\frac{1}{2}M_x\frac{\partial^2 w}{\partial x^2}dx\cdot dy
$$

Likewise, the work by M_y is

$$-\frac{1}{2}M_y\left(\partial^2 w/\partial y^2\right)dx\cdot dy$$

Summing up these two,

$$dU = -\frac{1}{2}\left(M_x\frac{\partial^2 w}{\partial x^2} + M_y\frac{\partial^2 w}{\partial y^2}\right)dx\cdot dy$$

$$= \frac{1}{2}D\left[\left(\frac{\partial^2 w}{\partial x^2}\right)^2 + \left(\frac{\partial^2 w}{\partial y^2}\right)^2 + 2\nu\frac{\partial^2 w}{\partial x^2}\cdot\frac{\partial^2 w}{\partial y^2}\right]dx\cdot dy$$

For the plate having area A, we have only to replace as $dU \to U$ and $dxdy \to A$. In the case of $M_{xy} \neq 0$, the work done by M_{xy} must be considered. The rotation angle of the x axis at $x=x$ is $\partial w/\partial y$, the rotation angle at $x+dx$ is

$$\frac{\partial w}{\partial y} + \frac{\partial}{\partial x}\left(\frac{\partial w}{\partial y}\right)dx$$

Therefore, the work done by M_{xy} is expressed as follows.

$$\frac{1}{2}M_{xy}dy\frac{\partial^2 w}{\partial x\cdot\partial y}dx = \frac{1}{2}M_{xy}\frac{\partial^2 w}{\partial x\cdot\partial y}dx\cdot dy = \frac{1}{2}D(1-\nu)\cdot\left(\frac{\partial^2 w}{\partial x\cdot\partial y}\right)^2 dx\cdot dy$$

The work done by M_{yx} can be expressed by the similar derivation and it is expressed as $D(1-\nu)\cdot\left(\partial^2 w/\partial x\cdot\partial y\right)^2 dx\cdot dy$.

Since the works done by bending moment and twisting moment can be independently additive, the summation of the above results agrees with Equation 10.22.

Problems of Chapter 10

1. Verify by the same way as Example problem 1.1 of the plane problem in Chapter 1 of Part I.

2. $\sigma_{max} = (\sigma_r)_{r=a} = 3qa^2/4h^2$.

4. $b \leq r \leq a$

$$w = \frac{P}{8\pi D}\left[(a^2-r^2)\frac{a^2+b^2}{2a^2} + (b^2+r^2)\log\frac{r}{a}\right]$$

$0 \leq r \leq b$

$$w = \frac{P}{8\pi D}\left[(b^2+r^2)\log\frac{b}{a} + \frac{(a^2+r^2)\cdot(a^2-b^2)}{2a^2}\right]$$

5. $\dfrac{P}{8\pi D}\left[b^2\log\dfrac{b}{a} + \dfrac{(3+\nu)}{2(1+\nu)}(a^2-b^2)\right]$

Chapter 11

Problems of Chapter 11

1. $w = P/8\beta^3 D, M_x = P/4\beta$

   $w = (P/4\beta^4 D)e^{-\beta x}(\cos\beta x + \sin\beta x) - P/4\beta^4 D$

2. Assume the displacement in the radial direction in the circular plate is 0.

$$\sigma_x = \frac{3[\beta^3 a^3 + 2(1+\nu)]}{\beta^2 h^2 [4\beta a + 2(1+\nu)]} p.$$

4. In the beam theory of the strength of materials,

$$EI(d^2 w/dx^2) = -M_x, EI(d^3 w/dx^3) = -(dM_x/dx) = -Q_x$$
$$EI \cdot (d^4 w/dx^4) = -(dQ_x/dx) = q \text{(distributed loading)}.$$

   The equation of this problem is

$$EI\frac{d^4 w}{dx^4} = p - kw$$

   Then,

$$EI\frac{d^4 w}{dx^4} + kw = p, \quad \text{or} \quad \frac{d^4 w}{dx^4} + \frac{k}{EI}w = \frac{p}{EI}$$

   This equation is the same form as Equation 11.12.

5. $\sigma = \dfrac{6M_0}{h^2}$,

   where

$$M_0 = \frac{E\delta l}{C_0}, \quad C_0 = \frac{bl^3}{3I} + \frac{2a^2\beta}{h} + \frac{2a^2\beta^2 l}{h} + \frac{2a^2\beta^2(1+2\beta l)l}{h}, \quad b = \frac{2\pi a}{n}, \quad I = \frac{bh^3}{12}.$$

Chapter 12

Problem 12.1.1 Stress for the case of Figure 12.1b is:

$$\sigma_x = -\alpha T_0 E\left(\frac{2y}{h}\right)^n$$

The stress for the case of Figure 12.1a is:

$$\sigma_x = \alpha T_0 E\left[\frac{1}{2(n+1)}\{1-(-1)^{n+1}\} + \frac{3}{n+2}\{1-(-1)^{n+2}\}\frac{y}{h} - \left(\frac{2y}{h}\right)^n\right]$$

Problem 12.1.2 After substituting $\{\sigma\} = [D](\{\varepsilon\} - \{\varepsilon_0\})$ into Equation 9.35, continue the calculation.

Problem 12.2.1 $\sigma_\theta = 65.3\,\text{MPa}$

Problem 12.2.2 $\sigma_\theta = \alpha T_0 E \cdot \dfrac{n}{(n+2)}$. σ_θ is larger for $n > 1$ than $n < 1$.

Problem 12.2.3

$$\sigma_r = \alpha E \left(\frac{1}{r_2{}^2 - r_1{}^2} \int_{r_1}^{r_2} Tr\,dr - \frac{1}{r^2} \cdot \frac{r_1{}^2}{r_2{}^2 - r_1{}^2} \int_{r_1}^{r_2} Tr\,dr - \frac{1}{r^2} \int_{r_1}^{r} Tr\,dr \right)$$

$$\sigma_\theta = \alpha E \left(\frac{1}{r_2{}^2 - r_1{}^2} \int_{r_1}^{r_2} Tr\,dr + \frac{1}{r^2} \cdot \frac{r_1{}^2}{r_2{}^2 - r_1{}^2} \int_{r_1}^{r_2} Tr\,dr - T + \frac{1}{r^2} \int_{r_1}^{r} Tr\,dr \right)$$

Problems of Chapter 12

1. 824 MPa.
2. $\sigma_{\theta A} = \sigma_{rA} = \sigma_{\varphi A} = -p$, $\sigma_{rB} = -p$, (θ : longitudinal direction, φ: meridian direction).

$$\sigma_{\theta B} = \sigma_{\varphi B} = -\frac{E_B}{E_A} \cdot \frac{1 - 2\nu_A + \nu_B \cdot \dfrac{E_A}{E_B}}{1 - \nu_B} p + \frac{E_B}{1 - \nu_B}(\alpha_A - \alpha_B)\Delta T.$$

3.

$$-\alpha E \left(T_1 \, \sin\frac{\pi y}{h} + T_2 \, \cos\frac{\pi y}{h} \right) + \frac{2\alpha E T_2}{\pi} + \frac{24\alpha E T_1 y}{\pi^2 h}$$

Chapter 13

Problems of Chapter 13

1. $K_I = 2\sigma_n \sqrt{a/\pi}$. (see Chapter 13 of Part II).
2. $K_I = \sigma_{net} \sqrt{\pi a}/2$, σ_{net}: nominal stress at the minimum section (see Chapter 13 of Part II).
3. $y = 0.786b$, $\tau_{45°\,max} = 0.300q_0$.

Appendix for Part I

A.1 Rule of Direction Cosines

$$\left.\begin{array}{l} l_1^2 + m_1^2 + n_1^2 = 1 \\ l_2^2 + m_2^2 + n_2^2 = 1 \\ l_3^2 + m_3^2 + n_3^2 = 1 \end{array}\right\} \qquad \left.\begin{array}{l} l_1^2 + l_2^2 + l_3^2 = 1 \\ m_1^2 + m_2^2 + m_3^2 = 1 \\ n_1^2 + n_2^2 + n_3^2 = 1 \end{array}\right\}$$

$$\left.\begin{array}{l} l_1 m_1 + l_2 m_2 + l_3 m_3 = 0 \\ m_1 n_1 + m_2 n_2 + m_3 n_3 = 0 \\ n_1 l_1 + n_2 l_2 + n_3 l_3 = 0 \end{array}\right\} \qquad \left.\begin{array}{l} l_1 l_2 + m_1 m_2 + n_1 n_2 = 0 \\ l_2 l_3 + m_2 m_3 + n_2 n_3 = 0 \\ l_3 l_1 + m_3 m_1 + n_3 n_1 = 0 \end{array}\right\}$$

A.2 Green's Theorem and Gauss' Divergence Theorem

A.2.1 2D Green's Theorem

$$\iint_s \frac{\partial f}{\partial x} dx \cdot dy$$

Figure A2.1 shows a transformation of the area integral over the region S into the line integral along Γ which envelopes the region S. $\partial f / \partial x$ is assumed continuous in the region S.

Theory of Elasticity and Stress Concentration, First Edition. Yukitaka Murakami.
© 2017 John Wiley & Sons, Ltd. Published 2017 by John Wiley & Sons, Ltd.

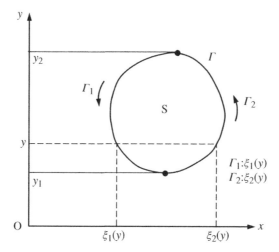

Figure A2.1

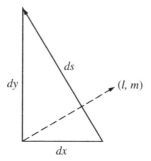

Figure A2.2

Referring to Figures A2.1 and A2.2,

$$\int_{\xi_1(y)}^{\xi_2(y)} \frac{\partial f}{\partial x} dx = f(\xi_2(y), y) - f(\xi_1(y), y)$$

$$\int_{y_1}^{y_2} [f(\xi_2(y), y) - f(\xi_1(y), y)] dy = \int_{y_1}^{y_2} f(\xi_2(y), y) dy + \int_{y_2}^{y_1} f(\xi_1(y), y) dy$$

$$= \int_{\Gamma_1} f(x, y) dy + \int_{\Gamma_1} f(x, y) dy$$

$$= \oint_{\Gamma} f(x, y) dy \qquad (A2.1)$$

$$= \oint_\Gamma f(x,y)\frac{dy}{ds}ds$$

$$= \oint_\Gamma f(x,y)l\,ds \qquad\qquad (A2.2)$$

Likewise, the following equation is transferred into the line integral:

$$\iint_S \frac{\partial g}{\partial y}dx\cdot dy$$

Assuming $\partial g/\partial y$ is continuous in S and referring to Figure A2.3,

$$\int_{\eta_1(x)}^{\eta_2(x)} \frac{\partial g}{\partial y}dy = g[x,\eta_2(x)] - g[x,\eta_1(x)]$$

$$\int_{x_1}^{x_2} \{g[x,\eta_2(x)] - g[x,\eta_1(x)]\}dx$$

$$= -\int_{x_2}^{x_1} g[x,\eta_2(x)]dx - \int_{x_1}^{x_2} g[x,\eta_1(x)]dx$$

$$= -\int_{\Gamma_2} g(x,y)dx - \int_{\Gamma_1} g(x,y)dx$$

$$= -\oint_\Gamma g(x,y)dx \qquad\qquad (A2.3)$$

$$= \oint_\Gamma g(x,y)m\,ds \qquad\qquad (A2.4)$$

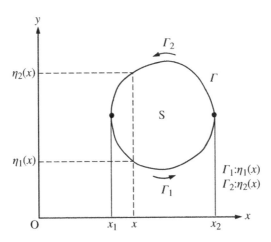

Figure A2.3

Then,

$$\iint_S \left(\frac{\partial f}{\partial x} + \frac{\partial g}{\partial y}\right) dx \cdot dy = \oint_\Gamma (f \cdot l + g \cdot m) ds \tag{A2.5}$$

Equation (A2.5) is called the 2D Green's theorem or Gauss' divergence theorem.

A.2.2   3D Green's Theorem

Like the 2D theorem, the 3D theorem is derived by a similar procedure. Although the calculation is skipped, the result is given as follows:

$$\iiint_V \left(\frac{\partial f}{\partial x} + \frac{\partial g}{\partial y} + \frac{\partial h}{\partial z}\right) dx \cdot dy \cdot dz = \iint_S (f \cdot l + g \cdot m + h \cdot n) dS \tag{A2.6}$$

where S is the area of the curved surface which envelopes the 3D domain of V.

Part II

Stress Concentration

1

Stress Concentration in Two Dimensional Problems

In this chapter, two dimensional (2D) elastic stress concentration problems are treated. Several useful ways of thinking and useful equations which are not shown in other literature will be introduced. Typical mistakes in the ways of thinking will be also explained.

1.1 Stress Concentration at a Circular Hole

Stress concentration at a circular hole is the most basic problem for understanding the way of thinking of stress concentration.

When a circular hole exists in a wide plate under remote uniform tensile stress $\sigma_x = \sigma_0$ (Figure 1.1), the stress distribution around the hole is given in the polar coordinates as follows (see Chapter 6 of Part I) [1]:

$$
\left.
\begin{aligned}
\sigma_r &= \frac{\sigma_0}{2}\left(1 - \frac{a^2}{r^2}\right) + \frac{\sigma_0}{2}\left(1 - \frac{4a^2}{r^2} + \frac{3a^4}{r^4}\right)\cos 2\theta, \\
\sigma_\theta &= \frac{\sigma_0}{2}\left(1 + \frac{a^2}{r^2}\right) - \frac{\sigma_0}{2}\left(1 + \frac{3a^4}{r^4}\right)\cos 2\theta, \\
\tau_{r\theta} &= -\frac{\sigma_0}{2}\left(1 + \frac{2a^2}{r^2} - \frac{3a^4}{r^4}\right)\sin 2\theta
\end{aligned}
\right\} \tag{1.1}
$$

Using Equation 1.1, σ_y along the y axis or at $\theta = \pi/2$ is given by the following equation:

$$
\sigma_x = \sigma_\theta|_{\theta=\frac{\pi}{2}} = \sigma_0\left(1 + \frac{a^2}{2r^2} + \frac{3a^4}{2r^4}\right) \tag{1.2}
$$

Theory of Elasticity and Stress Concentration, First Edition. Yukitaka Murakami.
© 2017 John Wiley & Sons, Ltd. Published 2017 by John Wiley & Sons, Ltd.

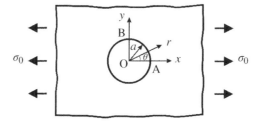

Figure 1.1

and σ_θ at $r = a$ (the boundary of the circular hole) is calculated as

$$\sigma_\theta = \sigma_0 - 2\sigma_0 \ \cos 2\theta \tag{1.3}$$

Therefore, the maximum tensile stress occurs at point B, irrespective of the size of hole. The value at the point B is three times higher than the remote stress as follows:

$$\sigma_{max} = 3\sigma_0 \tag{1.4}$$

From this nature of stress distribution, the *stress concentration factor* K_t for a circular hole in a wide plate under uniaxial stress is defined as follows:

$$K_t = 3 \tag{1.5}$$

Although, in general, the stress $\sigma_\theta \left(= \sigma_y \right)$ at point A is given no attention, it is interesting to note that

$$\sigma_\theta \left(= \sigma_y \right)$$

is compression as follows:

$$\sigma_\theta = \sigma_y = -\sigma_0 \tag{1.6}$$

It is useful to understand this nature for solving many practical problems by the application of the *principle of superposition* (see Part I). For example, the principle of superposition can be applied to evaluating the stress concentration under a biaxial stress field or under general remote stress $\left(\sigma_{x\infty}, \sigma_{y\infty}, \tau_{xy\infty} \right)$. Several simple, but very useful, examples will be shown in the following.

Example problem 1.1
Stress concentration at a circular hole under a remote stress $\left(\sigma_{x\infty}, \sigma_{y\infty}, \tau_{xy\infty} \right) = \left(2\sigma_0, \sigma_0, 0 \right)$
    The problem is shown in Figure 1.2.

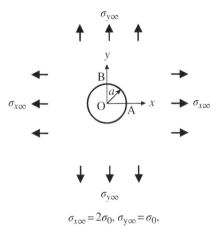

$$\sigma_{x\infty} = 2\sigma_0, \ \sigma_{y\infty} = \sigma_0,$$

Figure 1.2

The stress at the point B is influenced by $\sigma_{x\infty}$ as $3\sigma_{x\infty}$ and $\sigma_{y\infty}$ as $-\sigma_{y\infty}$, so that the total effect is

$$\sigma_{xmax} = 3\sigma_{x\infty} - \sigma_{y\infty} = 5\sigma_0 \ (\text{at point B}).$$

Example problem 1.2

Stress concentration at a circular hole under a remote stress $\left(\sigma_{x\infty}, \sigma_{y\infty}, \tau_{xy\infty}\right) = \left(0, 0, \tau_{xy\infty}\right)$
The problem is shown in Figure 1.3.

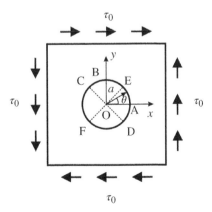

Figure 1.3

First of all, the principal stresses (see Section 1.3.4.1 in Part I) are calculated as:

$$\sigma_1 = \tau_0, \sigma_2 = -\tau_0$$

σ_1 acts in the direction $\theta = 45°$ and σ_2 acts in the direction $\theta_2 = -45°$.

The problem of Figure 1.3 can be regarded as equivalent to the superposition of two stresses in Figure 1.4 by inclining the plate by $\pm 45°$. Thus, the maximum stress σ_{max} occurs at point C $\left(\theta = \frac{3}{4}\pi\right)$ and D $\left(\theta = -\frac{\pi}{4}\right)$ and is given as

$$\sigma_{max} = 3\sigma_1 - \sigma_2$$
$$= 4\tau_0$$

The minimum stress σ_{min} occurs at point E $\left(\theta = \frac{\pi}{4}\right)$ and F $\left(\theta = -\frac{3}{4}\pi\right)$ and is given as

$$\sigma_{min} = -4\tau_0$$

This problem is important in fatigue problems, such as the *torsional fatigue* of an axle containing the oil supply hole (Figure 1.5) and nodular cast iron containing spherical graphites

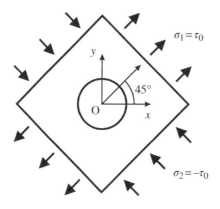

Figure 1.4

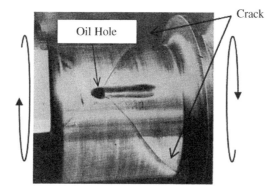

Figure 1.5 Fatigue cracks emanated in $\pm 45°$ direction from oil supply hole under cyclic torsion [2]

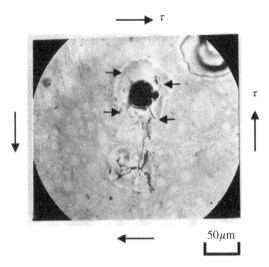

Figure 1.6 Fatigue cracks from graphites in nodular cast iron under cyclic torsion (iron emanated from graphite)

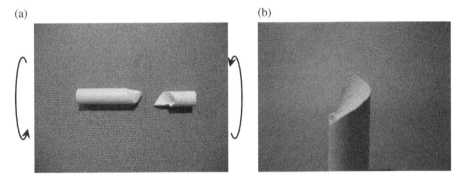

Figure 1.7 Stick of chalk fractured by torsion. (a) Fracture caused by torsion. (b) Magnification of fractured surface

(Figure 1.6) which have an approximately equivalent effect to fatigue strength as a circular hole.

If a stick of chalk is twisted, the fracture surface has a spiral shape, as shown in Figure 1.7. This morphology can be understood as the result of the combined effect of the shear stress in Figure 1.3 and small voids contained in chalk acting equivalently to a circular hole.

Example problem 1.3

Stress concentration at a circular hole in a wide plate under a general remote stresses $\left(\sigma_x, \sigma_y, \tau_{xy}\right) = \left(\sigma_{x\infty}, \sigma_{y\infty}, \tau_{xy\infty}\right)$

See Figure 1.8a.

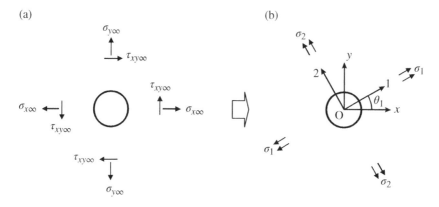

Figure 1.8 (a, b) The stress field (σ_1, σ_2) equivalent to the stress field $(\sigma_{x\infty}, \sigma_{y\infty}, \tau_{xy\infty})$

First of all, the principal stresses, σ_1 and σ_2, at a remote distance and the direction of the principal stresses, θ_1, are calculated as in Figure 1.8:

$$\sigma_1 = \frac{(\sigma_{x\infty} + \sigma_{y\infty}) + \sqrt{(\sigma_{x\infty} - \sigma_{y\infty})^2 + 4\tau_{xy\infty}^2}}{2}$$

$$\sigma_2 = \frac{(\sigma_{x\infty} + \sigma_{y\infty}) - \sqrt{(\sigma_{x\infty} - \sigma_{y\infty})^2 + 4\tau_{xy\infty}^2}}{2}$$

$$\theta_1 = \tan^{-1}\frac{\sigma_1 - \sigma_{x\infty}}{\tau_{xy\infty}}$$

Likewise, as in Example problem 1.2, the method of the superposition of individual problems of uniaxial stress, σ_1 and σ_2, can be applied. Thus, the maximum stress σ_{max} is given by

$$\sigma_{max} = 3\sigma_1 - \sigma_2$$

Example problem 1.4

The stress distribution around a circular hole under remote stress $(\sigma_{x\infty}, \sigma_{y\infty}, 0)$ is given by the following equations by the superposition of Equation 1.1 (Figure 1.9):

Along x axis

$$\left.\begin{aligned}
\sigma_x &= \sigma_{x\infty}\left(1 - \frac{5a^2}{2x^2} + \frac{3a^4}{2x^4}\right) + \sigma_{y\infty}\left(\frac{3a^2}{2x^2} - \frac{3a^4}{2x^4}\right), \\[2mm]
\sigma_y &= \sigma_{x\infty}\left(\frac{a^2}{2x^2} - \frac{3a^4}{2x^4}\right) + \sigma_{y\infty}\left(1 + \frac{a^2}{2x^2} + \frac{3a^4}{2x^4}\right)
\end{aligned}\right\} \tag{1.7}$$

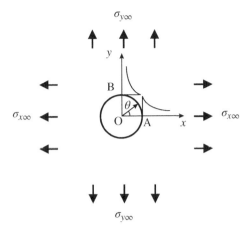

Figure 1.9 Stress distribution around a circular hole under biaxial remote stress

Along y axis

$$\left.\begin{aligned}
\sigma_x &= \sigma_{x\infty}\left(1+\frac{a^2}{2y^2}+\frac{3a^4}{2y^4}\right)+\sigma_{y\infty}\left(\frac{a^2}{2y^2}-\frac{3a^4}{2y^4}\right), \\
\sigma_y &= \sigma_{x\infty}\left(\frac{3a^2}{2y^2}-\frac{3a^4}{2y^4}\right)+\sigma_{y\infty}\left(1-\frac{5a^2}{2y^2}+\frac{3a^4}{2y^4}\right)
\end{aligned}\right\}$$

(1.8)

1.2 Stress Concentration at an Elliptical Hole

The solution of stress concentration for an elliptical hole is very useful for fatigue strength analysis, because the solution can be applied to simulate various defects by changing the *aspect ratio b/a* of the ellipse. The solution for an ellipse can be used as the solution for a *crack* by letting $b \to 0$. However, the solutions for an elliptical hole in most of literature and books [1] are given in the *elliptic coordinates* which are not convenient for fatigue analysis. In this chapter, the solution is expressed with the *Cartesian coordinates* (x, y) [3] (Figure 1.10).

1.2.1 Stress Distribution and Stress Concentration Under Remote Stress $\sigma_{y\infty}$ or $\sigma_{x\infty}$

The closed form solutions in the Cartesian coordinate (x, y) are expressed as follows.
Remote stress $(\sigma_x, \sigma_y, \tau_{xy}) = (0, \sigma_{y\infty}, 0)$
Along x axis (Figure 1.11):

$$\frac{\sigma_x}{\sigma_{y\infty}} = \frac{1}{\zeta^2-1}\cdot\frac{2\lambda+1}{\lambda-1}+F_1(\zeta)+F_2(\zeta)$$

(1.9)

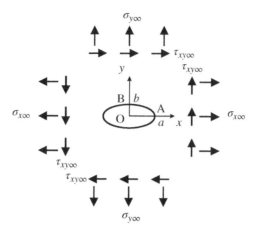

Figure 1.10

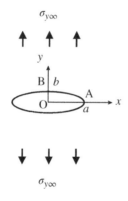

Figure 1.11

$$\frac{\sigma_y}{\sigma_{y\infty}} = \frac{1}{\zeta^2 - 1}\left(\zeta^2 + \frac{\lambda}{\lambda - 1}\right) - F_1(\zeta) - F_2(\zeta) \tag{1.10}$$

$$\frac{\tau_{\max}}{\sigma_{y\infty}} = \frac{\sigma_y - \sigma_x}{2}\Big/\sigma_{y\infty}$$
$$= \frac{1}{2(\zeta^2 - 1)}\left(\zeta^2 - \frac{\lambda + 1}{\lambda - 1}\right) - F_1(\zeta) - F_2(\zeta) \tag{1.11}$$

$$F_1(\zeta) = \frac{1}{(\zeta^2 - 1)^2}\left\{\frac{1}{2}\left(\frac{\lambda - 1}{\lambda + 1} - \frac{\lambda + 3}{\lambda - 1}\right)\zeta^2 - \frac{\lambda + 1}{(\lambda - 1)^2}\right\} \tag{1.12}$$

$$F_2(\zeta) = \frac{4\zeta^2}{(\zeta^2 - 1)^3}\left(\frac{1}{\lambda + 1}\zeta^2 - \frac{1}{\lambda - 1}\right)\frac{\lambda}{\lambda - 1} \tag{1.13}$$

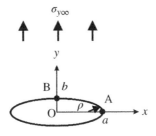

Figure 1.12

$$\text{where } \lambda = a/b, \; \zeta = \frac{x + \sqrt{x^2 - c^2}}{c}, \; c = \sqrt{a^2 - b^2} \tag{1.14}$$

Substituting $x = a$ into Equation 1.10 and determining σ_{max}, the *stress concentration factor* K_t under the remote stress $(\sigma_x, \sigma_y, \tau_{xy}) = (0, \sigma_{y\infty}, 0)$ is given by the following equation (Figure 1.12).

$$K_t = 1 + \frac{2a}{b} \tag{1.15}$$

Another expression of Equation 1.15 is also useful for various practical problems.

$$K_t = 1 + 2\sqrt{\frac{t}{\rho}} \tag{1.16}$$

where $t = a$, $\rho = b^2/a$. ρ is the *radius of the curvature* at the point A, that is the notch root radius at the point A (Figure 1.12).

It must be noted the stress σ_x at point B is

$$\sigma_x = -\sigma_{y\infty}$$

irrespective of the aspect ratio b/a.

The maximum shear stress τ_{max} expressed by Equation 1.11 is the value for a *plane stress problem*. If the plane stress problem is treated as a *3D problem* including the stress in the plate thickness direction, the maximum shear stress on the x axis is given by $\tau_{max} = \sigma_y/2$ for the position of $\sigma_x > 0$.

Example problem 1.5

Determine the location of the value of the maximum stress for an elliptical hole with the aspect ratio $b/a = 0.5$ under a remote stress condition $\sigma_x = 2\sigma_0$, $\sigma_y = \sigma_0$, as shown in Figure 1.13.

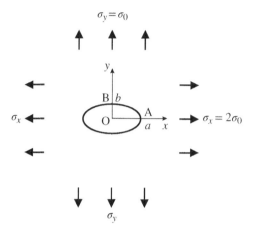

$$\sigma_y = \sigma_0$$

Figure 1.13

Solution

By using Equation 1.15, $\sigma_y = 5\sigma_0 - 2\sigma_0 = 3\sigma_0$ at point A and $\sigma_x = 2 \times 2\sigma_0 - \sigma_0 = 3\sigma_0$ at point B. It must be noted that when we apply Equation 1.15, the relationship of a and b in the equation is reversed for the remote stress σ_x and σ_y. In this problem, not only the stresses at point A and B, but also the stress is $3\sigma_0$ everywhere along the periphery of the elliptical hole (see Appendix A.4 of Part II).

Remote stress $(\sigma_x, \sigma_y, \tau_{xy}) = (\sigma_{x\infty}, 0, 0)$

Along the x axis (Figure 1.14):

$$\frac{\sigma_x}{\sigma_{x\infty}} = \frac{1}{\zeta^2 - 1}\left(\zeta^2 - \frac{1}{\lambda - 1}\right) - F_3(\zeta) + F_4(\zeta) \tag{1.17}$$

$$\frac{\sigma_y}{\sigma_{x\infty}} = -\frac{1}{\zeta^2 - 1} \cdot \frac{\lambda + 2}{\lambda - 1} + F_3(\zeta) - F_4(\zeta) \tag{1.18}$$

$$\frac{\tau_{max}}{\sigma_{x\infty}} = \frac{\sigma_y - \sigma_x}{2}\Big/\sigma_{x\infty}$$

$$= -\frac{1}{2(\zeta^2 - 1)}\left(\zeta^2 + \frac{\lambda + 1}{\lambda - 1}\right) + F_3(\zeta) - F_4(\zeta) \tag{1.19}$$

$$F_3(\zeta) = \frac{1}{(\zeta^2 - 1)^2}\left\{\frac{1}{2}\left(3\frac{\lambda + 3}{\lambda - 1} - \frac{\lambda - 1}{\lambda + 1}\right)\zeta^2 - \frac{(\lambda + 1)\lambda}{(\lambda - 1)^2}\right\} \tag{1.20}$$

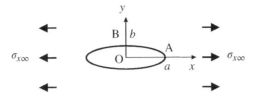

Figure 1.14

$$F_4(\zeta) = \frac{4\zeta^2}{(\zeta^2-1)^3}\left(\frac{\lambda}{\lambda+1}\zeta^2 - \frac{\lambda-2}{\lambda-1}\right)\frac{1}{\lambda-1} \tag{1.21}$$

Remote stress $(\sigma_x, \sigma_y, \tau_{xy}) = (\sigma_{x\infty}, \sigma_{y\infty}, 0)$

Using Equations 1.9 to 1.13 and 1.17 to 1.21, the stress invariant $(\sigma_x + \sigma_y)$ and the maximum shear stress along the x axis can be obtained easily.

$$\sigma_x + \sigma_y = \frac{\sigma_{x\infty}}{\zeta^2-1}\left(\zeta^2 - \frac{\lambda+3}{\lambda-1}\right) + \frac{\sigma_{y\infty}}{\zeta^2-1}\left(\zeta^2 + \frac{3\lambda+1}{\lambda-1}\right) \tag{1.22}$$

$\tau_{max} = (\sigma_y - \sigma_x)/2$

$$= \left\{\frac{\sigma_{y\infty}}{\zeta^2-1}\left(\zeta^2 - \frac{\lambda+1}{\lambda-1}\right) - \frac{\sigma_{x\infty}}{\zeta^2-1}\left(\zeta^2 + \frac{\lambda+1}{\lambda-1}\right) - 2\sigma_{y\infty}(F_1+F_2) + 2\sigma_{x\infty}(F_3-F_4)\right\}\bigg/2 \tag{1.23}$$

These equations are useful for the discussion of biaxial fatigue problems. The maximum shear stress expressed by Equation 1.23 is the value defined as the plane stress problem. If the maximum shear stress is defined including the stress in the thickness direction, the maximum shear stress along the x axis is calculated as $\tau_{max} = \sigma_x/2$ or $\tau_{max} = \sigma_y/2$ when the position of σ_y or σ_x is positive.

1.2.2 Stress Concentration and Stress Distribution Under Remote Shear Stress $\tau_{xy\infty}$

The problem of a circular hole under the remote stress condition $(\sigma_{x\infty} = 0, \sigma_{y\infty} = 0)$ can be solved easily by calculating the principal stresses σ_1 and σ_2, and then by using the superposition of two solutions for two principal stresses (see Example problem 1.2). However, the elliptical problem (Figure 1.15) cannot be solved by the same method, because the location of the maximum stress varies depending on the aspect ratio b/a.

The maximum stress $\sigma_{\eta max}$, the minimum stress $\sigma_{\eta min}$ and the locations of those stresses are expressed as follows (see Appendix A.4 of Part II).

$$\frac{\sigma_{\eta min}}{\tau_{xy\infty}} = -\frac{(a+b)^2}{ab} \text{ at } z = \frac{a^2}{\sqrt{a^2+b^2}} + i\frac{b^2}{\sqrt{a^2+b^2}} \tag{1.24}$$

$$\frac{\sigma_{\eta\max}}{\tau_{xy\infty}} = \frac{(a+b)^2}{ab} \text{ at } z = \frac{a^2}{\sqrt{a^2+b^2}} - i\frac{b^2}{\sqrt{a^2+b^2}} \qquad (1.25)$$

where z is the complex coordinate ($z = x + iy$) at the locations of the maximum and the minimum stress along the periphery of the elliptical boundary and σ_η is the normal stress along the tangential direction on the boundary of the elliptical hole. It is interesting to note that the locations of the maximum and the minimum stress are the contact point between the elliptical hole and the lines having the inclination of $\pm 45°$ against the x axis. Considering this nature, it is easy to predict the locations of a fatigue crack initiation from the elliptical hole.

On the other hand, τ_{xy} along the x axis in Figure 1.16 is expressed as follows:

$$\frac{\tau_{xy}}{\tau_{xy\infty}} = \frac{\zeta^2}{\zeta^2-1} + \frac{1}{(\zeta^2-1)^2}\left\{\frac{3\lambda+1}{\lambda-1}\zeta^2 - \left(\frac{\lambda+1}{\lambda-1}\right)^2\right\}$$
$$-\frac{2\zeta^2}{(\zeta^2-1)^3}\left\{\frac{\lambda-1}{\lambda+1}\zeta^2 - \frac{\lambda-3}{\lambda-1}\right\}\frac{\lambda+1}{\lambda-1} \qquad (1.26)$$

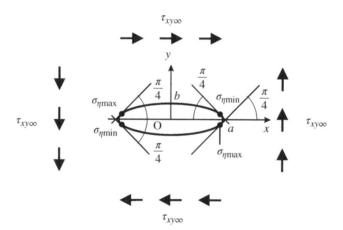

Figure 1.15

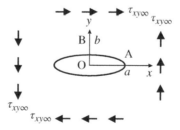

Figure 1.16

where

$$\lambda = a/b, \ \zeta = \frac{x + \sqrt{x^2 - c^2}}{c}, \quad c = \sqrt{a^2 - b^2} \tag{1.27}$$

In the case of an elliptical hole, the shear stress τ_{xy} at point A is always 0. Namely, the maximum shear stress τ_{max} occurs at some distance away from point A (Figure 1.17). The maximum shear stress τ_{max} is given by the following equation:

$$
\begin{aligned}
\frac{\tau_{xy\,max}}{\tau_{xy\,\infty}} &= \frac{\zeta^{*2}}{\zeta^{*2} - 1} + \frac{1}{\left(\zeta^{*2} - 1\right)^2} \left\{ \frac{3\lambda + 1}{\lambda - 1} \zeta^{*2} - \left(\frac{\lambda + 1}{\lambda - 1}\right)^2 \right\} \\
&\quad - \frac{2\zeta^{*2}}{\left(\zeta^{*2} - 1\right)^3} \left\{ \frac{\lambda - 1}{\lambda + 1} \zeta^{*2} - \frac{\lambda - 3}{\lambda - 1} \right\} \frac{\lambda + 1}{\lambda - 1}
\end{aligned}
\tag{1.28}
$$

where

$$\lambda = a/b, \ \zeta^* = \frac{x^* + \sqrt{x^{*2} - c^2}}{c}, \quad c = \sqrt{a^2 - b^2},$$

$$\left(x^* + \sqrt{x^{*2} - c^2}\right)^2 = a^2 \frac{1}{\lambda^2} \left\{ \lambda^2 + 5 + \sqrt{12\left(\lambda^2 + 2\right)} \right\} \tag{1.29}$$

x^* is the coordinate on the x axis where the maximum shear stress τ_{max} occurs. The value of τ_{max} can be calculated by Equation 1.29. Especially, in the case of a circular hole, $\lambda = 1 (a = b)$, see Figure 1.17, $x^* = \sqrt{3}a$. The location x^* of the maximum shear stress approaches the edge of the elliptical hole as the ellipse becomes slender. Figure 1.18 shows the experimental result of the torsional fatigue test related to the above discussion.

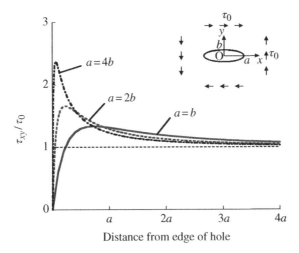

Figure 1.17   Shear stress distribution for elliptical hole along the x axis under shear stress

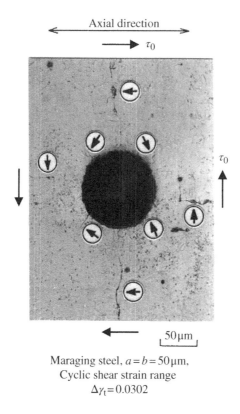

Maraging steel, $a = b = 50\,\mu\text{m}$,
Cyclic shear strain range
$\Delta\gamma_t = 0.0302$

Figure 1.18 Cracks initiated by cyclic torsion in a maraging steel [4]. Cracks initiate from four points located on the edge of the hole where the maximum normal stress occurs and also at four points along the x and y axes where the shear stress τ_{xy} reaches a maximum (diameter of hole = depth of hole = 100 μm)

Figure 1.18 shows the morphologies of crack initiation around an artificial small hole $(a = b)$ in low cycle torsional fatigue in a cylindrical maraging steel specimen [4]. Eight fatigue cracks initiate around the hole, four of which initiate from the edge of the hole at the points $\pm 45°$ and $\pm 135°$ from the x axis, and the remaining four cracks initiate further in the x and y directions at the points where the maximum shear stress occurs.

1.3 Stress Concentration at a Hole in a Finite Width Plate

The maximum stress at the edge of a circular hole, existing at the center of a finite width plate (Figure 1.19), is more than three times larger than the remote stress σ. Since the stress concentration factor K_t in these cases is based on the nominal stress σ_n at the minimum section where the circular hole exists, K_t is smaller than 3, namely the maximum stress is smaller than $3\sigma_n$. However, it must be noted that the actual maximum stress at the edge of the circular hole, σ_{max}, is of course larger than 3σ, because σ_n is larger than σ. The notations related to this problem are shown in the following and the *stress concentration factor* K_t is shown in Figure 1.19.

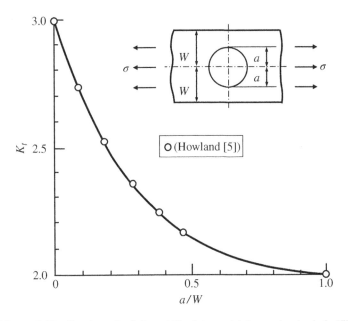

Figure 1.19 Tension of a finite width plate containing a circular hole [5]

Circular Hole

σ: Remote stress

σ_n: Nominal stress at the minimum section

$$\sigma_n = \frac{W}{W-a}\sigma \tag{1.30}$$

σ_{max}: Maximum stress at the edge of circular hole

$$K_t = \frac{\sigma_{max}}{\sigma_n} \tag{1.31}$$

$$\sigma_{max} = K_t \frac{W}{W-a}\sigma \tag{1.32}$$

Since the minimum section becomes very slender like Figure 1.20 as $a/W \rightarrow 1$, the typical misconception arising from this problem is that $K_t \rightarrow 1.0$ as $a/W \rightarrow 1$. However, the correct value is $K_t \rightarrow 2.0$ as $a/W \rightarrow 1$ (see the verification in Appendix A.3 of Part II).

Elliptical Hole

The definition of the *stress concentration factor* K_t for an elliptical hole existing at the center of a finite width plate (Figure 1.21) is the same as the definition in the case of a circular hole.

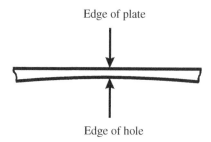

Edge of plate

Edge of hole

Figure 1.20   Condition of the minimum section as $a/W \rightarrow 1.0$

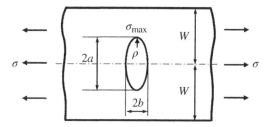

Figure 1.21   Tension of finite width plate containing an elliptical hole

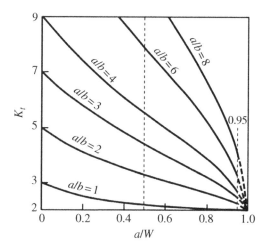

Figure 1.22   Stress concentration factor [6] for elliptical hole in finite width plate under tension (Figure 1.21) $K_t = \sigma_{\max}/\sigma_n$, $\sigma_n = \frac{W}{W-a}\sigma$

The solutions [6] for elliptical holes having various aspect ratios b/a are shown in Figure 1.22.

The reason why $K_t \rightarrow 2.0$ when $a/W \rightarrow 1.0$, irrespective of the value of b/a, is the same as the case of circular hole.

1.4   Concept of Equivalent Ellipse

1.4.1   Basic Concept

In real cases, most holes, notches and defects do not necessarily have a completely circular or elliptical shape. In order to obtain the exact stress concentration factor for holes and notches which do not have an ideal circular or elliptical shape, numerical analyses such as the finite element method (FEM), may become necessary.

However, if we apply the concept of equivalent ellipse [7], we could estimate the value with practically sufficient accuracy. We do not necessarily need exact solutions for various practical problems. It is not recommended to jump directly to time-consuming computer analyses such as FEM. It is a good engineering attitude and useful to try to seek an approximate solution at first without computer analyses.

Most engineering accidents are not caused by a mistake of precision in stress concentration analysis. Actual accidents are rather caused by a lack of attention to more important factors such as boundary conditions, service loadings and environments.

By relying too much on precise FEM analysis, design engineers lose the sense of imaging the physical reality of the structure in question.

The basic concept of equivalent ellipse and its applications to estimating the approximate stress concentration factor of various holes and notches will be explained in the following.

As shown in Figure 1.23, there is a wide plate under a remote stress $\sigma_y = \sigma_0$ which contains a rounded square hole with rounded corners.

In order to estimate the stress concentration at point A, we can imagine an elliptical hole (dotted line) which has the same radius of curvature as the radius of curvature ρ_A at point A of the square hole ρ_A and the same length $(2a)$ of the major axis in the x direction of the hole. By this approximation, the *stress concentration K_t* at point A can be estimated as

$$K_t \cong 1 + 2\sqrt{\frac{a}{\rho_A}} \tag{1.33}$$

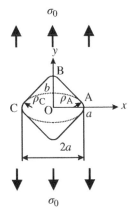

Figure 1.23   Application of the concept of equivalent ellipse

If we need to estimate the stress concentration at a point C in Figure 1.23, we have only to replace ρ_A by the radius of the curvature ρ_C at the point C. Even if the radius of the curvature ρ_C is different from ρ_A, this estimation is valid.

For example, as the application of the concept of the equivalent ellipse, the stress concentration factor K_t for the case of $\rho_A = 1$, $t = a = 4$ (Equation 1.16) is estimated as $K_t = 5$. The value of K_t for the same problem calculated by FEM is $K_t = 5.19$. The difference between these two values is only 4%. Thus, we do not always need FEM for design and structural integrity analyses. It is important to realize that using FEM for every problem is a waste of time and money.

Some other useful examples of the concept of the equivalent ellipse applications will be shown in the following.

Example 1   V-shaped notch
As shown later in Section 1.4.3, the open angle of a V-shaped notch (Figure 1.24) does not affect much to K_t. Thus, K_t can be estimated by the following equation with sufficient accuracy:

$$K_t \cong 1 + 2\sqrt{\frac{t}{\rho}}$$

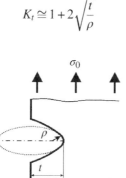

Figure 1.24   Application of the concept of equivalent ellipse in a V-shaped notch

The effect of the notch open angle will be discussed in detail in Section 1.4.3.

Example 2
The stress concentration for the holes shown in Figures 1.25 and 1.26 can also be estimated by the same concept. The stress concentration factors K_t estimated by the concept of the equivalent ellipse are shown in these figures.

The exact analytical solution for the case of $a_1 = a_2$ in Figure 1.25 is $K_t = 3.7 \sim 3.8$ (see Figure 4.3). On the other hand, K_t estimated by the concept of the equivalent ellipse is $K_t = 1 + 2\sqrt{2} \cong 3.83$, which is very close to the exact solution.

The problem of Figure 1.26 shows a circular hole in contact with the edge of a semi-infinite plate under a remote tension. Although an analytical solution related to this problem is shown in Figure 1.27, the value of K_t for the limiting case of $e/a \rightarrow 1.0$ is not given. However, applying the concept of the equivalent ellipse, we can easily estimate the stress concentration factor for the case of $e/a \rightarrow 1.0$ at point A′ in Figure 1.26 as $K_{tA'} \cong 3.83$.

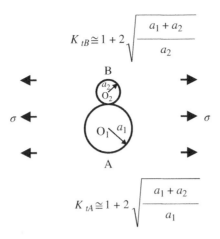

$$K_{tB} \cong 1 + 2\sqrt{\frac{a_1 + a_2}{a_2}}$$

$$K_{tA} \cong 1 + 2\sqrt{\frac{a_1 + a_2}{a_1}}$$

Figure 1.25 Application of the concept of the equivalent ellipse: two circular holes touching each other (the contact part may be regarded as having no strength)

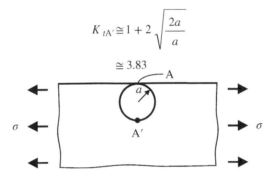

$$K_{tA'} \cong 1 + 2\sqrt{\frac{2a}{a}}$$

$$\cong 3.83$$

Figure 1.26 Application of the concept of the equivalent ellipse: a circular hole touching the edge of a semi-infinite plate (the contact part may be regarded having no strength)

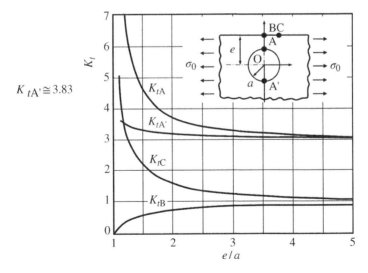

$$K_{tA'} \cong 3.83$$

Figure 1.27 Stress concentration for a circular hole near the edge of a semi-infinite plate [8]. The solution for Figure 1.26 is given by $e/a \rightarrow 1.0$

Example 3 Torsional fatigue of nodular cast iron

In the case of irregularly shaped defects in an infinite plate subjected to remote uniform shear stress $(0, 0, \tau)$, as shown in Figure 1.28, the application of Equations 1.24 and 1.25) is valid based on the concept of the equivalent ellipse. Figures 1.29 and 1.30 show cracks emanating from nodules in a torsional fatigue test.

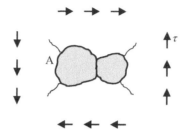

Figure 1.28 Application of the concept of equivalent ellipse: two nodules in contact to each other

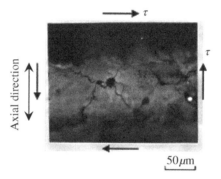

Figure 1.29 Cracks initiated from nodules of nodular cast iron in torsional fatigue

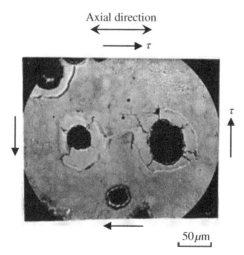

Figure 1.30 Cracks initiated from nodules of nodular cast iron in a torsional fatigue test

1.4.2 Difference of Stress Concentration at a Circular Hole in an Infinite Plate and at a Semi-Circular Hole on the Edge of a Semi-Infinite Plate

The stress concentration factor K_t for point B in Figure 1.31a is $K_t = 3$. But how much is the stress concentration factor K_t for point B in Figure 1.31b? Based on the concept of the equivalent ellipse, K_t for Figure 1.31b is $K_t = 3$. What is the exact value, $K_t = 3$ or not?

Maunsell [9] challenged this problem and calculated the value of K_t for the semi-circular notch (Figure 1.31b) up to two digits, obtaining $K_t \cong 3.0$. However, he was not confident with the non-zero value of the third digit and concluded that the exact solution would be $K_t = 3$ and the non-zero third digit appeared due to a numerical calculation error. Since the computing machine was manual at that time (1936), this conclusion was natural. According to an exact calculation done in 1953 by M. Ishida [10], it is

$$K_t = 3.065 \tag{1.34}$$

Since the difference between this value and $K_t = 3$ is only 2%, using $K_t = 3$ for actual cases does not cause any problem. However, even if the difference is small, it is important to understand the source of the difference between Figure 1.31a and b for an exact understanding of other similar problems.

The difference between Figure 1.31a and b is the difference in the stress state along the x axis. The stress state along the x axis in Figure 1.31a is $\sigma_y \neq 0$, $\tau_{xy} = 0$[1]. On the other hand, the boundary condition along the x axis in Figure 1.31b is $\sigma_y = 0$, $\tau_{xy} = 0$[2]. The stress state $\tau_{xy} = 0$, which Figure 1.31a and b have in common, is an important factor for closeness of Figure 1.31a and b from the viewpoint of the theory of elasticity. This consideration gives us the explanation that the stress concentration for Figure 1.31b is not equal to $K_t = 3$ but $K_t = 3.065$ due to differences in σ_y along the x axis in Figure 1.31a and b.

Figure 1.32 shows the imaginary stress state which is made by cutting the problem of Figure 1.31a along the x axis.

The dotted curve in Figure 1.32 shows the distribution of σ_y which acts along the x axis in Figure 1.31.

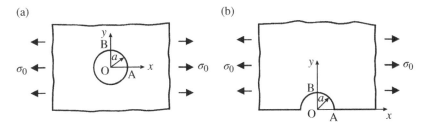

(a)   (b)

Figure 1.31 What is the difference between the stress concentrations for a circular hole and a semi-circular notch? (a) Circular hole in infinite plate. $K_t = 3$. (b) Semi-circular notch in semi-infinite plate. $K_t = ?$

---

[1] Although the reason for $\tau_{xy} = 0$ along the x axis can be understood from Equation 1.1, it is better to understand the reason from the nature of Hooke's law that shear strain γ_{xy} along the x axis is $\gamma_{xy} = 0$ from the symmetry of the problem, and accordingly $\tau_{xy} = 0$ from Hooke's law $\tau_{xy} = G\gamma_{xy}$ (see Chapter 3 in Part I).

[2] The reason for $\tau_{xy} = 0$ along the x axis is that the x axis is the edge of a stress-free semi-infinite plate.

As already described, $\sigma_y = -\sigma_0$ at point A. However, the resultant force, by integrating σ_y along the x axis, must be 0, because the infinite plate is subjected only to $\sigma_x = \sigma_0$ at remote distance. Therefore, σ_y along the x axis satisfies the following equation:

$$\int_a^\infty \sigma_y dx = 0 \tag{1.35}$$

Since σ_y is compressive near the point A, σ_y must tend to tensile stress as the coordinate becomes far from point A and reaches a maximum, finally decreasing to 0. The state of Figure 1.31b is equivalent to the state in which the distribution of σ_y is removed from Figure 1.32. In other words, the problem of Figure 1.31b is produced by superimposing the stress having the opposite sign of σ_y along the x axis in Figure 1.32 to vanish the stress σ_y along the x axis. How much effect on σ_x at point B is produced by vanishing σ_y which is expressed by the dotted curve in Figure 1.32? Since the stress σ_y to be vanished acts in the y direction, the influence of vanishing σ_y on σ_x at point B should be small. The amount analyzed later by M. Ishida is 0.065 of $K_t = 3.065$.

If shear stress τ_{xy} exists as the stress to be vanished along the x axis, the influence is naturally not small, because τ_{xy} acts in the x direction and the vanishing τ_{xy} generates additional force in the x direction, eventually influencing σ_x at point B in the same direction.

1.4.3 Limitation of Applicability of the Concept of the Equivalent Ellipse

The problem of Figure 1.24 is a typical example of an application of the concept of the equivalent ellipse. However, how valid is the application of the concept of the equivalent ellipse to the problems such as Figure 1.33 when the open angle θ of a V-shape notch changes?

Figure 1.34 and Table 1.1 compare the stress concentration factor estimated by the concept of the equivalent ellipse, $K_t \cong 1 + 2\sqrt{\frac{t}{\rho}}$, and the exact values by numerical analysis [11]. It is confirmed that K_t estimated by the concept of equivalent ellipse has practically sufficient accuracy up to $\theta = \sim 90°$.

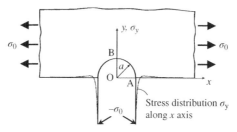

Figure 1.32 Stress state above the x axis of Figure 1.31a

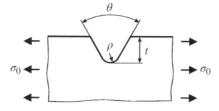

Figure 1.33 V-shaped notch at the edge of semi-infinite plate under remote tension σ_0

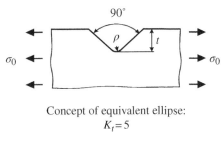

90°

Concept of equivalent ellipse:
$K_t = 5$

Exact numerical analysis:
$K_t = 5.274$

Figure 1.34 Application of the concept of the equivalent ellipse to V-shaped notched and the exact numerical value for $t/\rho = 4$

Table 1.1 Comparison of stress concentrations calculated by the numerical analysis [11] and the concept of the equivalent ellipse for V-shaped notches with different open angles

		t/ρ	1 θ	2 θ	4 θ	8 θ
		Concept of equivalent ellipse	3	3.83	5	6.66
K_t	Open angle of V-shaped notch θ	0°	3.065	3.997	5.334	7.824
		30°	3.065	3.997	5.339	7.264
		60°	3.065	3.995	5.331	7.243
		90°	3.062	3.976	5.274	7.097
		120°	3.016	3.839	4.941	6.409
		150°	2.654	3.725	3.725	4.413

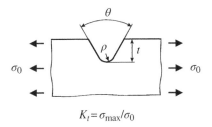

$K_t = \sigma_{max}/\sigma_0$

1.5   Stress Concentration at Inclusions

There are many misconceptions regarding the stress concentration at *inclusions*. A typical mis-conception is that the stress concentration is high at hard inclusions and low at soft inclusions. It must be noted that the hardness of inclusions expressed by the terms "*hard*" and "*soft*" has no relationship with Young's modulus E and Poisson's ratio ν which control stress concentrations. For example, the Young's modulus E and Poisson's ratio ν of a heat treated steel are almost the same as those of an annealed steel with the same chemical composition regardless of a large difference in the hardness due to heat treatment. Therefore, the basic subjects to be understood in this problem will be explained as follows.

Case A   Circular hole in a wide plate under remote uniaxial tensile stress

If the Young's modulus E_I of a 2D circular inclusion is 0, the inclusion is substantially equivalent to a circular hole. Namely, a hole can be regarded as the softest inclusion. In this case, the stress concentration factor K_{tA} and the maximum stress at point A of the matrix side are expressed as follows.

$$K_{tA} = 3 \tag{1.36}$$

$$\sigma_{yA} = 3\sigma_0 \tag{1.37}$$

As shown later, it will be understood that the stress concentration factor for a hole is always larger than for any inclusion having the same shape and various Young's modulus E and Poisson's ratio ν.

Case B   Rigid circular inclusion in a wide plate under remote uniaxial tensile stress

In the case of a rigid inclusion, the Young's modulus E_I is ∞ and the maximum stress occurs near point B of Figure 1.35. The maximum stress is lower than that at point A for a hole ($E_I = 0$). It must be noted that the stress σ_{yA} at the point A of the matrix side is always lower than the stress σ_{yB} at the point B, that is:

$$\sigma_{yA} < \sigma_0$$

Namely, the location of the maximum stress varies depending on $E_I < E_M$ or $E_I > E_M$ and the value of the maximum stress also varies. As shown in Table 1.2, the stresses σ_{yB} and σ_{yA} for $E_I = \infty$ have the following values:

$$\sigma_{yB} = 1.509\sigma_0 \tag{1.38}$$

$$\sigma_{yA} = 0.009\sigma_0 \tag{1.39}$$

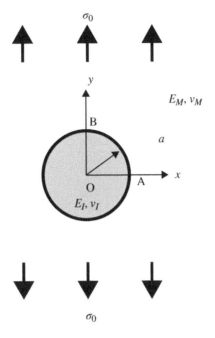

Figure 1.35   Circular inclusion

Although the strain $\varepsilon_{yA} = 0$ at the interface of the matrix side where the matrix and the inclusion are completely bonded, it does not necessarily mean that the stress $\sigma_{yA} = 0$. This nature holds also for rigid elliptic inclusions and rigid spherical inclusion. The reason can be understood from *Hooke's law* (see Chapter 3 of Part I and also Problem 5 of Chapter 5 of Part I.).

The nature of stress concentration described above agrees with the experimental results of fatigue crack initiation from inclusions and defects (see Figure 1.36). The crack initiation from spherical inclusions such as Al_2O_3 and $Al_2O_3(CaO)_x$ which have higher Young's moduli than matrices occurs at the polar point (C or C′ of Figure 1.37) and then the crack causes the delamination of the interface between the inclusion and the matrix. Eventually, a complete shape of an inclusion can be observed at the fracture origin.

Once delamination at the matrix–inclusion interface occurs, the inclusion cannot sustain tensile stress and the domain of inclusion becomes mechanically equivalent to a stress free hole or cavity (Figure 1.38). After delamination of the interface, the maximum stress does not occur at the polar point (point C) as it moves to the equator of the inclusion (point A) where a *fatigue crack* initiates and propagates into the matrix. However, as this problem is essentially a 3D problem, the detailed discussion will be made in Chapter 3 of Part II.

Table 1.2 Stress concentration factors for circular and elliptical inclusions ($E_M = 206$ GPa, $\nu_I = \nu_M = 0.3$, perfect bonding)

E_I/E_M	point	stress	b/a 0.1	0.5	1.0	2.0	10.0
0	A	σ_x/σ_0	0.000	0.000	0.000	0.000	0.000
		σ_y/σ_0	21.012	5.004	3.003	2.001	1.200
	B	σ_x/σ_0	−1.001	−1.003	−1.003	−1.002	−1.001
		σ_y/σ_0	0.000	0.000	0.000	0.000	0.000
0.5	A	σ_x/σ_0	0.113	0.031	−0.007	−0.025	−0.014
		σ_y/σ_0	1.889	1.671	1.498	1.320	1.079
	B	σ_x/σ_0	−0.061	−0.191	−0.239	−0.246	−0.189
		σ_y/σ_0	0.961	0.840	0.748	0.656	0.537
0.9	A	σ_x/σ_0	0.022	0.006	−0.001	−0.005	−0.003
		σ_y/σ_0	1.106	1.088	1.071	1.050	1.015
	B	σ_x/σ_0	−0.009	−0.026	−0.034	−0.037	−0.033
		σ_y/σ_0	0.996	0.979	0.964	0.945	0.913
1.1	A	σ_x/σ_0	−0.022	−0.005	0.001	0.004	0.003
		σ_y/σ_0	0.912	0.925	0.938	0.954	0.986
	B	σ_x/σ_0	0.008	0.023	0.029	0.033	0.032
		σ_y/σ_0	1.003	1.017	1.032	1.050	1.084
5	A	σ_x/σ_0	−0.565	−0.089	0.019	0.072	0.068
		σ_y/σ_0	0.068	0.213	0.279	0.360	0.642
	B	σ_x/σ_0	0.131	0.264	0.333	0.425	0.764
		σ_y/σ_0	1.018	1.174	1.370	1.712	3.130
10	A	σ_x/σ_0	−0.868	−0.113	0.024	0.091	0.105
		σ_y/σ_0	−0.133	0.089	0.150	0.212	0.454
	B	σ_x/σ_0	0.187	0.312	0.390	0.516	1.158
		σ_y/σ_0	1.013	1.199	1.436	1.878	4.255
∞	A	σ_x/σ_0	−1.509	−0.142	0.029	0.114	0.182
		σ_y/σ_0	−0.453	−0.043	0.009	0.034	0.0544
	B	σ_x/σ_0	0.299	0.367	0.453	0.624	1.989
		σ_y/σ_0	0.997	1.225	1.509	2.079	6.637

Element: 4 sides, 8 nodes
Error. Estimated less than 1% from the exact solution $K_t = 1 + 2a/b$ for elliptical hole.

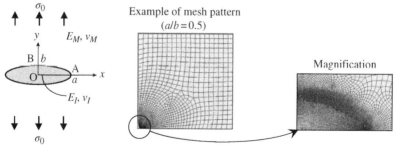

Example of mesh pattern
($a/b = 0.5$)

Magnification

Matrix: Young's modulus E_M. Poisson's ratio ν_M
Inclusion: Young's modulus E_I. Poisson's ratio ν_I

- Element mesh at the interface between matrix and inclusion: 200.
- Cercular inclusion: uniform meshing.
- Eilliptical inclusion: Node distance at radius of curvature is adjusted to be approximately equal to the case of circular inclusion.

Note: σ_x inside the inclusion is equal to σ_x at A of the matrix; and σ_y inside the inclusion is equal to σ_y at B of the matrix [12].

(a)   (b)

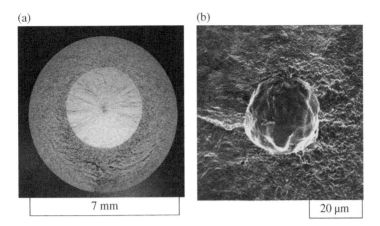

| 7 mm | 20 μm |

Figure 1.36 Nonmetallic inclusion observed at fatigue fracture. Material: JIS SCM435. Stress amplitude: $\sigma = 560\,$MPa. Number of cycles to failure: $N_f = 1.11 \times 10^8$. (a) Fatigue fracture surface. (b) Inclusion at fracture origin. Chemical composition: $Al_2O_3 \cdot (CaO)_x$

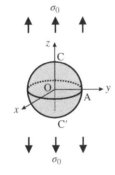

Figure 1.37   Spherical inclusion

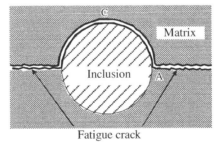

Figure 1.38   Mechanism of fatigue crack initiation and growth from nonmetallic inclusion

Problems of Chapter 1

1. The stress distribution for the problem of Figure 1.39 is given by Equation 1.1. Subtract σ_0 from the distribution of the stress σ_x along the y axis and integrate $(\sigma_x - \sigma_0)$ from $y = a$ to ∞. The integral is the partial resultant force of σ_x above σ_0. Calculate the integral and explain the physical meaning of the integral value. Assume plate thickness unity.

$$\int_a^\infty (\sigma_x - \sigma_0)\,dy = ?$$

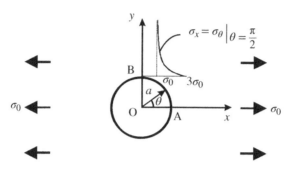

Figure 1.39

2. Integral the distribution of stress σ_y for the problem of Figure 1.39 along the x axis from $x = a$ to ∞ and explain the meaning of the value of the integral.

$$\int_a^\infty \sigma_y\,dx = ?$$

3. (1) The stress distribution for the problem of Figure 1.40 is given by Equation 1.10. Subtract σ_0 from the distribution of the stress σ_y along the x axis and integrate $(\sigma_y - \sigma_0)$ from $x = a$ to ∞. The integral is the partial resultant force of σ_y above σ_0. Calculate the integral and explain the physical meaning of the integral value. Assume plate thickness unity.

$$\int_a^\infty (\sigma_y - \sigma_0)\,dx = ?$$

(2) Integrate the distribution of stress σ_x along the y axis in Figure 1.40 from $y = b$ to ∞ and explain the meaning of the value of the integral.

$$\int_b^\infty \sigma_x\,dy = ?$$

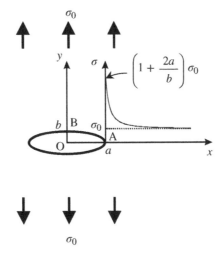

Figure 1.40   Elliptical hole in a plate under remote tension $\sigma_y = \sigma_0$ (unit thickness)

4.   The stress concentration factor K_t for a circular hole in Figure 1.1 is $K_t = 3$. However, machined circular holes are not necessarily perfectly circular and they always contain surface roughness and burrs produced by machining (Figure 1.41). The stress concentration factor K_t for a circular hole having a rough surface caused by rough machining becomes larger than $K_t = 3$. Explain the reason and the estimation method of stress concentration factor.

Thus, in the strength design, the stress concentration factor must not be estimated simply based on the design drawing as $K_t = 3$. Rather, a stress concentration factor higher than $K_t = 3$ must always be taken into consideration.

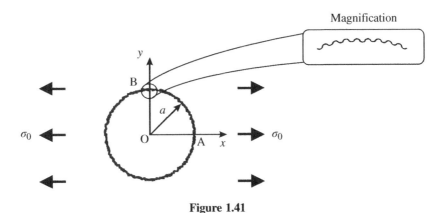

Figure 1.41

5.   Calculate the exact shape of an elliptical inclusion after deformation in the problem of Figure 1.42. Use the values in Table 1.2.

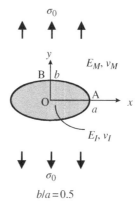

$$b/a = 0.5$$

Matrix: Young's modulus E_M, Poisson's ratio v_M
Inclusion: Young's modulus E_I, Poisson's ratio v_I
$E_I/E_M = 0.5$, $v_M = v_I = 0.3$

Figure 1.42

6. A thin sheet of material B containing a circular hole of radius a is bonded onto the surface of thick material A (Figure 1.43). The thickness h of the sheet is very small compared to the radius a of the circular hole and $h < a/100$. When tensile stress σ_0 is applied to material A, determine the stress σ_x in the x direction at the edge of circular hole B in sheet material B.

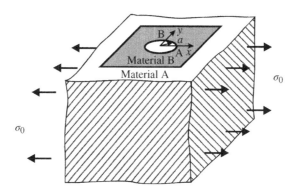

Figure 1.43 Material A: Young's modulus E_A, Poisson's ratio v_A. Material B: Young's modulus E_B, Poisson's ratio v_B

7. The stress concentration of a notch like Figure 1.44a is caused by introducing the notch and relieving the force which sustained the force applied to the part before introduction of the notch (note Problem 1). Figure 1.44b shows a structure which contains a convex part at the straight edge. The stress concentration at the corner points A or B of the convex part occurs for a reason different from Figure 1.44a. Explain the reason.

Figure 1.44c is a structure which is made by bonding or welding a different material onto the straight edge. In this case too, stress concentration occurs for the same reason as Figure 1.44b.

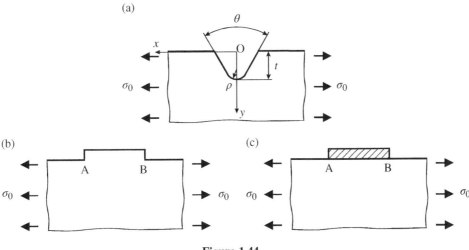

Figure 1.44

8. Figure 1.45a shows a circular hole in an infinite plate under a remote biaxial tensile stress σ_0. Figure 1.45b shows a rigid inclusion in an infinite plate under the same remote biaxial tensile stress σ_0. Before the application of the remote stress in Figure 1.45b, the rigid inclusion is inserted into a circular hole without clearance and is perfectly bonded with an infinite plate at the interface without any initial stress.

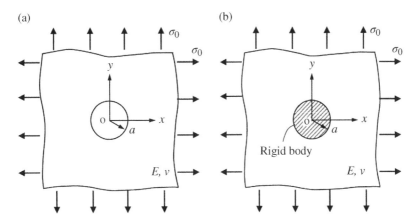

Figure 1.45

Calculate the stress σ_θ^* in the circumferential direction at the edge of the circular hole in Figure 1.45a and the stress σ_r^* in the radial direction in Figure 1.45b. Compare which is larger, σ_θ^* or σ_r^*.

Young's modulus and Poisson's ratio of the infinite plate are E and ν, respectively.

References

[1] S. P. Timoshenko and J. N. Goodier (1970) *Theory of Elasticity*, 3rd edn, McGraw–Hill, New York.

[2] R. Cazaud, G. Pomey, P. Rabee and C. Janssen (1969) *La Fatigues des Metaux*, Dunod, Paris.

[3] T. Kanezaki, K. Nagata and Y. Murakami (2007) *J. Solid Mech. Mat. Eng., Japan Soc., Mech. Engrs.*, **1**:232–243.

[4] M. Abe, H. Ito and Y. Murakami (1990) Tension–compression and torsional low-cycle fatigue of maraging steel, *Proc. 4th Intl Conf. Fatigue*, vol. **3**, eds, H. Kitagawa and T. Tanaka, pp. 1661–1666.

[5] R. C. Howland (1930) *Phil. Trans. R. Soc. A*, **229**:48–86.

[6] M. Isida (1952) *Appl. Mech.*, (in Japanese) **5**:131–140.

[7] F. Hirano (1950) *Trans. Jpn Soc. Mech. Eng.*, **16**:52–58; F. Hirano (1950) *Trans. Jpn Soc. Mech. Eng.*, **17**:12–16.

[8] T. Udoguchi (1947) *Trans. Jpn Soc. Mech. Eng.*, **13**:17–40.

[9] F. G. Maunsell (1936) *Phil. Mag.*, **21**:765–773.

[10] M. Isida (1953) *Form-Factors of a Strip with Semicircular Notches in Tension and Bending*, Research Report of Faculty of Engineering, Tokushima University, **4**:67–69.

[11] H. Nisitani (1978) *Mech. Fract.*, Noordhoff Int. Pub. **5**:1–68.

[12] J. D. Eshelby (1957) *Proc. R. Soc. Lond. Ser. A*, **241**:376–396.

2

Stress Concentration at Cracks

2.1 Singular Stress Distribution in the Neighborhood of a Crack Tip

A crack always exists prior to a fracture. Extremely high stress concentration occurs at the crack tip. The mechanics model which has been accepted for a crack in the history of strength of materials is defined as the limiting state of an elliptic hole. This is called the *Griffith Crack* [1].

In order to analyze the stress distribution in the vicinity of a crack tip, the crack is commonly modelled as a linear slit in plane [2, 3]. However, in this chapter the solution is derived by using the solution for an elliptical hole (Equations 1.9 to 1.13 and 1.17 to 1.21).

In the solution for the elliptical hole of Figure 2.1, σ_y along the x axis is expressed as Equation 1.10. Equation 1.10 can be rewritten as follows:

$$\sigma_y = \sigma_0 \left[\frac{1}{\zeta^2 - 1}\left(\zeta^2 + \frac{a}{a-b}\right) - \frac{1}{(\zeta^2-1)^2}\left\{\frac{1}{2}\left(\frac{a-b}{a+b} - \frac{a+3b}{a-b}\right)\zeta^2 - \frac{(a+b)b}{(a-b)^2}\right\} \right.$$
$$\left. - \frac{4\zeta^2}{(\zeta^2-1)^3}\left(\frac{b}{a+b}\zeta^2 - \frac{b}{a-b}\right)\frac{a}{a-b} \right] \tag{2.1}$$

where

$$\zeta = \frac{x + \sqrt{x^2 - c^2}}{c}, \quad c = \sqrt{a^2 - b^2}$$

Putting $b \to 0$ for a crack, Equation 2.1 is reduced to the following equation:

$$\sigma_y = \frac{\sigma_0 x}{\sqrt{x^2 - a^2}} \quad (x > a) \tag{2.2}$$

Theory of Elasticity and Stress Concentration, First Edition. Yukitaka Murakami.
© 2017 John Wiley & Sons, Ltd. Published 2017 by John Wiley & Sons, Ltd.

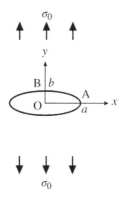

Figure 2.1

Therefore, the stress distribution in the vicinity of a crack tip, that is at $x = a + r (r/a \ll 1)$ is approximately expressed by the following equation:

$$\sigma_y = \frac{\sigma_0 \sqrt{a}}{\sqrt{2r}} \tag{2.3}$$

After the concept of the *stress intensity factor* has been introduced [4, 5], Equation 2.3 has been modified from the viewpoint of energy to the following traditional equation which includes π in the expression.

$$\sigma_y = \frac{\sigma_0 \sqrt{\pi a}}{\sqrt{2 \pi r}} = \frac{K_{\mathrm{I}}}{\sqrt{2 \pi r}} \tag{2.4}$$

K_{I} is called the *Stress Intensity Factor* of *Mode I*. As $b \to 0$ in Figure 2.1, the ellipse is reduced to a crack and the stress intensity factor K_{I} is $K_{\mathrm{I}} = \sigma_0 \sqrt{\pi a}$. Thus, K_{I} is an important factor which prescribes the stress field in the vicinity of the crack tip. Equations 2.3 and 2.4 indicate that the stress distribution at a crack tip has the *singularity* of $r^{-0.5}$. It is known that 3D cracks also have the singularity of $r^{-0.5}$ at the crack tip [3].

Equation 2.4 shows the approximate stress distribution near a crack tip. The exact stress distribution is expressed by Equation 2.2 with $x = a + r$. The difference between Equations 2.2 and 2.4 is illustrated as Figure 2.2.

Even if the remote stress $\sigma_x = \sigma_{x \infty}$ is applied in the x direction, it does not influence the singularity $r^{-0.5}$ of Equation 2.4; and accordingly the remote stress $\sigma_{x \infty}$ does not influence the stress intensity factor K_{I}. However, for the discussion of multiaxial fatigue problems, it must be noted that $\sigma_{x \infty}$ influences the stress distribution and yield condition along the x axis.

On the other hand, it must be noted that the remote stress $\sigma_y = \sigma_0$ produces not only the singularity of the stress distribution σ_y near the crack tip but also the singularity of σ_x ahead of the

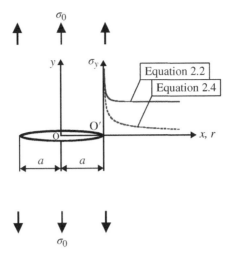

Figure 2.2   Crack in a wide plate under remote stress $\sigma_y = \sigma_0$

crack tip along the x axis. The distribution of σ_x near a crack tip is given by the following equation:

$$\sigma_x = \sigma_y - \sigma_0$$

$$= \frac{\sigma_0 x}{\sqrt{x^2 - a^2}} - \sigma_0 \tag{2.5}$$

2.2   Stress Distribution Near Crack Tip Under Biaxial Stress Field

Using Equations 1.9 to 1.13 and 1.17 to 1.21, the stress invariant $(\sigma_x + \sigma_y)$ and the maximum shear stress τ_{max} on the x axis for the problem of Figure 2.3 are given by the following equation:

$$\sigma_x + \sigma_y = \sigma_1 + \sigma_2$$

$$= \left(\frac{2x}{\sqrt{x^2 - a^2}} - 1\right)\sigma_{y\infty} - \sigma_{x\infty} \tag{2.6}$$

$$\tau_{max} = (\sigma_y - \sigma_x)/2$$

$$= (\sigma_{y\infty} - \sigma_{x\infty})/2 \tag{2.7}$$

The maximum shear stress τ_{max} expressed by Equation 2.7 is the one defined in a plane stress problem and is constant regardless of the value of x. However, if the maximum shear stress is defined as a 3D problem considering the thickness direction, τ_{max} is expressed as follows:

$$\tau_{max} = \frac{\sigma_y - \sigma_z}{2}$$

$$= \frac{\sigma_y}{2} \tag{2.8}$$

$$= \frac{\sigma_{y\infty} x}{2\sqrt{x^2 - a^2}}$$

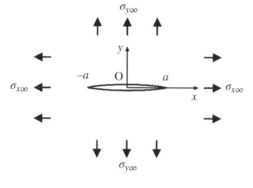

Figure 2.3

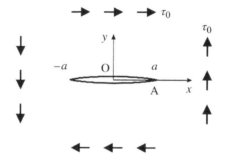

Figure 2.4   Crack in a wide plate under shear stress

The plastic zone size of the so-called the *Dugdale model* [6] (see Section 2.5) is determined by the yielding phenomenon caused by the shear stress based on Equation 2.8.

2.3   Distribution of Shear Stress Near Crack Tip

First, let us consider an elliptical hole under remote shear stress τ_0. By letting $b \to 0$, the elliptical hole becomes increasingly slender and is finally reduced to a crack, as shown in Figure 2.4. As $b \to 0$, the location of τ_{max} approaches the crack tip A (actually reaching point A in the limit, see Figure 1.17). As a result, the shear stress near the crack tip is expressed from Equation 1.26 in the same manner as tensile mode (opening mode, Mode I) by the following equation:

$$\tau_{xy} = \frac{\tau_0 x}{\sqrt{x^2 - a^2}} \tag{2.9}$$

Therefore, the shear stress for $x = a + r (r/a \ll 1)$ is

$$\tau_{xy} \cong \frac{\tau_0 \sqrt{a}}{\sqrt{2r}} = \frac{K_{\mathrm{II}}}{\sqrt{2\pi r}} \tag{2.10}$$

where K_{II} is the stress intensity factor of Mode II, as expressed by

$$K_{\mathrm{II}} = \tau_0 \sqrt{\pi a}.$$

2.4 Short Cracks and Long Cracks

Fracture mechanics was born as the science which treats crack problems with infinite stress concentration at the crack tip based on a new concept of mechanics [4, 5]. As the stress field near the crack tip is prescribed by the *stress intensity factor*, it has been thought that the fracture criterion can be quantitatively given by focusing the stress intensity factor regardless of crack size. On the other hand, it has been also reported that the *fatigue crack growth rate* da/dN (N = number of cycles) and the *threshold stress intensity factor range* for fatigue crack growth ΔK_{th} for cracks shorter than 1 mm have values different from those for long or large cracks [7]. To understand this problem, it is necessary not to simply use the approximate solution with the stress intensity factor as the stress distribution near crack tip but to return to the exact solution expressed by Equation 2.2.

Let us investigate the cases of 2D cracks in a wide plate, where five kinds of combinations of crack length $2a$ and remote stress σ_0 are compared.

1. $a_1 = 10\ \mu\mathrm{m}$, $\sigma_0 = 1000\ \mathrm{MPa}$
2. $a_2 = 100\ \mu\mathrm{m}$, $\sigma_0 = 100\ \sqrt{10}\ \mathrm{MPa}$
3. $a_3 = 1000\ \mu\mathrm{m}$ (=1 mm), $\sigma_0 = 100\ \mathrm{MPa}$
4. $a_4 = 10\ \mathrm{mm}$, $\sigma_0 = 10\ \sqrt{10}\ \mathrm{MPa}$
5. $a_5 = 100\ \mathrm{mm}$, $\sigma_0 = 10\ \mathrm{MPa}$

All these cracks have the same value of K_{I}, that is $K_{\mathrm{I}} = \sigma_0\sqrt{\pi a} = \sqrt{10\pi}$ MPa$\cdot\sqrt{\mathrm{m}}$. If the stress distribution ahead of the crack tip is expressed by K_{I}, we have $\sigma_y = K_{\mathrm{I}}/\sqrt{2\pi r}$ (Equation 2.4) and all the stress distributions for cases 1–5 become the same. However, if the exact solution Equation 2.2 is used, the stress distributions near the crack tip are as illustrated as Figure 2.5.

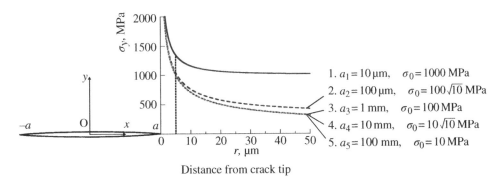

Distance from crack tip

Figure 2.5 Difference in stress distribution near the crack tip for cracks with the same value of K_I and different lengths

As shown in Figure 2.5, although the stress distributions near crack tip for cracks with $a > 1$ mm are almost close, the stress distribution for cracks with $2a_1 = 2 \times 10$ μm is clearly different from that of cracks with $2a_5 = 2 \times 100$ mm (e.g., see the big difference at the crack tip $r \cong 5$ μm).

These differences in stress distribution at 5–10 μm ahead of the crack tip naturally cause a difference in the fatigue crack growth rate da/dN and the fatigue threshold stress intensity factor ΔK_{th}. Based on fracture mechanics analysis and experimental studies, it is well accepted that ΔK_{th} for long or large cracks is a material constant under constant stress ratio. However, for short 2D cracks and small 3D cracks, ΔK_{th} is not a material constant and is smaller than the values for long or large cracks. Namely, the values of ΔK_{th} depend on the size of the crack.

This is fundamental for studying strength problems related to cracks.

2.5 Plastic Zone Ahead of Crack Tip: Dugdale Model

The stress intensity factor for a crack in a wide plate under tensile stress as shown in Figure 2.6 is $K_I = \sigma_0 \sqrt{\pi a}$. The elastic stress distribution is extremely high near a crack tip which is the singular point of stress and the stress at the crack tip becomes infinite.

However, the stress at the crack tip in real materials never becomes infinite, because *yielding* occurs at the crack tip and a *plastic zone* is produced. The situation can be modelled as Figure 2.7. After the plastic zone is produced, the stress singularity vanishes and the stress becomes constant (yield stress σ_Y) inside the plastic zone. The stress outside the plastic zone decreases with distance from the edge of the plastic zone.

The model with a constant yield stress of σ_Y inside the plastic zone, such as Figure 2.7, is called the *Dugdale model* [6]. The yield condition of this model considers the 3D stress condition, including stress in the thickness direction. Namely, σ_y is the maximum among all stresses compared to stresses in other directions, that is $\sigma_x \neq 0$, σ_z (thickness direction) $= 0$. Therefore, the maximum shear stress τ_{max} is $\tau_{max} = \frac{1}{2}\sigma_y$ and the yielding of ductile materials occurs by sliding in the 45° direction against the plate surface (x–y plane).

The equation to determine the plastic zone size R is given by the following equation [6].

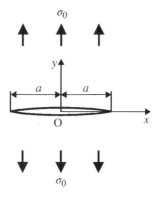

Figure 2.6

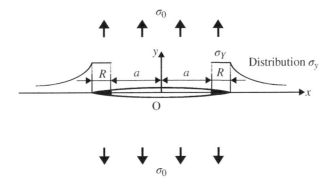

Figure 2.7 Dugdale model

$$\frac{R}{a} = \sec\left(\frac{\pi\sigma_0}{2\sigma_Y}\right) - 1 \tag{2.11}$$

If the remote tensile stress σ_0 is sufficiently small compared to σ_Y, Equation 2.11 can be approximately expressed by the following equation:

$$\frac{R}{a} \cong \frac{1}{2}\left(\frac{\pi}{2}\cdot\frac{\sigma_0}{\sigma_Y}\right)^2 \tag{2.12}$$

The above equation can be rewritten with the stress intensity factor K_I as follows:

$$R \cong \frac{\pi}{8\sigma_Y^2}\cdot K_I^2 \tag{2.13}$$

Equation 2.13 means that, if the applied σ_0 is sufficiently smaller than the yield stress σ_Y, the plastic zone size R is proportional to the square of the stress intensity factor K_I. When the concept of the stress intensity factor was first introduced, the physical meaning of the stress intensity factor was not smoothly accepted due to the unrealistic elastic singular stress distribution near the crack tip. Nevertheless, Equation 2.13 shows the possibility of a correlation between the stress intensity factor and the reality of the phenomena which occur ahead of the crack tip, such as the fatigue crack growth problem.

However, as the applied stress increases and σ_0 becomes close to σ_Y, the approximate approach introduced above does not hold. The importance of this problem is pronouncedly revealed in the case of constant stress intensity factor K_I for different crack sizes, as explained in the previous section (Figure 2.5).

Although Figure 2.5 compares the stress fields near the crack tip for five cracks with different length, Figure 2.8 compares the difference in the plastic zone size R for the identical stress intensity factor and different crack length. It is notable that, as the size of crack becomes smaller with the constant K_I, the plastic zone size increases. Large plastic zone size means large crack tip opening displacement. From such a viewpoint too, the difference between small cracks and large cracks under $K_I = a$ *constant value* can be understood. In other words, the stress state near the crack tip and crack tip opening displacement are not the same for small cracks and large cracks even with the same K_I.

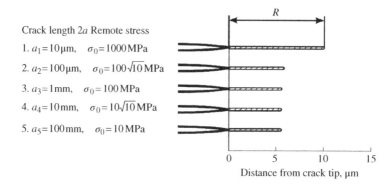

Crack length 2a Remote stress

1. $a_1 = 10\,\mu m$, $\sigma_0 = 1000\,MPa$

2. $a_2 = 100\,\mu m$, $\sigma_0 = 100\sqrt{10}\,MPa$

3. $a_3 = 1\,mm$, $\sigma_0 = 100\,MPa$

4. $a_4 = 10\,mm$, $\sigma_0 = 10\sqrt{10}\,MPa$

5. $a_5 = 100\,mm$, $\sigma_0 = 10\,MPa$

Distance from crack tip, μm

Figure 2.8   Plastic zone size R for $K_I = \sqrt{10\pi}\ MPa\cdot\sqrt{m}$, Yield stress $\sigma_Y = 1500\,MPa$

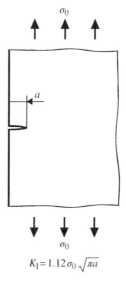

$$K_I = 1.12\sigma_0\sqrt{\pi a}$$

Figure 2.9   Edge crack in a semi-infinite plate [8]

2.6   Approximate Estimation Method of K_I

If we understand the fundamental nature of the stress concentration of holes or notches, it is possible to estimate the stress intensity factor with sufficient accuracy without detailed numerical analysis. Before performing a detailed and troublesome calculation, it is recommended to have a good engineering sense of this kind of estimation. Several basic examples and their applications are shown in the following.

Basic solutions

Basic examples are given in Figures 2.9 to 2.12.

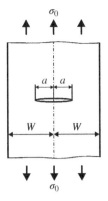

$$K_I = F(\lambda)\sigma_0\sqrt{\pi a}$$
$$F(\lambda) = (1 - 0.025\lambda^2 + 0.06\lambda^4)\sqrt{\sec(\pi\lambda/2)}$$
$$\lambda = a/W$$

Figure 2.10 Center crack in a strip [9]

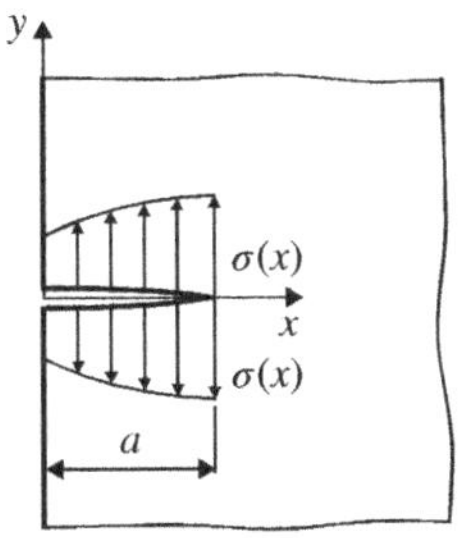

$$\sigma(x) = -\sigma\sum_{n=0}^{3}C_n\left(\frac{x}{a}\right)^n$$

$$K_I = \sigma\sqrt{\pi a}(1.1215C_0 + 0.6829C_1 + 0.5255C_2 + 0.4410C_3)$$

Figure 2.11 Edge crack in a semi-infinite plate under nonlinear pressure distribution [10]

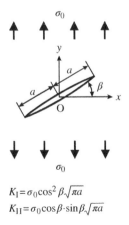

$$K_I = \sigma_0\cos^2\beta\sqrt{\pi a}$$
$$K_{II} = \sigma_0\cos\beta\cdot\sin\beta\sqrt{\pi a}$$

Figure 2.12 Inclined crack

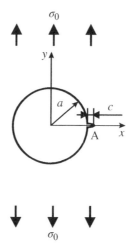

Figure 2.13   A short crack emanating from a circular hole $(c \ll a)$. [Answer: $K_{\mathrm{I}} \cong 1.12 \times 3\sigma_0 \sqrt{\pi c}$]

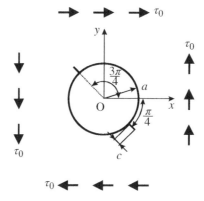

Figure 2.14   A short crack emanating from a circular hole under shear stress $(c \ll a)$. [Answer: $K_{\mathrm{I}} \cong 1.12 \times 4\tau_0 \sqrt{\pi c}$]

Application problems

Estimate the stress intensity factors for cracks in Figures 2.13 to 2.16.

2.7   Crack Propagation Path

In which direction does a crack propagate in structures under external loading? It is most natural to think that the new crack will start propagation from the tip of the original crack. However, the direction of the new crack is not necessarily self-evident. The stress distribution in the vicinity of a crack tip and the material properties influence the direction of crack propagation. Based on

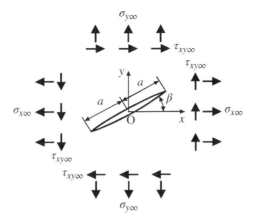

Figure 2.15 An inclined crack in a wide plate under remote stres, $\sigma_{x\infty}$, $\sigma_{y\infty}$, $\tau_{xy\infty}$

$$\text{Answer :} \begin{bmatrix} K_{\mathrm{I}} = \left(\sigma_{x\infty}\sin^2\beta + \sigma_{y\infty}\cos^2\beta - 2\tau_{xy\infty}\cos\beta\cdot\sin\beta\right)\sqrt{\pi a} \\ K_{\mathrm{II}} = \left(\sigma_{y\infty} - \sigma_{x\infty}\right)\cos\beta\cdot\sin\beta + \tau_{xy\infty}\left(\cos^2\beta - \sin^2\beta\right)\sqrt{\pi a} \end{bmatrix}$$

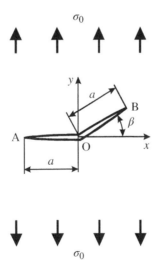

Figure 2.16 A kinked crack in a wide plate under remote stress σ_0 in the y direction

$$\text{Answer :} \begin{bmatrix} K_{\mathrm{IA}} \cong \sigma_0\sqrt{\pi a(1+\cos\beta)/2},\ K_{\mathrm{IIA}} \cong 0 \\ K_{\mathrm{IB}} \cong \sigma_0\cos^2\beta\sqrt{\pi a(1+\cos\beta)/2},\ K_{\mathrm{IIB}} \cong \sigma_0\cos\beta\cdot\sin\beta\sqrt{\pi a(1+\cos\beta)/2} \end{bmatrix}$$

experience obtained by many experiments, a crack propagates mostly by *Mode I* (opening mode), rarely by *Mode II* (in-plane shear mode) or *Mode III* (out of plane shear mode).

The way of thinking about the crack propagation direction will be discussed in the following. The basic equations to be used for the calculation are summarized in Appendix A.2 of Part II.

2.7.1 Propagation Direction of Mode I Crack

Several criteria have been proposed to predict the *direction of crack propagation*. If the material in question can be assumed to be homogeneous, isotropic and brittle, it is reasonable to presume that the Mode I fracture occurs by maximum tensile stress at the crack tip. In other words, the new crack propagates from the tip of the initial crack to a direction perpendicular to the direction of the *maximum tangential stress* $\sigma_{\theta max}$ in the vicinity of the initial crack [11] (Figure 2.17). This criterion may not be strictly applied to a fatigue crack in which more microscopic analysis on the scale of atoms or dislocations is necessary. Nevertheless, if we look at the direction of crack propagation on a more macroscopic scale, the eventual result is more or less not different from the conclusion derived from the same criterion.

In general, the direction of crack propagation is discussed in terms of the angle observed on the surface of material. If this problem is discussed from the 3D viewpoint, we need to pay attention to the *special singularity of the stress distribution* near the point where the crack front meets the material free surface (Figure 2.18). This special point is called the *corner point*. It is known that the singularity of the stress distribution near the corner point is different from that inside material [2–19]. When a 3D crack has K_{II} and a singularity of Mode II exists near crack tip, a singularity of Mode III is generated near the corner point in a coupled way together with a Mode II singularity.

The problem of the *corner point singularity* is ignored in most discussions of crack propagation direction. Concluding this problem first, various detailed discussions on the direction of crack propagation based on various criteria do not create fruitful results for the purpose of practical applications. Thus, the maximum tangential stress $\sigma_{\theta max}$ criterion [11] as the simplest one

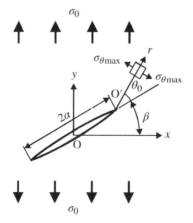

Figure 2.17   Direction of crack propagation by the $\sigma_{\theta max}$ criterion

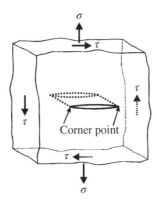

Figure 2.18 Corner point (cross point with crack front and free surface). K_{III} is generated in a coupled way with K_{II}

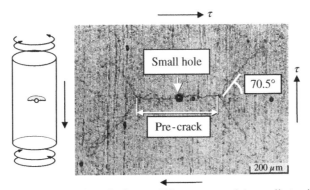

Branched cracks from a 400 μm pre-crack by cyclic torsion
0.45% C steel, $\tau = 152$ MPa, $N = 4.0 \times 10^5$

Figure 2.19 Crack branching from a semi-elliptic surface crack under torsional fatigue. (Due to $K_I = 0$ and $K_{II} \neq 0$, the branch angle is calculated as $\theta_0 = \pm 70.5°$ by Equation 2.14. Although the loading is torsion (shear stress), the crack propagation type is Mode I.)

and the $K_{II} = 0$ criterion [20–27] are intuitively clear and practical for the prediction of crack propagation direction for Mode I (opening mode) fatigue crack. According to the $\sigma_{\theta max}$ criterion [11], the direction θ_0 of crack propagation against the direction of the initial crack is given by the following equation (see Figure 2.17):

$$\tan \frac{\theta_0}{2} = \frac{1 \pm \sqrt{1 + 8\gamma^2}}{4\gamma} \tag{2.14}$$

where $\gamma = K_{II}/K_I$.

Crack propagation by Mode II must be carefully discussed, because even if the loading mode is Mode II, crack propagation is mostly Mode I. In such cases, although the value of K_{II} is discussed, the fracture mode itself is Mode I as shown in Figure 2.19.

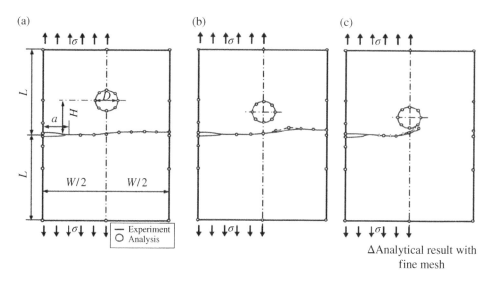

Figure 2.20   Effect of a circular hole on the crack propagation path [28]. $a/W = 0.2$, $L/W = 0.75$, $D/a = 1.0$. (a) $H/a = 1.5$. (b) $H/a = 1.0$. (c) $H/a = 0.75$. Source: T. Norikura and Y. Murakami 1983 [28]. Reproduced with permission of The Japan Society of Mechanical Engineers

$H = 20$ mm, $a = 19.24$ mm,
$W = 100$ mm, $L = 75$ mm,
$D = 20$ mm, $t = 5$ mm

Figure 2.21   Experimental result of Figure 2.20b [28]. Material: acryl plate. Crack propagates as if it is attracted by the hole. Source: T. Norikura and Y. Murakami 1983 [28]. Reproduced with permission of The Japan Society of Mechanical Engineers

Figures 2.20 to 2.23 show the analytical and experimental studies of the effects of hole and rigid inclusion on the crack propagation path in a brittle material [28]. Two general characteristics of the crack propagation direction are summarized as follows.

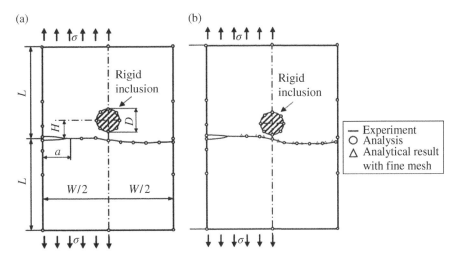

Figure 2.22 Effect of rigid inclusion on crack propagation path [28]. Experiment and analysis. $a/W = 0.2$, $L/W = 0.75$, $D/a = 1.0$, $n = 0.3$. (a) $H/a = 0.75$. (b) $H/a = 0.5$. Source: T. Norikura and Y. Murakami 1983 [28]. Reproduced with permission of The Japan Society of Mechanical Engineers

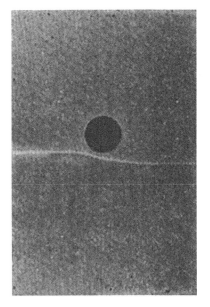

$H = 10\,\text{mm}$, $a = 19.70\,\text{mm}$, $W = 100\,\text{mm}$,
$L = 75\,\text{mm}$, $D = 20\,\text{mm}$, $t = 5\,\text{mm}$

Figure 2.23 Experimental result of Figure 2.22c [28]. Material: acryl plate. Inclusion: 0.45% C circular steel plate. Crack propagates by avoiding the inclusion with high rigidity. Source: T. Norikura and Y. Murakami 1983 [28]. Reproduced with permission of The Japan Society of Mechanical Engineers

Mode II crack propagation   Branding by Mode I cracks

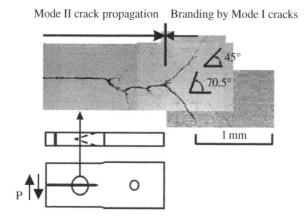

Figure 2.24   Mode II crack growth test. JIS S45C, ΔP = 11.8 kN. Mode II crack and branching to ±70.5° by Mode I [29]. Source: Y. Murakami, K. Takahashi and R. Kusumoto 2003 [29]. Reproduced with permission of The Japan Society of Mechanical Engineers

1. Crack propagates into the domain with lower rigidity such as hole.
2. Crack propagates by avoiding the domain with higher rigidity in structures.

2.7.2   *Propagation Direction of Mode II Crack*

If the loading mode is Mode II and the crack propagation mode is also Mode II, the crack propagates in the direction of the maximum shear stress. Although it is very difficult to achieve Mode II fatigue crack growth by laboratory test (see Figure 2.24) [29], many actual failure cases by Mode II fatigue have been experienced in railroad rails, steel making rolls, bearings, gears and so on. Even if K_I = 0 and only the stress intensity factor of Mode II, K_{II}, exists, there is a possibility of crack branching by Mode I in the direction of ±70.5°, as calculated by Equation 2.14. In Figures 2.24 and 2.25, as the result of competition between Mode II and Mode I crack growth, Mode II crack growth prevails before final branching by Mode I.

2.7.3   *Propagation Direction of Mode III Crack*

A Mode III fatigue fracture is very rare even in actual cases. Axles and circumferentially notched cylinders which are subjected to even Mode III loading finally fracture by zigzag growth of a Mode I crack.

Figure 2.26 shows the failure case of a ship component fractured by Mode III fatigue[1].

Figure 2.27 shows the fracture surface of a cylindrical specimen with an initial circumferential crack failed by torsional fatigue. At the beginning, the crack grows by Mode III from an initial pre-crack and afterwards the crack grows by Mode I by forming a *factory roof* morphology. Knowledge about the mechanism of forming this kind of fracture morphology will be useful for failure analysis.

---

[1] Courtesy of Nippon Kaiji Kyokai.

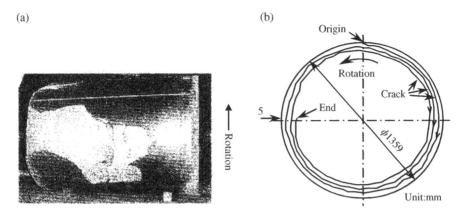

Figure 2.25 Mode II crack propagated almost three circles below surface by cyclic contact load [30]. (a) Fractured back up roll. Subsurface crack propagation. Source: Y. Ohkomori, C. Sakae and Y. Murakami 2001 [30]. Reproduced with permission of The Society of Materials Science, Japan

Figure 2.26 Ship component fractured by Mode III crack propagation. Courtsey of Nippon Kaiji Kyokai

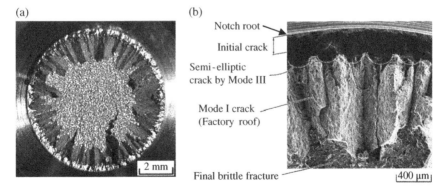

Figure 2.27 Mode III fatigue crack growth and the formation of factory roof morphology by torsional fatigue of a circumferentially notched specimen [29]. 0.45% C steel, $\tau_a = 132$ MPa, $\Delta K_{III} = 11.5$ MPa $\sqrt{m}$ (a) Fatigue fracture surface of pre-cracked specimen. (b) Magnification of (a). Source: Y. Murakami, K. Takahashi and R. Kusumoto 2003 [29]. Reproduced with permission of John Wiley & Sons

Problems of Chapter 2

1. Explain the reason why the remote stress $\sigma_x = \sigma_{x\infty}$ in Figure 2.2 does not influence the stress intensity factor for the crack.

2. (1) Calculate the following integral in terms of Figure 2.28 and explain the meaning of the integral.

$$\int_a^\infty \left[\frac{\sigma_0 x}{\sqrt{x^2 - a^2}} - \sigma_0 \right] dx = ?$$

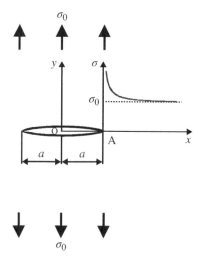

Figure 2.28   Crack in an infinite plate under remote stress $\sigma_y = \sigma_0$ (plate thickness: 1)

(2) Calculate the resultant force corresponding to the integral on σ_x from $y = 0$ to ∞ along the y axis and explain the meaning of the integral.

$$\int_0^\infty \sigma_x dy = ?$$

3. When we denote the stress concentration factor for a notch having the same notch root radius as a part of an elliptic hole by K_t, the notch is reduced to a crack by letting the notch root radius $\rho \to 0$. It is known that the stress intensity factor in this case can be determined by the following equation [31]:

$$K_I = \frac{1}{2} \sqrt{\pi} \lim_{\rho \to 0} \sqrt{\rho} \sigma_{max}$$

$$= \frac{1}{2} \sqrt{\pi} \lim_{\rho \to 0} \sqrt{\rho} K_t \sigma_0$$

Using this equation and the stress concentration factor for a V-notch given in Table 1.1 of Chapter 1, estimate the approximate stress intensity factor for the edge crack of the semi-infinite plate of Figure 2.9 and compare the estimated value with the exact value.

4. If the stress concentration factor K_t for a notch is expressed as a function of the notch root radius ρ, the stress intensity factor K_I can be determined by the following equation [31]:

$$K_I = \frac{1}{2}\sqrt{\pi} \lim_{\rho \to 0} \sqrt{\rho} K_t \sigma_n$$

where ρ_n is the nominal stress defined at the net section. Namely,

$$\sigma_n = \frac{W}{W-a}\sigma$$

In Figure 2.29a, we have (see Figure 1.22):

$K_t = 5.55$ for $a/W = 0.5$, $a/\rho = 16$
$K_t = 10.27$ for $a/W = 0.5$, $a/\rho = 64$

Using these values, estimate K_I for the crack of Figure 2.29b.

(a)   (b)

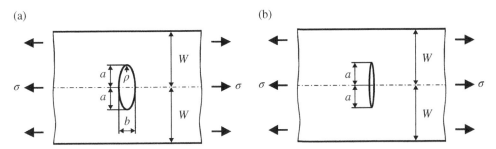

Figure 2.29   (a) Elliptic hole in a strip under tensile stress. (b) Crack in a strip under tensile stress ($a/W = 0.5$)

5. Estimate the stress intensity factor K_I for the crack in Figure 2.30.

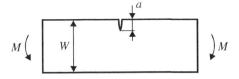

Figure 2.30   Bending of a plate containing a short edge crack. $a \ll W$, plate thickness $= t$

6. When an elliptical hole exists in a pure shear stress field, as in Figure 2.31, the maximum stress occurs at the contact point between the elliptical hole and straight lines with a $\pm 45°$ gradient. In this condition, cracks initiate and grow in the $\mp 45°$ direction (see Section 1.2.2). However, in the case of a crack in a pure shear stress field, as in

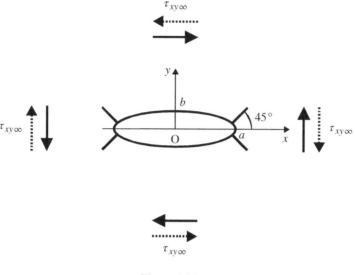

Figure 2.31

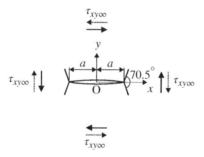

Figure 2.32

Figure 2.32, the crack first branches from the initial crack tip to the $\pm 70.5°$ direction, as shown in Figure 2.19, and then grows. As the minor axis b of the ellipse is infinitesimally reduced to 0 ($b \to 0$), the shape of the ellipse approaches to a crack. It seems that there is a contradiction for the directions of crack initiation, branching and growth in these two cases. Explain the reason for the apparent contradiction and then explain the reason why there is actually no contradiction.

7. Figure 2.20a–c and Figure 2.21 show the analytical and experimental results of a crack propagation path in a plexiglass plate which contains a circular hole and an edge crack. By tensile loads applied at the upper and lower ends, the crack grows along a path influenced by the presence of the hole. Explain the reason why the crack path is curved by the presence of the hole and is likely attracted by the hole.

Figure 2.22a, b and Figure 2.23 show the analytical and experimental results of a crack propagation path in a plexiglass plate which contains a steel disk fitted and bonded in the plate and an edge crack. By tensile loads applied at the upper and lower ends, the crack grows along a path influenced by the presence of the steel disk. Explain the reason why the crack path is curved by the presence of the steel disk and likely is avoiding the steel disk.

References

[1] A. A. Griffith (1920) The phenomena of rupture and flaw in solids, *Phil. Trans. R. Soc. Lond. A* **221**:163–198.

[2] H. M. Westergaad (1939) Bearing pressures and cracks, *Trans. ASME, J. Appl. Mech. A* **66**:49.

[3] I. N. Sneddon (1946) The distribution of stress in the neighbourhood of a crack in an elastic solid, *Proc. R. Soc. Lond. A* **187**:229–260.

[4] G. R. Irwin (1957) Analysis of stresses and strains near the end of a crack traversing a plate, *Trans. ASME, J. Appl. Mech.*, **24**:361–364.

[5] G. R. Irwin (1958) Fracture, in *Handbuch der Physik*, vol. **6**, Springer, Berlin, pp.551–590.

[6] D. S. Dugdale (1960) Yielding of steel sheets containing slits, *J. Mech. Phys. Solids*, **8**:100–104.

[7] H. Kitagawa and S. Takahashi (1979) Fracture mechanics approach to very small fatigue crack growth and to the threshold condition, *Trans. Jpn Soc. Mech. Eng. A*, **45**:1289–1303.

[8] W. T. Koiter (1965) Rectangular tensile sheet with symmetric edge cracks, *Trans. ASME, J. Appl. Mech. Ser. E*, **32**:237.

[9] H. Tada (1971) A note on the finite width corrections to the stress intensity factor, *Eng. Fract. Mech.*, **3**:345–347.

[10] M. P. Stallybrass (1970) A crack perpendicular to an elastic half plane, *Int. J. Eng. Sci.*, **8**351–362.

[11] F. Erdogan and G. C. Sih (1963) On the crack extension in plates under plate loading and transverse shear, *J. Basic Eng. Trans. ASME Ser. D*, **85**:519–527.

[12] J. P. Benthem (1977) State of stress at the vertex of a quarter-infinite crack in a half-space, *Int. J. Solids Struct.*, **13**:479–492.

[13] J. P. Benthem (1980) The quarter-infinite crack in a half space: alternative and additional solutions, *Int. J. Solids Struct.*, **16**:119–130.

[14] L. P. Pook (1992) A note on corner point singularities, *Int. J. Fract.*, **53**:R3–R8.

[15] L. P. Pook (1994) Some implications of corner point singularities, *Eng. Fract. Mech.*, **48**:367–378.

[16] G. Dhondt (1998) On corner point singularities along a quarter circular crack subject to shear loading, *Int. J. Fract.*, **89**:L33–L38.

[17] Z. P. Bazant and L. F. Estenssoro (1979) Surface singularity and crack propagation, *Int. J. Solids Struct.*, **15**:405–426.

[18] Z. P. Bazant (1974) Three-dimensional harmonic functions near termination or intersection or gradient singularity lines: a general numerical method, *Int. J. Eng. Sci.*, **12**:221–243.

[19] Y. Murakami and H. Natsume (2002) Stress singularity at the corner point of 3-D surface crack under mode II loading, *Int. J. Jpn. Soc. Mech. Eng. A* **45**:161–169.

[20] N.V. Banichuk (1970) *Izv. An. SSR MTT*, **7**:130–137.

[21] R.V. Goldstein and R.L. Salganik (1970) *Izv. An SSR MTT* **7**:69–82.

[22] R.V. Goldstein and R.L. Salganik (1974) *Int. J. Fract.* **10**:507–523.

[23] H. Kitagawa, R. Yuuki, K. Togho, M. Tanabe (1985) *ASTM STP* **853**:164–183.

[24] B. Cotterell and J.R. Rice (1980) Slightly curved or kinked cracks, *Int. J. Fract.*, **16**:155–169.

[25] Y. Sumi, S. Nemat-Nasser, L. M. Keer (1983) On crack branching and curving in a finite body, *Int. J. Fract.*, **21**67–79; Y. Sumi, S. Nemat-Nasser, L. M. Keer (1984) Erratum, *Int. J. Fract.*, **24**:159.

[26] Y. Sumi, S. Nemat-Nasser, L. M. Keer (1985) On crack path stability in a finite body, *Eng. Fract. Mech.*, **22**:759–771.

[27] Y. Sumi (1985) Computational crack path prediction, *Theor. Appl. Fract. Mech.*, **4**:149–156.

[28] T. Norikura and Y. Murakami (1983) Solution of two dimensional mixed mode boundary problems by the body force method and applications to crack problems, *Trans. Jpn Soc. Mech. Eng. A*, **49**:818–828.

[29] Y. Murakami, K. Takahashi, R. Kusumoto (2003) Threshold and growth mechanism of fatigue cracks under mode II and III loadings, *Fatigue Fract. Eng Mat. Struct.*, **26**:523–531.

[30] Y. Ohkomori, C. Sakae, Y. Murakami (2001) Analysis of mode II crack growth behavior in spalling failure of backup roll, *J. Soc. Mat. Sci. Jpn*, **50**:249–254.

[31] M. Isida (1976) Elastic stress analysis of cracks (in Japanese), Vol. 2 of Lecture Series of *Fracture Mechanics and Strength of Materials*, Baifukan, 56, Tokyo.

3

Stress Concentration in Three Dimensional Problems

Finding analytical solutions for three dimensional (3D) problems is much more difficult compared to 2D problems. Existing exact solutions are extremely limited to a few special problems. Although nowadays numerical solutions with practically sufficient accuracy can be obtained by the finite element method (FEM), we can learn useful important factors from these exact solutions.

3.1 Stress Concentration at a Spherical Cavity

We assume a *spherical cavity* which exists in an infinite body under remote uniaxial tensile stress. In the problem of Figure 3.1, stress σ_z on the x–y plane $(z = 0)$ is given by the following equation [1]:

$$\sigma_z = \sigma_0 \left[1 + \frac{4-5\nu}{2(7-5\nu)} \frac{a^3}{r^3} + \frac{9}{2(7-5\nu)} \frac{a^5}{r^5} \right] \tag{3.1}$$

If we compare the above equation with stress σ_x on the y axis (Equation 1.2 and Figure 1.1), we note that the constant stress terms are σ_0 both in 2D and 3D problems, but the terms related to the distance from the center of the circular hole and spherical cavity which control the shape of the stress distribution are different. The stress distribution in the 2D problem decays in terms of $1/r^2$ and $1/r^4$. In contrast, the stress distribution of the 3D problem decays in terms of $1/r^3$ and $1/r^5$. This nature of stress distribution means that the influence of 3D stress concentration decays faster with distance than that of 2D problems. In other words, the interaction effect

Theory of Elasticity and Stress Concentration, First Edition. Yukitaka Murakami.
© 2017 John Wiley & Sons, Ltd. Published 2017 by John Wiley & Sons, Ltd.

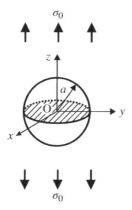

Figure 3.1

in 3D stress concentration is weaker than that of 2D stress concentration. For example, when another cavity with the same size exists in the neighborhood of the cavity of Figure 3.1, the interaction effect is smaller than the interaction effect for the case of two 2D circular holes (see Chapter 4 of Part II).

Assuming a spherical cavity as a globe, maximum stress σ_{max} occurs on the equator and the value and the stress concentration factor K_t are expressed as follows [1]:

$$\sigma_{max} = \frac{27 - 15\nu}{2(7 - 5\nu)}\sigma_0 \tag{3.2}$$

$$K_t = \frac{27 - 15\nu}{2(7 - 5\nu)} \tag{3.3}$$

Depending on Poisson's ratio, σ_{max} has the following values:

$\sigma_{max} = 1.929\sigma_0$ for $\nu = 0$.
$\sigma_{max} = 2.045\sigma_0$ for $\nu = 0.3$.

These values are much smaller than $K_t = 3$ for a 2D circular hole. However, it must be noted that the stress concentration in 3D problems is influenced by Poisson's ratio, though the effect is small.

3.2   Stress Concentration at a Spherical Inclusion

The stress concentration of this problem is influenced by Young's modulus E_M and Poisson's ratio ν_M of the microstructure and also E_I and ν_I of the inclusion (Figure 3.2). The location of the maximum stress is dependent on the combination of (E_M, ν_M) and (E_I, ν_I).

In the case of $E_I < E_M$, the maximum stress occurs at the interface of the equator (at the side of the matrix).

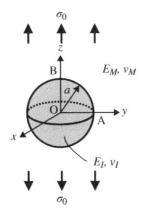

Figure 3.2   Spherical inclusion

But, in the case of $E_I > E_M$, the maximum stress occurs near the pole (Point B). For example, in the case of a rigid inclusion ($E_I/E_M = \infty$), the stress σ_z at Point B at the interface is $\sigma_z = 1.938\sigma_0$ for $\nu_I = 0.3$ and $\nu_M = 0.3$. The value of σ_z at point B is the same for both the matrix and the inclusion at the interface and a little smaller than the maximum stress $2.046\,\sigma_0$ at the equator for a spherical cavity.

Since the solutions for inclusion problems are not given in the closed form, the results of the numerical analysis by FEM are given for reference in Table 3.1. These solutions will be useful for the discussion of crack initiation and growth from a *spherical inclusion* (Figures 1.36–1.38).

3.3   Stress Concentration at an Axially Symmetric Ellipsoidal Inclusion

Since the solution for the problem of an *axially symmetric ellipsoidal inclusion* of Figure 3.3 is not given in an easily useful closed form, the numerical solutions by FEM are shown in Table 3.1.

$\sigma_x = \sigma_y$ inside the inclusion and σ_y at point A of the matrix is equal to σ_x at point C. Another nature to be noted is that σ_z in the inclusion is equal to σ_z at the matrix [2].

Naturally, the stress concentration is smaller for the case that Young's modulus E_I of the inclusion is closer to the Young's modulus E_M of the matrix.

3.4   Stress Concentration at a Surface Pit

The stress concentration due to surface *defects* in machine components and structures such as corrosion pits, surface defects of cast irons and other machining defects can be discussed as the problem of a small surface pit. If the size of pit is *sufficiently small* compared to the size of components and structures, the stress concentration can be approximated by that of a surface pit existing on the surface of a semi-infinite body. The definition of sufficiently small may be regarded to be the order of 1/10 of the representative size of the structure where the pit exists.

Several examples are summarized in Figures 3.4 to 3.7.

Table 3.1 Stress concentration factor of spherical and ellipsoidal inclusion (the shape of inclusion is axially symmetric with respect to the z axis). (E_M = 206 GPa, $\nu_I = \nu_M = 0.3$, perfect bonding)

E_I/E_M	Location	Stress	b/a				
			0.1	0.5	1.0	2.0	10.0
0	Equator A	σ_x/σ_0	3.499	0.470	0.136	0.018	-0.005
		σ_y/σ_0	0.000	0.000	0.000	0.000	0.000
		σ_z/σ_0	13.520	3.313	2.046	1.440	1.044
	Pole B	σ_x/σ_0	-0.795	-0.747	-0.681	-0.591	-0.454
		σ_y/σ_0	-0.795	-0.747	-0.681	-0.591	-0.454
		σ_z/σ_0	0.000	0.000	0.000	0.000	0.000
0.5	Equator A	σ_x/σ_0	0.260	0.100	0.036	0.005	-0.002
		σ_y/σ_0	0.166	0.063	0.023	0.003	-0.001
		σ_z/σ_0	1.823	1.544	1.348	1.181	1.022
	Pole B	σ_x/σ_0	-0.074	-0.209	-0.246	-0.247	-0.221
		σ_y/σ_0	-0.074	-0.209	-0.246	-0.247	-0.221
		σ_z/σ_0	0.947	0.785	0.679	0.591	0.516
0.9	Equator A	σ_x/σ_0	0.035	0.014	0.005	0.001	0.000
		σ_y/σ_0	0.033	0.013	0.005	0.001	0.000
		σ_z/σ_0	1.103	1.078	1.055	1.032	1.004
	Pole B	σ_x/σ_0	-0.011	-0.032	-0.040	-0.043	-0.043
		σ_y/σ_0	-0.011	-0.032	-0.040	-0.043	-0.43
		σ_z/σ_0	0.994	0.971	0.950	0.929	0.914
1.1	Equator A	σ_x/σ_0	-0.031	-0.012	-0.004	-0.001	0.000
		σ_y/σ_0	-0.032	-0.012	-0.005	-0.001	0.000
		σ_z/σ_0	0.912	0.932	0.950	0.970	0.996
	Pole B	σ_x/σ_0	0.010	0.029	0.036	0.041	0.044
		σ_y/σ_0	0.010	0.029	0.036	0.041	0.044
		σ_z/σ_0	1.005	1.026	1.045	1.067	1.108

Matrix: Young's modulus E_M
Poisson's ratio ν_M
Inclusion: Young's modulus E_I
Poisson's ratio ν_I

The shape made by rotation of an ellipse with major axis $2a$ and minor axis $2b$ with respect to z-axis.

Example of mesh pattern for analysis (b/a = 0.5)

Magnification

5	Equator A	σ_x/σ_0	-0.459	-0.128	-0.051	-0.010	0.006
		σ_y/σ_0	-0.845	-0.235	-0.094	-0.019	0.010
		σ_z/σ_0	-0.085	0.175	0.293	0.444	0.856
	Pole B	σ_x/σ_0	0.182	0.392	0.539	0.768	1.483
		σ_y/σ_0	0.182	0.392	0.539	0.768	1.483
		σ_z/σ_0	1.023	1.281	1.627	2.250	4.320
10	Equator A	σ_x/σ_0	-0.628	-0.149	-0.060	-0.013	0.011
		σ_y/σ_0	-1.294	-0.306	-0.123	-0.026	0.019
		σ_z/σ_0	-0.398	0.015	0.129	0.257	0.727
	Pole B	σ_x/σ_0	0.261	0.481	0.670	1.027	2.821
		σ_y/σ_0	0.262	0.481	0.670	1.027	2.813
		σ_z/σ_0	1.014	1.328	1.768	2.669	7.287
∞	Equator A	σ_x/σ_0	-0.958	-0.171	-0.069	-0.016	0.037
		σ_y/σ_0	-2.236	-0.398	-0.162	-0.037	0.072
		σ_z/σ_0	-0.958	-0.171	-0.069	-0.016	0.038
	Pole B	σ_x/σ_0	0.423	0.591	0.831	1.405	9.932
		σ_y/σ_0	0.423	0.591	0.831	1.406	9.933
		σ_z/σ_0	0.987	1.378	1.938	3.280	23.174

- Element: 8 nodes, a quarter axisymmetric mesh model.
- Error: less than 1%
- 200 elements at the inclusion–matrix interface
- Circular inclusion: uniformly divided elements.
- Elliptic inclusion: similar to a circular inclusion with respect to root radius ρ.

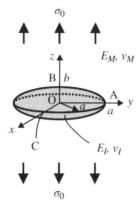

Figure 3.3   Ellipsoidal inclusion. The shape of the inclusion is axially symmetric with respect to the z axis

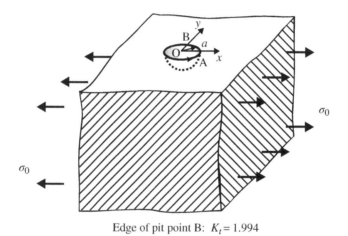

Edge of pit point B: $K_t = 1.994$

Figure 3.4   Semi-spherical pit [3] ($\nu = 0.3$)

In the case of the *spherical pit* of Figure 3.4, the fatigue crack in tension compression or rotating/bending initiates mostly at the bottom of the pit. In some rare cases, cracks initiate at the edge of pit due to the scatter of microstructure and slight irregularities of shape of the edge of pit.

Tension compression and torsional fatigue cracks at a *drilled hole* like Figure 3.6 mostly initiate at the edge of hole. If the bottom of the drilled hole is very sharp and the root radius ρ is 0, the bottom becomes a singular point of stress and the stress at the bottom of the hole becomes unbounded. However, a fatigue crack does not initiate from the bottom of a hole with an aspect ratio larger than $h = a$. The reason will be

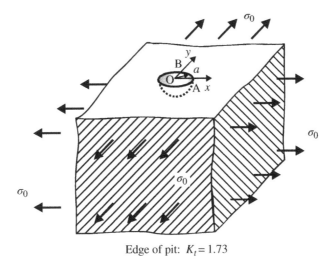

Edge of pit: $K_t = 1.73$

Figure 3.5 Semi-spherical pit [3, 4] under biaxial tension $(\nu = 0.3)$

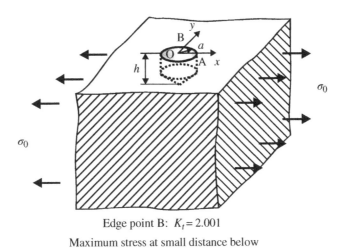

Edge point B: $K_t = 2.001$

Maximum stress at small distance below

Figure 3.6 Drilled hole under uniaxial tension [5] $(\nu = 0.3)$. $h/a = 1.0$, top angle of pit $= 120°$

explained in detail in Chapter 5 of Part II. The simple explanation is that the region of high stress is limited to a very small region of the bottom of the hole. On the other hand, the region of stress of the order of $K_t \cong 2.0$ at the edge of the hole is relatively larger than that of the bottom of the hole.

The stress concentration factor at the edge of pits and drilled holes is mostly about $K_t = 2.0$ regardless of the size. However, since the size of the region exposed to high stress influences

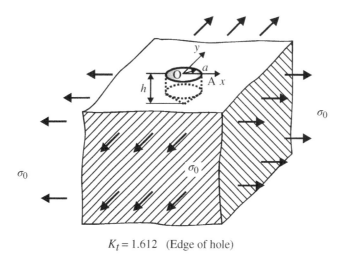

$K_t = 1.612$ (Edge of hole)

Figure 3.7   Drilled hole under biaxial tension [4] ($\nu = 0.3$). $h/a = 1.0$, top angle of pit = $120°$

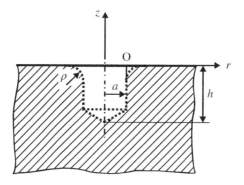

Figure 3.8   Pit with rounded corner edge

the fatigue strength, the size of a pit or drilled hole influences the fatigue strength through the size of the region of high stress around the pit or hole (see Chapter 5 of Part II).

The edge of a hole (e.g. Figures 3.6 and 3.7) is usually perpendicular to the surface of a semi-infinite body. If the corner of the edge is rounded by machining, the stress concentration at the edge of the hole changes. It is possible to decrease the stress concentration by *rounding* or *coning* the edge of the hole, as in Figures 3.8 and 3.9.

Although rounding or coning decreases the stress concentration at the edge of a hole, the stress a little below the surface does not decrease and a large improvement in fatigue strength cannot be expected simply by rounding or coning. From the viewpoint of fatigue strength, it is worth noting that removing small *burrs* at the edge of a hole has a large effect of improvement (Figures 3.10 and 3.11).

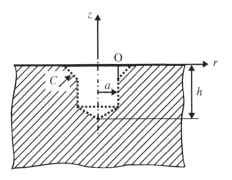

Figure 3.9 Pit with cone forming

(a)   (b)   (c)

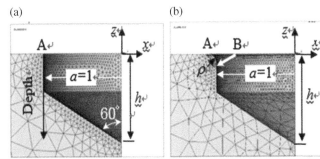

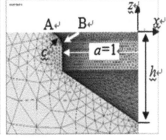

Point A at edge: $K_t = 1.990$.
Numerical analysis 2.001.
Maximum value occurs at a
small distance below edge:
2.029. Numerical analysis: 2.03.

Point A at edge: $K_t = 1.656$
Point B: 2.054.
Maximum value occurs at a
small distance below edge:
2.054.

Point A at edge: $K_t = 1.680$.
Point B: 2.058.
Maximum value occurs at a
small distance below edge:
2.071.

(d)

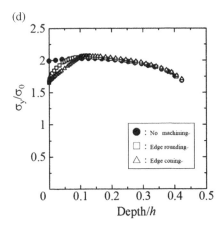

Figure 3.10 Comparison of stress distribution and concentration for different machining. Stress acts in the y axis direction. (a) Without forming; $h/a = 1.0$. (b) Round forming; $\rho/a = 0.1$, $h/a = 1.0$. (c) Cone forming $c/a = 0.1$, $h/a = 1.0$. (d) Stress distribution along the depth of hole

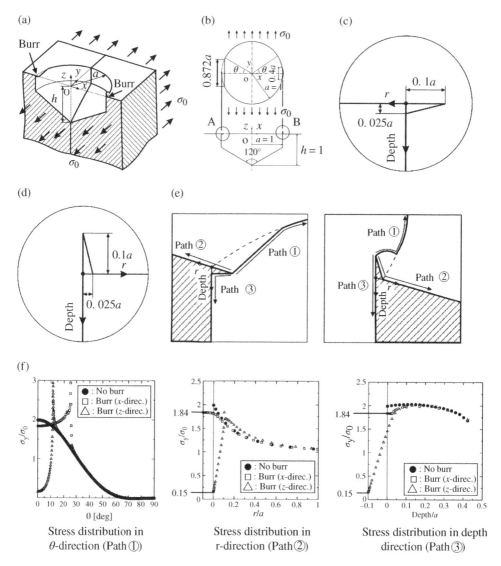

Figure 3.11   Comparison of stress concentration factors with and without burr. (a) Analytical model (1/2 model). (b) Shape of drilled hole. (c) Magnification of the burr (x direction) of A in (b). (d) Magnification of the burr (z direction) of B in (b). (e) The arrowed paths where stress distribution was presented. (f) FEM analysis

3.5   Stress Concentration at a 3D Crack

3.5.1   Basic Problems

The *singularity of the stress distribution* near the crack tip for 3D cracks is also $r^{-0.5}$, the same as 2D cracks. Therefore, the stress field near crack tip can be prescribed by analysis of the *stress intensity factor*.

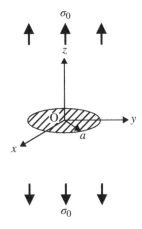

Figure 3.12   Penny shaped crack

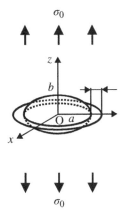

Figure 3.13   Crack emanating from axisymmetric ellipsoidal cavity

Several examples of stress intensity factors for 3D cracks will be shown in the following. K_I for a *penny shaped crack* of Figure 3.12 is given by the following equation [6]:

$$K_I = \frac{2}{\pi}\sigma_0\sqrt{\pi a} \tag{3.4}$$

This value is smaller than a 2D crack with length $2a$ by $2/\pi$ (63.7%).

K_I for a very small crack with length $c(c \ll a)$ emanating from a axially symmetric ellipsoidal cavity (like Figure 3.13) has the following value (see Figure 2.13):

$$K_I = 1.12 K_t \sigma_0 \sqrt{\pi c} \tag{3.5}$$

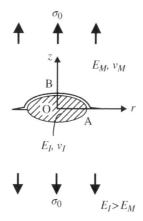

Figure 3.14   Fatigue crack emanating from interface at ellipsoidal inclusion

where K_t is the stress concentration factor for the axially symmetric ellipsoidal cavity. If the crack length c satisfies the condition $c > a/4$, the value of K_I can be approximately estimated by the following equation [7]:

$$K_I \cong \frac{2}{\pi}\sigma_0\sqrt{\pi(a+c)} \tag{3.6}$$

In the case of hard inclusions, fatigue cracks initiate at the pole B (Figure 3.14) followed by debonding of the interface (Figure 1.36) and finally become equivalent to a penny shaped crack.

Although the analysis of 3D cracks is far more difficult, nowadays many solutions are summarized in the handbooks [8–12]. Simple and useful solutions are proposed for cracks in an infinite body and a semi-infinite body. Especially, the following solutions will be practically useful.

3.5.2   *Approximate Equation for the Stress Intensity Factor K_I for a Crack of Arbitrary Shape in an Infinite Body*

We denote the *area of crack* by *area* and assume the remote stress σ acting in the direction perpendicular (z direction) to the crack plane (x–y plane). Although the value of stress intensity factor varies along the crack front, the maximum value $K_{I\max}$ is approximately estimated by the following equation (Figure 3.15) [13]:

$$K_{I\max} \cong 0.5\sigma\sqrt{\pi\sqrt{area}} \tag{3.7}$$

Regarding the application of Equation 3.7, the definition of *area* for a crack with irregular convex and concave shape such as Figure 3.16 must be modified by rounding the shape as the dotted curve. This modification is reasonable, because the stress intensity factor at the points such as concave crack front A, B and C becomes extremely large, and fracture occurs at low stress first until the crack grows to the dotted area where the local stress intensity factor decreases from the beginning of fracture by forming a stable round crack shape.

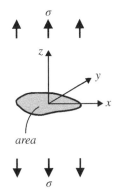

Figure 3.15   3D crack of arbitrary shape in infinite body

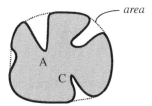

Figure 3.16   3D crack having irregular shape and definition of area

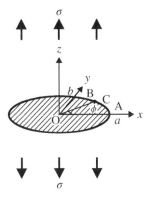

Figure 3.17   Elliptical crack

Equation 3.7 was originally obtained from the nature of the stress intensity factor for an elliptical crack whose stress intensity factor is expressed with Equation 3.8 in terms of the aspect ratio b/a and the maximum value $K_{I\text{max}}$ occurs at point B [13].

The stress intensity factor at crack front C for an *elliptic crack* (Figure 3.17) is given by the following equation [14].

$$K_{\mathrm{I}} = \frac{\sigma\sqrt{\pi b}}{E(k)} \left[\frac{\sin^2\phi + \dfrac{b^4}{a^4}\cos^2\phi}{\sin^2\phi + \dfrac{b^2}{a^2}\cos^2\phi} \right]^{\frac{1}{4}} \tag{3.8}$$

where

$$\phi = \tan^{-1}\frac{y}{x},$$

$$E(k) = \int_0^{\frac{\pi}{2}} \left[1 - k^2\sin^2\phi\right]^{\frac{1}{2}} d\phi$$

$$k^2 = \frac{a^2 - b^2}{a^2}$$

For $a > b$, K_I has a maximum value at point B ($\phi = \pi/2$) and the value is expressed as follows:

$$K_{\mathrm{IB}} = \frac{\sigma\sqrt{\pi b}}{E(k)} \tag{3.9}$$

Thus, investigating the relationship between Equation 3.9 and the area of ellipse ($area = \pi ab$), Equation 3.7 was made as an approximate equation for a crack with arbitrary shape.

However, since the crack in Figure 3.17 is actually reduced to a 2D crack by putting $b = $ constant and $a \to \infty$, the stress intensity factor for such a case is obviously expressed as:

$$K_{\mathrm{Imax}} = K_{\mathrm{IB}} = \sigma\sqrt{\pi b}$$

If we simply substitute the value of $\sqrt{area}$ ($= \infty$) for $b = $ constant and $a \to \infty$ into Equation 3.7, we have $K_{\mathrm{Imax}} \to \infty$. This does not agree with the correct value. In order to apply Equation 3.7 to such a slender crack, it is necessary to consider the saturation of K_I at $b/a \cong 0.2$. It follows that we need to use the effective $\sqrt{area}$ instead of using the actual total value of $\sqrt{area}$ as $\sqrt{area}$ for Equation 3.7 [15].

3.5.3 Approximate Formula of the Stress Intensity Factor K_I for a Crack of Arbitrary Shape at the Surface of a Semi-Infinite Body

For a 3D surface crack, as shown in Figure 3.18, the following approximate equation is proposed [16].

$$K_{\mathrm{Imax}} \cong 0.65\sigma\sqrt{\pi\sqrt{area}} \tag{3.10}$$

Like the previous problem, K_I varies along the crack front, and the effective $area$ must be used to estimate K_{Imax} for very slender cracks.

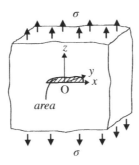

Figure 3.18   Surface crack of arbitrary shape

Problems of Chapter 3

1. Equation 3.1 shows the distribution of stress σ_z around a spherical cavity. Calculate the integral of the quantity made by subtracting σ_0 from σ_z over the region outside the spherical cavity. Namely, calculate the integral of the following equation and explain the meaning of the value.

$$\int_a^\infty (\sigma_z - \sigma_0)\cdot 2\pi r dr = ?$$

2. Indicate the location of the maximum value of K_I for cracks having the shapes shown in Figure 3.19a–c in an infinite body. Assume that the remote stress is applied in the z axis direction.

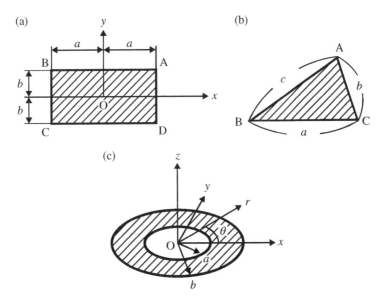

Figure 3.19   (a) Rectangular crack ($a > b$). (b) Triangular crack (on x–y plane: $a > c > b$). (c) Annular crack

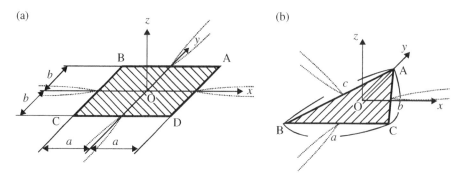

Figure 3.20 (a) Crack outside rectangular domain. (b) Crack is outside triangular domain $(c > a > b)$

3. Indicate the location of the maximum value of K_I for the cracks having the shapes as Figure 3.20a and b in an infinite body. The upper and lower parts ($z>0$ and $z<0$) of the infinite body are connected by the shaded areas of Figure 3.20a and b and the crack occupies the outside of the shaded area ($z = 0$). Namely, the crack tip is located at the boundary of the shaded area. Assume that the remote stress is applied in the z axis direction.

4. The exact solution of the stress intensity factor K_{IB} for an elliptical crack in Figure 3.17 is given by Equation 3.8 and the maximum value is given by K_{IB} of Equation 3.9. The approximate maximum value K_{Imax} of the stress intensity factor for 3D internal cracks of arbitrary shape can be estimated by Equation 3.7. When we have elliptical cracks with the aspect ratios of Case (1) $a = b$ and Case (2) $a = 2b$, compare the maximum values K_{Imax} of the stress intensity factors calculated by the exact solution and the values K_{Imax} estimated by Equation 3.7 for these two cases. The applied stress is a remote tensile stress σ. The values of $E(k)$ are given as $E(0) = \pi/2$ for $a = b$ and $E(0.8660) = 1.211$ for $a = 2b$.

References

[1] S. P. Timoshenko and J. N. Goodier (1970) *Theory of Elasticity*, 3rd edn, McGraw-Hill, New York.

[2] J. D. Eshelby (1957) The determination of the elastic field of an ellipsoidal inclusion, and related problems, *Proc. R. Soc. Lond. Ser. A* **241**:376–396.

[3] H. Noguchi, H. Nisitani, H. Goto, K. Mori (1987) stress concentration for a axially symmetric surface pit contained in a semi-infinite body under uni-axial tension, *Trans. Jpn Soc. Mech. Eng. A*, **53**:820–826.

[4] Y. Murakami, H. Tani-ishi, H. Nisitani (1982) Stress concentration for a axially symmetric surface pit contained in a semi-infinite body under biaxial tension, *Trans. Jpn Soc. Mech. Eng. A*, **48**:150–159.

[5] H. Noguchi, H. Nisitani, H. Goto (1988) Stress concentration for a drilled hole contained in a semi-infinite body under uni-axial tension, *Trans. Jpn Soc. Mech. Eng. A*, **54**:977–982.

[6] I. N. Sneddon (1946) The distribution of stress in the neighbourhood of a crack in an elastic solid, *Proc. R. Soc. Lond. A* **187**:229–260.

[7] Y. Murakami, M. Uchida (1992) Effects of cooling residual stress and applied stress on stress intensity factors for crack emanating from an axially symmetric nonmetallic inclusion, *Proc. Jpn Soc. Mech. Eng.*, **920–78**, B:239–241.

[8] H. Tada et al. (1973) *The Stress Intensity Factors Handbook*, DEL Research Corporation, New York.

[9] G. C. Sih (1974) *Handbook of Stress Intensity Factors for Researchers and Engineers*, vols **1**, **2**, Inst. Frac. Solids Mech., Lehigh University.

[10] D. P. Rooke and D. J. Cartwright (1976) *Compendium of Stress Intensity Factors*, Hillingdon,London.

[11] M. Isida (1976) *Elastic Analysis of Cracks and Stress Intensity Factors*, Baifukan, Tokyo.

[12] Y. Murakami (ed.): (1987) *Stress Intensity Factors Handbook*, vol. 1, Pergamon, London ; (1987) *Stress Intensity Factors Handbook*, vol. 2, Pergamon, London; (1993) *Stress Intensity Factors Handbook*, vol. **3**, Pergamon, London.

[13] Y. Murakami, S. Kodama, S. Konuma (1988) Quantitative evaluation of effects of nonmetallic inclusions on fatigue strength steel, *Trans. Jpn. Soc. Mech. Eng. A*, **54**:688–696.

[14] G. R. Irwin (1962) Crack extension force for a part through crack in a plate, *Trans. J. Appl. Mech.*, **29**:651–654.

[15] Y. Murakami, M. Endo (1986) Effects of hardness and crack geometry on ΔK_{th} of small cracks, *J. Soc. Mat. Sci. Jpn*, **35**:911–917.

[16] Y. Murakami, M. Isida (1985) Analysis of surface crack of arbitrary shape and stress distribution near surface, *Trans. Jpn Soc. Mech. Eng. A*, **51**:1050–1056.

4

Interaction Effects of Stress Concentration

The *interaction effect of stress concentration* occurs when holes and notches exist close to each other. Stress concentration increases or decreases depending on the mutual configuration of holes and notches. The interaction effect can be used positively to decrease stress concentration by considering the nature of stress distribution due to holes and notches.

4.1 Interaction Effect which Enhances Stress Concentration

Stress concentration in the case of Figure 4.1 increases due to the mutual interaction of holes compared to the case of a single hole. This nature can be understood by referring to Figure 4.2.

If we assume that the hole having its center at O_2 (the second hole) does not exist, the stress $\sigma_x{}^*$ in the neighborhood of O_2 is higher than σ_0 due to the influence of the hole having its center at O_1 (the first hole). The value $\sigma_x{}^*$ can be estimated by Equation 1.1. If the second hole exists in a stress field higher than σ_0, the stress concentration should be higher than $K_t = 3$. On the other hand, as the existence of the second hole influences the stress concentration of the first hole, the value of $\sigma_x{}^*$ should be higher than the value calculated simply by Equation 1.1. Thus, an interaction like this case increases the stress concentration. Figure 4.3 shows the analytical solution for this problem [1].

The notch having an additional small notch ahead of the tip, as in Figure 4.4, is called a *double notch*. The double notch of Figure 4.4 is composed of two semicircular notches. The stress concentration factor in this case can be estimated by the following idea.

Solution 1 If the smaller notch does not exist, as in Figure 4.5, the stress concentration factor for the larger notch K_{t1} is $K_{t1} \cong 3$ (the exact solution is $K_{t1} = 3.065$; see Section 1.4.2). We describe a smaller semicircular notch existing in the stress field $K_{t1}\sigma_0$ of the notch tip of a larger

---

Theory of Elasticity and Stress Concentration, First Edition. Yukitaka Murakami.
© 2017 John Wiley & Sons, Ltd. Published 2017 by John Wiley & Sons, Ltd.

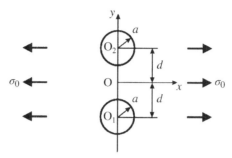

Figure 4.1   Interaction effect enhancing stress concentration

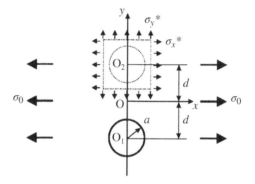

Figure 4.2

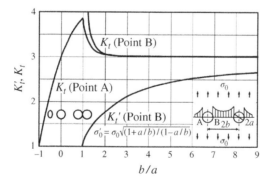

Figure 4.3   Analytical solution for interacting two circular holes [1]

notch, as in Figure 4.6. As the stress concentration factor K_{t2} of the smaller notch is $K_{t2} \cong 3$ for a single notch, the approximate stress concentration factor K_t for the double notch is estimated as follows:

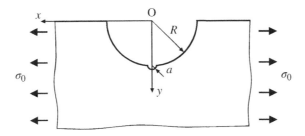

Figure 4.4 Double notch ($a = R/10$)

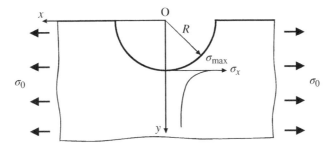

Figure 4.5 Single larger notch

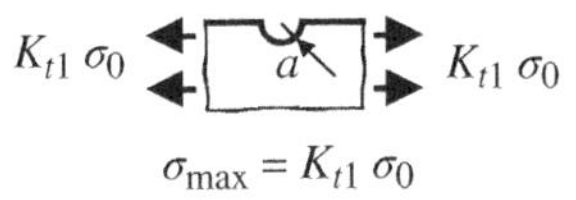

$$K_{t1}\,\sigma_0 \qquad a \qquad K_{t1}\,\sigma_0$$

$$\sigma_{max} = K_{t1}\,\sigma_0$$

Figure 4.6 Smaller semi-circular notch existing at the tip of a larger notch

$$K_t \cong K_{t1} \times K_{t2}$$
$$= 3 \times 3$$
$$= 9$$

However, this estimation is thought to be a little higher than the exact value, because the stress distribution in the neighborhood of the bottom of the larger notch decays quickly from the maximum value to smaller values into the interior of the plate, as shown in Figure 4.5. Considering $a = R/10$ in this case, it may be more reasonable to assume that the smaller notch exists in the stress field at $y \cong R + 0.5a \cong R + R/20$ on the y axis. In order to apply this idea, we substitute $r = a + a/20$ and $\theta = \pi/2$ into σ_θ of Equation 1.1 and can estimate the value as $\sigma_\theta \cong 2.72\sigma_0$. Thus, a more probable value than $K_t = 9$ is estimated as follows:

$$K_t \cong 2.72 \times 3 \cong 8.16$$

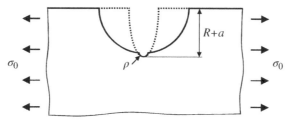

Figure 4.7

Solution 2 Application of the equivalent ellipse is another method. Considering $\rho = a$, $t = R + a$ in Figure 4.7,

$$K_t \cong 1 + 2\sqrt{\frac{R+a}{a}}$$
$$= 1 + 2\sqrt{11}$$
$$\cong 7.63$$

If we use FEM for the analysis of the problem of Figure 4.4, we can obtain $K_t \cong 7.37$.

As estimated in the above discussion, the stress concentration K_t of a double notch does not exceed the product of the stress concentrations for two notches, $K_{t1} \times K_{t2}$. It is useful to note $K_{t1} \times K_{t2}$ as the upper bound of the estimated value.

Figure 4.4 is only one example of double notches. Although other various combinations of double notches such as a V-notch + semi-circular notch, or a notch + crack can be assumed, the basic way of thinking is the same and must be understood correctly.

It is useful to check the results of FEM analyses every time from the viewpoint of this way of thinking, because it will be helpful to avoid simple mistakes by computer analyses.

4.2 Interaction Effect which Mitigates Stress Concentration

In the configuration of two holes, as in Figure 4.8, the stress concentration by interaction effect is lower than that for a single hole. This nature can be understood by referring to Figure 4.9. If we assume that the hole having a center at O_2 (the second hole) does not exist, the stress σ_x^* in the neighborhood O_2 is lower than σ_0 due to the influence of the hole having a center at O_1 (the first hole). Especially, the stress at the contact point with the first hole is 0. This stress can be estimated by Equation 1.1. If the second hole exists in a stress field lower than σ_0, the stress concentration should be lower than $K_t = 3$. On the other hand, as the existence of the second hole influences the stress concentration of the first hole, the value of σ_x^* should be lower than the value calculated simply by Equation 1.1. Thus, an interaction like this case decreases the stress concentration. Figure 4.10 shows the analytical solution for this problem [1].

Other similar examples of the interaction effect are shown in Figures 4.11 and 4.12.

Regarding the influence of *surface roughness* on fatigue strength, a rough surface has in general a higher stress concentration and is more detrimental. Nevertheless, it must be noted that there are some exceptional cases, if we pay attention to the interaction effect of roughness.

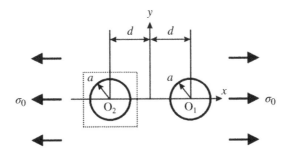

Figure 4.8 Interaction mitigating stress concentration

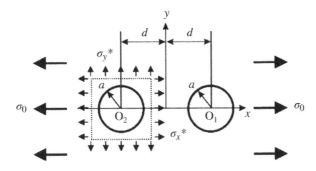

Figure 4.9

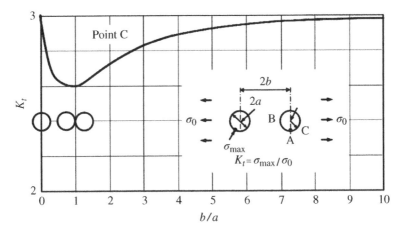

Figure 4.10 Analytical solution for interacting two circular holes [1]

When the roughness is extensive, the pitch of each rough part is dense and the stress concentration is small. On the other hand, when the roughness is limited, the isolated roughness causes a large stress concentration. Therefore, a large isolated *scratch* [2, 3] among a few areas of roughness results in a large influence on fatigue strength. Table 4.1 and Figures 4.13 and 4.14 show experimental data for understanding these phenomena.

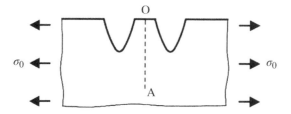

Figure 4.11   Interacting two edge notches

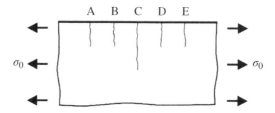

Figure 4.12   Growth of thermal cracks

Table 4.1   Roughness of artificially introduced surface roughness, notch pitch and hardness of materials [4]. Source: Murakami 2002 [4]. Reproduced with permission of Elsevier

Specimen[1]	Max. height R_y (μm)	Average depth a_{mean} (μm)	Notch pitch $2b$ (μm)	Vickers hardness HV (kgf/mm^2)
100A	27.3	14.5	100	180
150A	66.4	37.5	150	180
200A	74.0	53.3	200	180
150QT	20.5	13.0	150	650

[1]A = annealed specimen, QT = quenched and tempered specimen

As explained, we must understand the nature of the interaction effect in stress concentration for the prevention of fatigue failure.

For reference in understanding the effect of surface roughness, analytical examples of the relationship among a number of surface notches, pitches and stress concentrations are shown in Figures 4.15 and 4.16 and Tables 4.1 to 4.5 [5].

4.3   Stress Concentration at Bolt Threads

Bolts have spiral threads. As the groove of the thread turn around once, a notch having the same shape appears every time on the same section. This situation is analogous to the

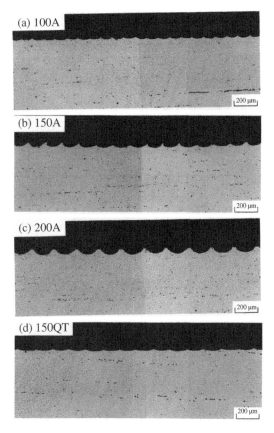

(a) 100A

200 μm

(b) 150A

200 μm

(c) 200A

200 μm

(d) 150QT

200 μm

Figure 4.13 Periodic artificial surface notches to imitate surface roughness: transverse section of cylindrical specimens [4]. Source: Murakami 2002 [4]. Reproduced with permission of Elsevier

interaction problem of an array of notches in Figure 4.15. The difference between Figure 4.15 and the bolt problem is that the loading in the case of bolts is not a remote tensile stress but a tension generated by fastening a bolt with a nut. As the first approximation, the FEM analysis for the stress concentration of a bolt was carried out as follows. In the analysis, the spiral grooves were approximated by an array of V-shaped notches. Figure 4.17a shows the bolt–nut fastened unit. Figure 4.17b shows the tension of a bolt as a single unit. Figure 4.17c shows the tension of a cylindrical bar which contains periodical V-shaped notches with the same geometry as the thread of the bolt. The bottom of each individual thread has its own number to identify the location. The thread is M16 and the shape of the thread is expressed with the following parameters of the root radius ρ at the bottom of groove, the notch depth t and the notch open angle 2α:

$$\rho = 0.25\,\text{mm}, t = 1.27\,\text{mm}, 2\alpha = 60°$$

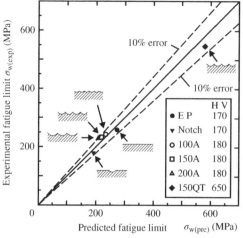

E P: Electro-polished after emery paper finish
Notch: Machining notch to E P specimen

Figure 4.14 Relationship between predicted fatigue limit and experimental fatigue limit [4]. The fatigue limit for a constant hardness is lowest for the specimen with single notch. However, residual stress must be always considered in the surface roughness effect on fatigue. Source: Murakami 2002 [4]. Reproduced with permission of Elsevier

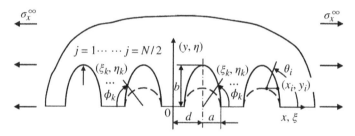

Figure 4.15 Interaction effect of semi-elliptic notches in an semi-infinite plate [5]. N: Number of notches. Source: Murakami 1996 [5]. Reproduced with permission of The Japan Society of Mechanical Engineers

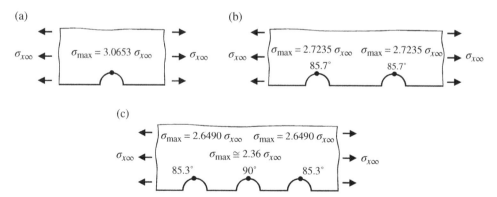

Figure 4.16 Maximum stress and its location for plural semi-circular notches existing at a semi-infinite plate under tension ($b/a = 1$, $a/d = 0.6$) [5]. (a) A single notch. (b) Twin notches. (c) Three notches; the stress concentration for the central notch is fairly decreased compared with the case of a single notch. Angles in this figure show the location of the maximum stress. Source: Murakami 1996 [5]. Reproduced with permission of The Japan Society of Mechanical Engineers

Table 4.2 Maximum stress and its location for an array of semi-circular notches in a semi-infinite plate and circular notches in an infinite plate. The maximum stress occurs at the notch or hole at the outer end of the array. N: Number of notches, $S_{max} = \max(S_{jmax})$. j: notch at the outer end of the array. $S_{jmax} = \sigma_{jmax}/\sigma_0$, $\sigma_0 = \sigma_{max} = 3.0653\sigma_{x\infty}$ for a single semi-circular notch ($N=1$); or $\sigma_0 = \sigma_{max} = 3\sigma_{x\infty}$ for a single circular hole ($N=1$). The maximum stress σ_{max} occurs at a location a little deviated (80–90°) from the notch tip (90°) of the outermost notch. Array of semi-circular notches: $\sigma_{max} = 3.0653S_{max}\cdot\sigma_{x\infty}$. Array of circular holes: $\sigma_{max} = 3S_{max}\cdot\sigma_{x\infty}$.

a/d	0.0		0.0		0.2		0.4		0.6		0.8	
N	(deg.)	σ_0	(deg.)	S_{max}	(deg.)	S_{max}	(deg.)	S_{max}	(deg.)	S_{max}	(deg.)	S_{max}
Reference [5]												
2	90.0	3.0653	90.0	1.0000	89.5	0.9718	87.6	0.9203	85.7	0.8885	84.5	0.8740
3	90.0	3.0653	90.0	1.0000	89.5	0.9652	87.4	0.9032	85.3	0.8642	84.0	0.8427
4	90.0	3.0653	90.0	1.0000	89.5	0.9623	87.3	0.8956	85.2	0.8536	83.8	0.8332
5	90.0	3.0653	90.0	1.0000	89.5	0.9606	87.3	0.8912	85.1	0.8470	83.7	0.8242
6	90.0	3.0653	90.0	1.0000	89.5	0.9596	87.2	0.8885	85.1	0.8431	83.7	0.8203
7	90.0	3.0653	90.0	1.0000	89.5	0.9588	87.2	0.8864	85.1	0.8401	83.7	0.8161
8	90.0	3.0653	90.0	1.0000	89.5	0.9583	87.2	0.8850	85.0	0.8379	83.7	0.8137
9	90.0	3.0653	90.0	1.0000	89.5	0.9578	87.2	0.8839	85.0	0.8360	83.7	0.8123
10	90.0	3.0653	90.0	1.0000	89.5	0.9576	87.2	0.8829	85.0	0.8348	83.6	0.8097
11	90.0	3.0653	90.0	1.0000	89.5	0.9573	87.2	0.8824	85.0	0.8337	83.6	0.8082
12	90.0	3.0653	90.0	1.0000	89.5	0.9571	87.2	0.8817	85.0	0.8328	83.6	0.8071
∞	90.0	3.0653	90.0	1.0000		0.9549		0.8740		0.8229		0.7950
Reference [6, 7]												
2	90.0	3.000	90.0	1.000	89.7	0.976	88.0	0.926	85.9	0.890	84.2	0.874
3	90.0	3.000	90.0	1.000	89.6	0.970	87.7	0.911	85.6	0.869	84.2	0.849
4	90.0	3.000	90.0	1.000	89.6	0.967	87.7	0.904	85.4	0.859	84.1	0.837
5	90.0	3.000	90.0	1.000	89.6	0.966	87.7	0.900	85.4	0.853	84.0	0.830
6	90.0	3.000	90.0	1.000	89.6	0.965	87.7	0.897	85.4	0.850	84.0	0.826
7	90.0	3.000	90.0	1.000	89.6	0.964	87.7	0.896	85.4	0.847	83.9	0.822
8	90.0	3.000	90.0	1.000	89.6	0.964	87.7	0.893	85.4	0.844	83.9	0.819
∞	90.0	3.000	90.0	1.000		0.961		0.886		0.832		0.806

$\sigma_{x\infty}$

Table 4.3 Maximum stress and its location for array of elliptic notches at the edge of a semi-infinite plate [5]. The maximum stress occurs at the notch at the outer end of array. N: Number of notches, deg. = degrees. $S_{max} = \max (S_{j\,max})$. j: notch at the outer end of array. $S_{j\,max} = \sigma_{j\,max}/\sigma_0$. $\sigma_0 = \sigma_{max}$ for a single elliptic notch ($N=1$). Source: Murakami 1996 [5]. Reproduced with permission of The Japan Society of Mechanical Engineers

ρ/b	b/d		0.0			0.2		0.4		0.6		0.8	
	N		σ_0	(deg.)	S_{max}	(deg.)	S_{max}	(deg.)	S_{max}	(deg.)	S_{max}	(deg.)	S_{max}
0.2	10		5.7353	90.0	1.0000	89.9	0.9626	89.4	0.8894	88.6	0.8301	88.0	0.7930
	11		5.7353	90.0	1.0000	89.9	0.9624	89.4	0.8887	88.6	0.8290	88.0	0.7916
	12		5.7353	90.0	1.0000	89.9	0.9622	89.4	0.8882	88.6	0.8281	88.0	0.7905
	13		5.7353	90.0	1.0000	89.9	0.9620	89.4	0.8877	88.6	0.8274	88.0	0.7895
	∞		5.7353	90.0	1.0000		0.9596		0.8817		0.8190		0.7775
0.4	10		4.3120	90.0	1.0000	89.8	0.9609	88.8	0.8870	87.6	0.8311	86.7	0.7980
	11		4.3120	90.0	1.0000	89.8	0.9607	88.8	0.8863	87.6	0.8300	86.7	0.7966
	12		4.3120	90.0	1.0000	89.8	0.9605	88.8	0.8858	87.6	0.8291	86.7	0.7955
	13		4.3120	90.0	1.0000	89.8	0.9603	88.8	0.8853	87.6	0.8284	86.7	0.7945
	∞		4.3120	90.0	1.0000		0.9579		0.8793		0.8200		0.7825
0.6	10		3.6867	90.0	1.0000	89.7	0.9597	88.3	0.8854	86.7	0.8325	85.6	0.8024
	11		3.6867	90.0	1.0000	89.7	0.9594	88.3	0.8847	86.7	0.8313	85.6	0.8009
	12		3.6867	90.0	1.0000	89.7	0.9592	88.3	0.8841	86.7	0.8305	85.6	0.7998
	13		3.6867	90.0	1.0000	89.7	0.9591	88.3	0.8836	86.7	0.8297	85.6	0.7987
	∞		3.6867	90.0	1.0000		0.9571		0.8776		0.8201		0.7855
0.8	10		3.3166	90.0	1.0000	89.6	0.9586	87.8	0.8841	85.8	0.8339	84.6	0.8062
	11		3.3166	90.0	1.0000	89.6	0.9583	87.8	0.8833	85.8	0.8327	84.6	0.8046
	12		3.3166	90.0	1.0000	89.6	0.9580	87.8	0.8827	85.8	0.8318	84.6	0.8036
	13		3.3166	90.0	1.0000	89.6	0.9579	87.8	0.8822	85.8	0.8310	84.6	0.8025
	∞		3.3166	90.0	1.0000		0.9567		0.8762		0.8214		0.7893

Table 4.4 Maximum stress for an infinite array of elliptic notches at the edge of a semi-infinite plate or elliptic holes in an infinite plate [5, 6].

$N = \infty$, $S_{Jmax} = \sigma_{max}/\sigma_0$, $\sigma_0 = \sigma_{max}$ for a single elliptic notch or elliptic hole

$p/b = (a/b)^2$	b/a	σ_0	S_{max}				
		$b/d=0.0$	$b/d=0.0$	$b/d=0.2$	$b/d=0.4$	$b/d=0.6$	$b/d=0.8$
Reference [5]							
0.2	≐2.24	5.7353	1.0000	0.9214	0.7717	0.6535	0.5785
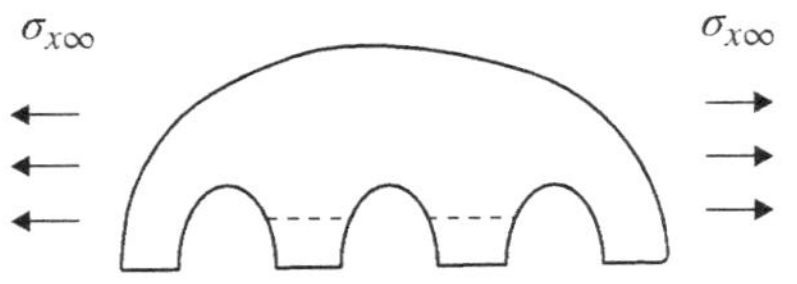 0.25	2.0	5.2202	1.0000	0.9205	0.7707	0.6533	0.5798
0.4	≐1.58	4.3120	1.0000	0.9179	0.7666	0.6525	0.5831
0.6	≐1.29	3.6867	1.0000	0.9152	0.7623	0.6521	0.5869
0.8	≐1.12	3.3166	1.0000	0.9128	0.7586	0.6518	0.5903
1.0	1.0	3.0653	1.0000	0.9106	0.7560	0.6529	0.5948
Reference [6]							
0.2	≐2.24	5.4721	1.0000	0.939	0.810	0.691	0.608
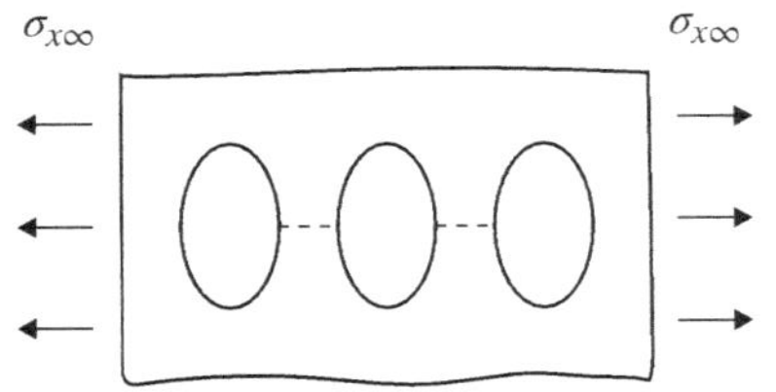 0.4	≐1.58	4.1623	1.0000	0.933	0.797	0.680	0.602
0.6	≐1.29	3.5820	1.0000	0.929	0.788	0.673	0.600
0.8	≐1.12	3.2361	1.0000	0.926	0.781	0.668	0.600
1.0	1.0	3.0000	1.0000	0.923	0.775	0.665	0.601

$\sigma_{x\infty}$

$\sigma_{x\infty}$

Table 4.5 Maximum stress at the central semi-circular notch of an array of semi-circular notches at the edge of a semi-infinite plate [5]. N: Number of notches, $S_{max} = \max(S_{/max})$, $p = 2c$, $c = \sqrt{2t\rho - t^2}$. Source: Murakami 1996 [5]. Reproduced with permission of The Japan Society of Mechanical Engineers

t/ρ	0.1	0.2	0.3	0.4	0.5
N					
3	1.4388	1.6120	1.7250	1.8262	1.9128
5	1.4102	1.5612	1.6574	1.7389	1.8076
7	1.3960	1.5372	1.6256	1.6994	1.7603
9	1.3878	1.5239	1.6080	1.6772	1.7336
11	1.3826	1.5154	1.5966	1.6627	1.7168
13	1.3790	1.5095	1.5885	1.6531	1.7046
15	1.3763	1.5051	1.5823	1.6454	1.6955
∞	1.359	1.477	1.542	1.595	1.636

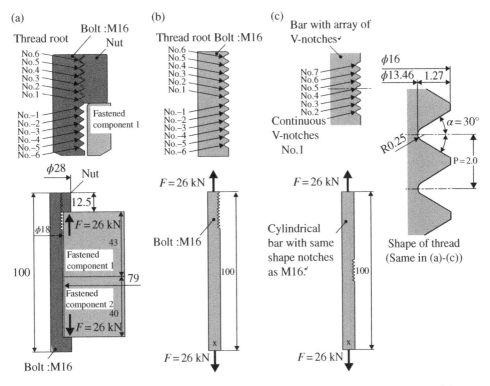

Figure 4.17 (a) Bolt-nut fastened unit. (b) Single bolt unit. (c) Cylindrical bar with array of the same V-shaped notches as an M16 bolt

Therefore, the basic stress concentration factor K_t for a single thread groove is estimated by the concept of an equivalent ellipse as follows:

$$K_t \cong 1 + 2\sqrt{\frac{1.27}{0.25}} \cong 5.5$$

Although the interaction effect of an array of thread grooves may act to decrease the stress concentration in a case of simple tension, the situation in the bolt–nut fastened unit is very different from a case of simple tension. The stress concentration for a bolt–nut fastened unit reaches $K_t = 4.75$ at the bottom of the first groove (No. 1) from the mating end between the nut and bolt. The value is not sufficiently decreased in comparison with the stress concentration factor $K_t = 5.5$ for a single V-shaped notch (Figure 4.18). Thus, in the case of bolt–nut units, the mitigation of stress concentration simply due to an interaction effect cannot be expected.

It must be noted that fatigue failures in bolt–nut units often occur due to an increase in stress amplitude caused by loosening the fastening force rather than the simple mistake of estimating the stress concentration.

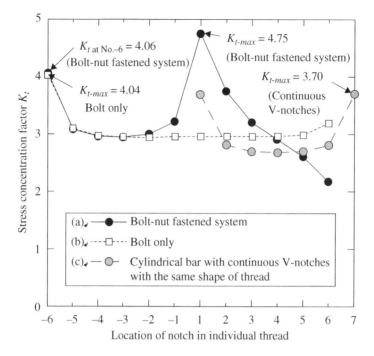

Figure 4.18 Stress concentration factors K_t at the bottom of a groove for a bolt–nut fastened unit, a single bolt and a periodical array of V-shaped notches

Problems of Chapter 4

1. Determine the stress intensity factors for the following cases.

    (1) Figure 4.19a:
        (a) K_{IB} at point B and K_{ID} at point D for $b/a = \infty$.
        (b) K_{IB} at point B and K_{ID} at point D for $b/a \to 0$.

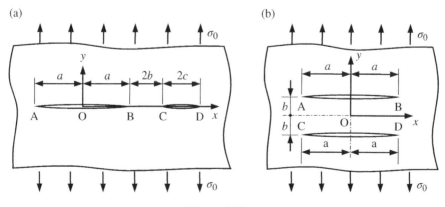

Figure 4.19

(2) Figure 4.19b:
    (a) K_{IB} at point B for $b/a = \infty$.
    (b) K_{IB} at point B for $b/a \rightarrow 0$.

2. Figure 4.20a and b shows two kinds of specimens which were actually used for a tension–compression fatigue test. The specimen in Figure 4.20a often fractured from the corner at the grip. Then, the shape was modified as the specimen in Figure 4.20b, which contains a shallow notch beside the grip corner and fatigue fracture from the corner at the grip was avoided. Explain the reason why fatigue fracture was avoided by the shape of the specimen in Figure 4.20b.

(a)   (b)

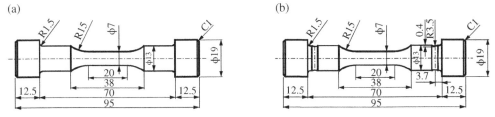

Figure 4.20 Shape of two types of tension–compression fatigue specimen (unit: mm). (a) Specimen shape before modification. (b) Specimen with a shallow circumferential notch beside the corner of the grip

3. Figure 4.21 shows an array of three circular holes along the x axis of x–y coordinates in an infinite plate. Answer the following questions.

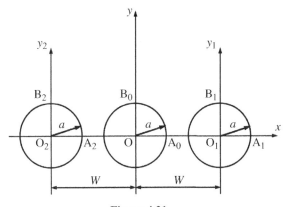

Figure 4.21

(1) When the plate is subjected to remote stress ($\sigma_x = \sigma_0 > 0$, $\sigma_y = 0$, $\tau_{xy} = 0$), judge which σ_x at point B_0 or B_1 is larger and explain the reason.

(2) When the plate is subjected to remote tensile stress ($\sigma_x = 0$, $\sigma_y = \sigma_0 > 0$, $\tau_{xy} = 0$), judge which σ_y at point A_0 or A_1 is larger and explain the reason.

(3) When $W \rightarrow 2a$ and the plate is subjected to remote tensile stress ($\sigma_x = \sigma_0 > 0$, $\sigma_y = \sigma_0$, $\tau_{xy} = 0$), estimate the stress σ_y at point A_1.

References

[1] T. Udoguchi (1947) Plane stress problem of plate containing two circular holes, *Appl. Math. Mech.* (in Japanese), **1**:14, 61–81.

[2] National Space Development Agency of Japan (2000) *Analysis of Launching Accident of No. 8 H-II Rocket*, http://www.nasda.go.jp/press.

[3] A. Konno, N. Sakazume (2001) *Turbomachinery* **29**:139–146 (in Japanese).

[4] Y. Murakami (2002) *Metal Fatigue: Effects of Small Defects and Nonmetallic Inclusion*, Elsevier, New York.

[5] T. Matsuo, N. Noda, Y. Murakami, S. Harada (1996) Tension of a semi-infinite plate containing a row of elliptical and circular-arc notches, *Trans. Japan Soc. Mech. Eng. A*, **62**:2276–2282.

[6] M. Ishida, H. Igawa (1992) Tension of a wide plate containing a row of elliptical holes and cracks (some asymptotic behavior and formulae of the stresses), *Jpn Soc. Mech. Eng. A*, **58**:1642–1649.

[7] N. Noda, T. Matsuo, H. Ishii (1995) Tension of a wide plate containing a row of elliptical inclusions, *Jpn Soc. Mech. Eng. A*, **61**:106–113.

5

Notch Effect and Size Effect in Fatigue: Viewpoint from Stress Concentration

Fatigue cracks mostly initiate at sites of stress concentration. Notches and holes decrease the fatigue strength of structures. This phenomenon is called the *fatigue notch effect*. However, it is difficult to avoid notches and holes, which are in general necessary for the function of structures. Thus, evaluation of the notch effect is very important for the prevention of fatigue fracture. Although many theories have been proposed on the fatigue notch effect [1–4], the notch effect and the size effect in fatigue are regarded as difficult problems for most strength design engineers to understand. The fundamental aspects of this problem will be explained clearly in this chapter.

The *size effect* in fatigue means the difference in fatigue strength depending on the size of the components. Geometrically similar structures have an identical stress concentration factor. It follows that the maximum stress σ_{max} becomes the same value under the same nominal stress. However, geometrically similar structures in general have different fatigue strengths, depending on the size of these structures. Thus, it must be noted that the difference in fatigue strength cannot be explained only by the stress concentration factor K_t, namely by the maximum stress σ_{max} at the tip of notches and edge of holes. In order to understand the notch effect and the size effect in fatigue, we must pay attention not only to the maximum stress σ_{max} at the notch root but also to the stress distribution in the neighborhood of the notch root.

First, let us compare the maximum stress σ_{max} and the stress distribution σ_y for two circular holes (Figure 5.1). The size of the circular hole of Figure 5.1a is smaller than that of Figure 5.1b. The maximum stresses for the circular holes in Figure 5.1a and b are identical, that is $3\,\sigma_{y\infty}$ under the identical remote stress $\sigma_{y\infty}$. However, the stress distributions along the x axis are completely different. The stress distributions for Figure 5.1a and b are compared in Figure 5.2. Although the maximum stresses are $3\,\sigma_{y\infty}$ for both cases, the stress distribution is higher for Figure 5.1b than for Figure 5.1a.

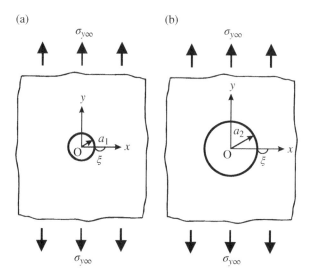

Figure 5.1 Two geometrically similar circular holes. (a) Stress distribution σ_y for a circular hole. (b) $a_2 = 2a_1$

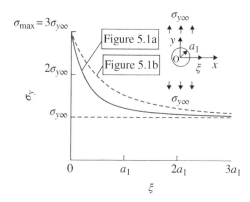

Figure 5.2 Distribution of σ_y along the x axis

The same situation occurs also in elliptic holes, as shown in Figures 5.3 and 5.4. The two elliptic holes in Figure 5.3a and b have the identical aspect ratio $b/a = 0.5$ and are geometrically similar. Based on the aspect ratio $b/a = 0.5$, the maximum stress σ_{max} at the end of the major axis is identical regardless of the absolute size of the elliptic holes (a, b) and is expressed by the following equation:

$$\sigma_{ymax} = \left(1 + \frac{2a}{b}\right)\sigma_{y\infty} = 5\sigma_{y\infty} \tag{5.1}$$

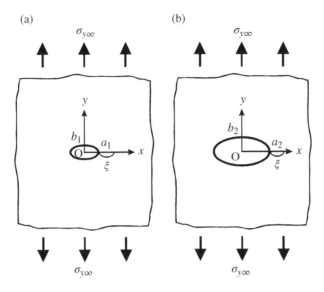

Figure 5.3   Two geometrically similar elliptical holes, $K_t = 5$. (a) $a_1 = 2b_1$. (b) $a_2 = 2a_1$, $b_2 = 2b_1$

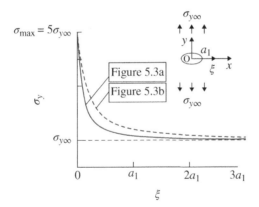

Figure 5.4   Distribution of σ_y along the x axis

However, the stress distributions for the two elliptic holes are completely different, as shown in Figure 5.4.

Let us pick up another problem related to notch effect. As shown in Figure 5.5, we compare two notched components (A and B) with different notch shapes, notch sizes and remote stresses. We assume that the stress concentration factors K_{tA} and K_{tB} are different and the remote stresses $\sigma_{y\infty A}$ and $\sigma_{y\infty B}$ are also different but these values satisfy the relationship $K_{tA}\sigma_{y\infty A} = K_{tB}\sigma_{y\infty B}$. In this case, the stress distributions in the neighborhood of the notch root are different for these two notches, as shown Figure 5.5.

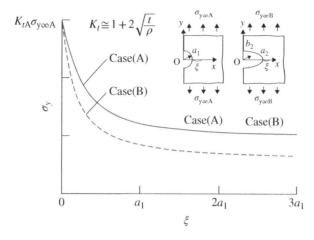

Figure 5.5   Two notches having an identical maximum stress at their notch tip

In all three cases (Figures 5.2, 5.4 and 5.5, as explained above), the maximum stresses σ_{max} at the notch root are identical but the stress distributions are different. It is easy to judge which stress distribution decreases more fatigue strength. Naturally, the stress distribution having a higher value near the notch root is in a more severe condition. Thus, the discussion from the viewpoint of relative and qualitative comparison is simple. However, it is not necessarily easy to predict fatigue strength quantitatively and accurately. The following two points are at least to be noted as the fundamental factors for the notch and size effects.

1. Regarding geometrically similar components containing notches, the fatigue strength of larger components is lower than that of smaller components.
2. Regarding geometrically dissimilar components containing notches with identical maximum stress σ_{max} at the notch root under different external loading, the fatigue strength is lower for the notch having a higher stress distribution near the notch tip.

Figure 5.6 shows the relationship between the fatigue limit σ_w and diameter d of a small hole introduced on the surface of specimens [5, 6]. All the small holes are geometrically similar and the sizes are sufficiently smaller than the diameter of the specimen (10 mm). Thus, the stress concentration factor K_t for each hole is equal regardless of hole diameter, $K_t \cong 2.0$ (see Figure 3.6). Nevertheless, as shown in Figure 5.6, the fatigue limit σ_w decreases with increasing hole diameter. This result agrees with the content of item 1 listed above. In conclusion, it must be noted that the design method of fatigue limit based only on stress concentration leads to simple mistakes. In this context, it is necessary from the above discussion to find the solution for cases in which the maximum stresses σ_{max} at the notch root are different and at the same time the stress distributions are also different, as shown in Figure 5.7. Figure 5.7 shows that the maximum stress σ_{max} at notch B is higher than that at notch A. However, the stress σ_y at some distance from the notch root is higher for notch A than for notch B. In these case, the situation is more complicated and it cannot be simply concluded 'Notch B is more detrimental than

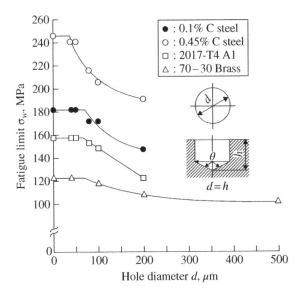

Figure 5.6   Hole diameter and rotating bending fatigue limit [5, 6]

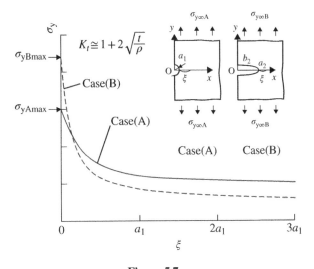

Figure 5.7

notch A.' In order to solve this problem, not only the analysis of stress distribution but also 'the experimental data on the material in question' is necessary.

If only two mechanics factors of stress concentration and stress distribution are the major factors to be considered, the *size effect* includes the same substantial content as the *notch effect*. However, if the materials contain defects and the defects have a statistical scatter, the size effect becomes more complicated. The reason is that large structures have a higher probability than small structures of containing large defects [7, 8].

(a)

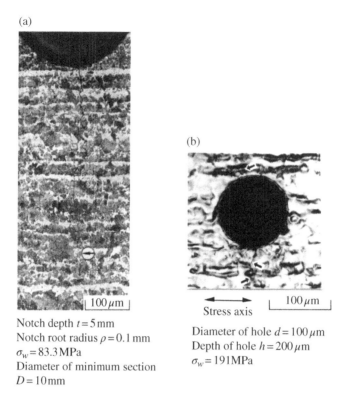

(b)

Stress axis

Notch depth $t = 5$ mm
Notch root radius $\rho = 0.1$ mm
$\sigma_w = 83.3$ MPa
Diameter of minimum section
$D = 10$ mm

Diameter of hole $d = 100\,\mu$m
Depth of hole $h = 200\,\mu$m
$\sigma_w = 191$ MPa

Figure 5.8 (a) Nonpropagating crack observed at the root of sharp notch after 10^7 cycles of rotating bending fatigue (0.45%C steel) [10]. Source: Murakami 1986 [10]. Reproduced with permission of The Japan Society of Mechanical Engineers. (b) Nonpropagating cracks observed at the edge of drilled small hole (annealed 0.45%C steel) [11]. These cracks are called *nonpropagating cracks* and they do not grow by additional stress cycles. Source: Murakami 1983 [11]. Reproduced with permission of The Japan Society of Mechanical Engineers

In the case of a sharp notch, the critical condition of the fatigue limit is determined by the nonpropagation of a fatigue crack emanated from a notch (see Figure 5.8). These cracks are called *nonpropagating cracks*. This peculiar phenomenon makes the discussion on notch effect and size effect more complicated. However, the existence of a nonpropagating crack at the fatigue limit is an experimental fact not to be denied and the nature of *nonpropagating* cracks must be understood for safe fatigue design. When fatigue strength is determined by the condition of fatigue crack nonpropagation, one quantitative solution is to pay attention to the threshold stress intensity factor range ΔK_{th} [12, 13].

Problem of Chapter 5

1. When the fatigue limit σ_w for the case of Figure 5.9a is $\sigma_w \cong 300$ MPa, estimate the fatigue limit $\sigma_w{}'$ for the case of Figure 5.9b, assuming that the materials are identical for Figure 5.9a and b.

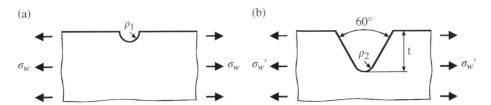

Figure 5.9   Notch at the edge of a semi-infinite plate. (a) Semi-circular notch (notch root radius $\rho_1 = 5$ mm). (b) V-shaped notch ($\rho_2 = \rho_1$, $t = 4\rho_1$)

References

[1] T. Isibasi (1967) *Fatigue of Metals and Prevention of Fracture* (in Japanese), Yokendo, Tokyo.

[2] E. Siebel and M. Stieler (1955) Ungleichformige Spannungsverteilung bei schwingender Beanspruchung, *VDI-Z.*, **97**:121–126.

[3] H. Nisitani (1968) Branch point in rotating bending fatigue test and size effect of fatigue limit (investigation by small-sized specimens), *Trans. Jpn Soc. Mech. Eng.*, **34**:371–382.

[4] D. Taylor (2004) Applications of the theory of critical distances to the prediction of brittle fracture in metals and non-metals, *Proc. 15th Eur. Conf. Fract.*, CD-ROM.

[5] Y. Murakami, Y. Tazunoki, T. Endo (1981) Existence of coaxing effect and effect of small artificial holes of 40 ~ 200 μm diameter on fatigue strength in 2017S-T4 Al alloy and 7:3 brass, *Jpn Soc. Mech. Eng. A*, **47**:1293–1300.

[6] Y. Murakami, H. Kawano, T. Endo (1979) Effect of micro-hole on fatigue strength (2nd report, effect of micro-hole of 40 μm–200 μm in diameter on the fatigue strength of quenched or quenched and tempered 0.46% carbon steel), *Jpn Soc. Mech. Eng. A*, **45**:1479–1486.

[7] Y. Murakami (1993) *Metal Fatigue: Effects of Small Defects and Nonmetallic Inclusion*, Yokendo, Tokyo.

[8] Y. Murakami (2002) *Metal Fatigue : Effects of Small Defects and Nonmetallic Inclusion*, Elsevier, New York.

[9] For example, Y. Murakami (2002) *Metal Fatigue : Effects of Small Defects and Nonmetallic Inclusion*, Elsevier, New York.

[10] Y. Murakami and K. Matsuda (1986) Dependence of threshold stress intensity factor range ΔK_{th} on crack size and geometry, and material properties, *Trans. Jpn Soc. Mech. Eng. A*, **54**:1492–1499.

[11] Y. Murakami, M. Endo (1983) A geometrical parameter for the quantitative estimation of the effects of small defects on fatigue strength of metals, *Jpn Soc. Mech. Eng. A*, **49**:127–136.

[12] R. A. Smith and K. J. Miller (1977) Fatigue cracks at notches, *Int. J. Mech. Sci.*, **19**:11–22.

[13] R. A. Smith and K. J. Miller (1978) Prediction of fatigue regimes in notched components, *Int. J. Mech. Sci.*, **20**:201–206.

6

Stress Concentration in Plate Bending

In this chapter, the stress concentration at a hole contained in a thin plate is considered. The expression of *thin plate* has a relative meaning. If the thickness of a plate is *sufficiently thin* compared to the size of the radius of a hole or notch root, the *classical theory* (or the *Kirchhoff theory* [1]) can be applied with practically sufficient accuracy. Although the expression of *sufficiently thin* is ambiguous, in general the difference in one order of the dimension may be assumed. Namely, if the plate thickness h is about the order of $1/10$ of the radius R of a hole, the accuracy of the solution by the classical theory is practically sufficient. However, even if so, the question that the solution for the case $h/R = 1/5$ is useless, is not appropriate. Such a solution will be valid as a useful estimation. If h and R is comparable, such problems should be treated as a 3D problem. Thus, it must be noted that *the absolute dimension of plate thickness and the absolute size of holes and notches cannot be the criterion for the application of the theory.*

6.1 Stress Concentration at a Circular Hole in a Wide Plate Under Bending

This problem can be solved by the classical theory (or the Kirchhoff theory) as 2D plane stress problems.

Let us pick up a problem like Figure 6.1 in which a plate is subjected to remote bending moment M_0. M_0 is the bending moment applied per unit length of the plate. Therefore, the stress concentration is defined by the maximum bending moment M_{max} at the edge of hole on the basis of the remote bending moment M_0.

The maximum bending moment M_{max} in the problem of Figure 6.1 occurs at $\theta = \pm \pi/2$ of the edge of the circular hole. The moment concentration factor $K_t (= M_{max}/M_0)$ at point B is given by the following equation.

Theory of Elasticity and Stress Concentration, First Edition. Yukitaka Murakami.
© 2017 John Wiley & Sons, Ltd. Published 2017 by John Wiley & Sons, Ltd.

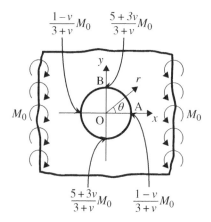

Figure 6.1   Stress concentration at circular hole in plate bending

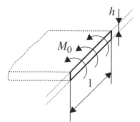

Figure 6.2   Definition of bending moment

$$K_t = \frac{5+3\nu}{3+\nu} \tag{6.1}$$

The maximum tensile stress at point B occurs on the bottom surface and the maximum compressive stress occurs on the upper surface. As the bending moment is defined as the quantity acting on the section with thickness h and width 1, as shown in Figure 6.2, the maximum stress is calculated by dividing the moment with the section modulus $h^2/6$. Thus, the maximum tensile stress at point B σ_{max} is expressed as

$$\sigma_{max} = K_t \frac{M_0}{\dfrac{h^2}{6}} = K_t \frac{6M_0}{h^2} \tag{6.2}$$

Therefore, in the plate bending problem of Figure 6.1 with bending moment M_0, the stress at the central section (neutral section) is 0, the stress on the bottom surface is tensile and the stress on the upper surface is compressive. The bending moment M at points $\theta = 0, \pi$ of the edge of the circular hole is given as:

$$M = \frac{1-\nu}{3+\nu} M_0 \tag{6.3}$$

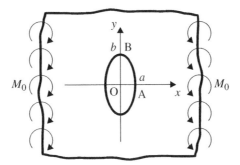

Figure 6.3   Stress concentration at elliptical hole in plate bending

Likewise of 2D plane stress problems, this value can be used extensively by the application of the principle of superposition.

6.2   Stress Concentration at an Elliptical Hole in a Wide Plate Under Bending

The moment concentration factors at points A and B for the problem of Figure 6.3 are given as follows [2, 3].

$$K_{tB} = 1 + \frac{2(1+\nu)b}{3+\nu}\frac{b}{a} \tag{6.4}$$

$$K_{tA} = \frac{1-\nu}{3+\nu} \tag{6.5}$$

For $a = b$, the values of Equation 6.4 naturally coincide with those of Equation 6.1. As seen in Equations 6.3 and 6.5, the bending moment M at point A is independent of a and b. It is interesting to note that this nature of stress concentration is the same as 2D plane stress problems. As shown in various examples of 2D plane stress problems, Equations 6.1–6.5 can be applied to various practical problems of plate bending. Naturally, the principle of the concept of equivalent ellipse can be used.

Problems of Chapter 6

1. Determine the maximum stress at the edge of a circular hole existing at the center of a wide square plate which is subjected to pairs of concentrated forces P at the four corners of the plate, as shown in Figure 6.4.
2. Determine the stress at the edge of a circular hole existing at the center of a wide circular plate which is subjected to bending moment M per unit length of the periphery, as shown in Figure 6.5.

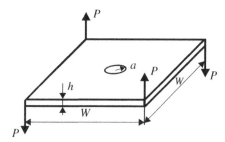

Figure 6.4   Circular hole at the center of a square plate ($a \ll W$, $h \ll a$)

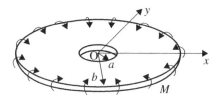

Figure 6.5   Bending of a wide circular plate containing a circular hole at the center ($a \ll b$, $h \ll a$)

References

[1] S. P. Timoshenko and S. Woinowsky-Krieger (1959) *Theory of Plates and Shells*, 2nd edn, McGraw-Hill, New York.

[2] J. N. Goodier (1936) The influence of circular and elliptical holes on the tranverse flexure of elastic plates, *Phil. Mag.*, **22**:69–80.

[3] K. Washizu (1952) On the bending of isotropic plates, *Jpn Soc. Mech. Eng.*, **18**:41–47.

7

Relevant Usage of Finite Element Method

7.1 Fundamentals for Element Meshing

Nowadays it is very common to use the *finite element method* (FEM) for the analysis of stress concentration. However, to the author's knowledge, no basic way of thinking about the relevant *element meshing* for strength analysis has been shown. Element meshing must be carried out by considering the basic nature of stress and strain concentration at notches. Otherwise, misjudgments based on wrong FEM analytical results are made on strength evaluation.

This chapter provides a guide for the accurate analysis of stress concentration and explains the basic method for element meshing related to strength evaluation, such as size effect and notch effect. The way of interpretation of the analytical results will be also explained.

First of all, let us study the analysis of the stress concentration for a circular hole in a wide plate (Figure 7.1a). Figure 7.1b shows an example of mesh pattern at the area of stress concentration.

In Figure 7.1, the size of one side of the triangular element at maximum stress is assumed by s. The basic way of thinking explained in the following is the same also for other types of elements such as quadratic elements or elements with a higher freedom. In FEM analysis, a concrete value such as a and s is given for the radius of circular hole and the minimum mesh size. However, it does not have a special meaning, because even if values 10 times larger or one-tenth smaller are given for a and s, the stresses obtained by FEM are identical. Such a basic nature of FEM is not necessarily understood by many students and researchers when they begin to use FEM. Although it is natural that the result of FEM analysis is influenced by the way of meshing, the result is not influenced by the absolute value of the element size. In the example of Figure 7.1b, the results are determined by the *relative size* of s to a. Namely, the results for $a = 10$ mm and $s = 1$ mm are the same as those for $a = 1$ m and $s = 10$ cm.

In the case of elliptical holes and notches too, as in Figures 5.1 to 5.7, the maximum stress is influenced by the mesh pattern at the notch root. The important point is to understand that the

Theory of Elasticity and Stress Concentration, First Edition. Yukitaka Murakami.
© 2017 John Wiley & Sons, Ltd. Published 2017 by John Wiley & Sons, Ltd.

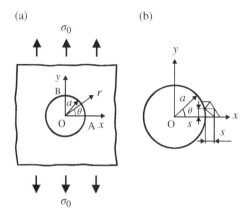

Figure 7.1 FEM analysis of stress concentration at circular hole and mesh pattern. (a) Problem of stress concentration. (b) Mesh pattern at stress concentration

value is determined not by the absolute value of s but the relative size compared to the notch root radius ρ.

It follows that, when we compare the results of FEM analyses of stress concentration for variously sized holes and notches, we need to 'keep the minimum size of element s at the notch root as an identical value relative to the notch root radius ρ'. If we need to compare accurate values by FEM analysis for several problems, it is necessary to decrease s/ρ while keeping the ratio s/ρ identical for individual problems.

This problem is explained from the nature of the analytical solution as follows. If the distribution of stress σ_θ (namely σ_y) along the x axis in the problem of Figure 7.1 is expressed by normalizing the distance x by a, the normalized curves become identical for any circular hole with different sizes (Equation 1.2). As the stress concentration factor K_t for general circular holes and notches can be approximately expressed by this form:

$$K_t \cong 1 + 2\sqrt{\frac{t}{\rho}},$$

the most influential geometrical factors which control stress concentration factors are the notch depth t and the notch root radius ρ. As we studied the *concept of the equivalent ellipse* in Chapter 1 of Part I, K_t is scarcely influenced by small differences in the local shape of a notch in the depth direction under the condition of identical notch root radius ρ. Therefore, it is important in general to perform FEM analysis of notches by paying attention notch root radius ρ in meshing elements.

7.2 Elastic–Plastic Analysis

How should we treat the *elastic–plastic analysis* of geometrically similar circular holes having different sizes, as shown in Figure 7.2a and b? In elastic–plastic analysis, in general the stress–strain relationship (so-called *constitutive equation*) like Figure 7.3 is assumed. We assume that

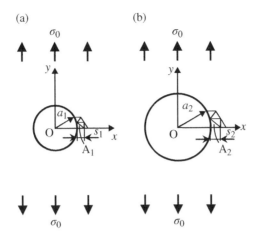

Figure 7.2   (a) $a_1 = a$. (b) $a_2 = 10a$, $s_2 = 10s_1$

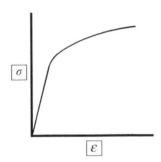

Figure 7.3   Stress–strain relationship

Figure 7.2b has a size 10 times larger than Figure 7.1a and the *mesh sizes are geometrically similar* ($s_1/a_1 = s_2/a_2$) for Figure 7.2a and b and the remote stresses σ_y are σ_0, identical for Figure 7.2a and b. We assume also that several elements in the area near the maximum stress satisfies the yielding condition and are in the elastic–plastic region. In this state, if we compare the results of FEM analysis for the minimum elements A_1 and A_2, the stresses and strains become identical for Figure 7.2a and b. This nature is the same as explained about elastic analysis. However, can we say that if the elastic stress concentrations factor are identical, the stress and strain concentrations in elastic–plastic condition are also identical for both the circular holes? Actually this is not the case, because the yielding phenomenon and plastic deformation are influenced by the absolute size of the area in question.

Even if yielding does not occur in the case of Figure 7.2a, yielding and extension of plastic zone can occur in the case of Figure 7.2b. The reason for the difference is that even if the elastic maximum stresses are identical for Figure 7.2a and b, the stress gradients and stress distributions are different (see Chapter 5 of Part II). This phenomenon is related to the problems of the notch effect and the size effect in fatigue.

How can we compare the difference of actual plastic deformation behaviors for two problems such as Figure 7.2a and b? One method is to change the element meshing in Figure 7.2b
not with a geometrically similar relationship in the minimum size of the element of Figure 7.2b
as in Figure 7.2a but with an identical element size for Figure 7.2b with Figure 7.2a. For
example, even if the minimum size of the element is the order of one grain size and yielding
does not occur in Figure 7.2a, yielding over the area of one to several grains may occur in
Figure 7.2b. In order to make such a comparison of yielding, it is important to determine
the size of the minimum element so that the accurate stress concentration can be obtained
by the elastic analysis of Figure 7.2a. In general, the size of the minimum element should
be about 1/10 of the notch root radius ρ or less. Such a fundamental knowledge of the
elastic–plastic analysis is very important for the application of FEM to strength analysis.

Although it may seem that the way of thinking described above is contradictory to the basic
policy of meshing described in Section 7.1, there is no contradiction because the basic rule in
elastic analysis is applied to comparison of stress concentrations for several notches and the
basic method of meshing element for elastic–plastic analysis is applied after the accurate elastic
analysis is attained.

7.3   Element Meshing for Cracks and Very Sharp Notches

Proceeding with the way of thinking described in the previous section to the limiting cases, we
become aware of the absence of a basic size parameter for the minimum element size in the case
of notch root radius $\rho = 0$ and cracks. For example, in the case of notch root radius $\rho = 0$ at sharp
notches and corners, such as Figure. 7.4a–c, higher stress can be obtained by refining the mesh
size at the notch root. In these problems, it is a typical misunderstanding to think that the calculation with a sufficiently small mesh will guarantee an accurate result. As the tip of these
sharp notches is a *singular point*, unlimited mesh refining results in divergence of the exact
elastic solution and the stress at tip of notch becomes unbounded. Therefore, a special method
is necessary to consider the *singularity* of stress distribution as well as the FEM analysis of
cracks, as shown in Figure 7.5. However, in actual machine components and structures, the
notch root radius ρ produced by machining is not $\rho = 0$, even in the notches and corners of

(a)   (b)   (c)

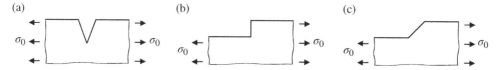

Figure 7.4   Sharp notches and corner

Figure 7.5   Crack as the limit of sharp notch (stress singularity: $r^{-0.5}$)

Figure 7.4a–c, and has a finite value. In such cases too, in general the stress concentration factor K_t can be expressed in the following form:

$$K_t = C_1 + C_2\sqrt{\frac{t}{\rho}} \tag{7.1}$$

Thus, we have to consider how large an influence a very small ρ gives to K_t. If K_t in these cases exceeds 10, cracks will initiate very easily at the stress concentration site even under low remote stress. Accordingly, it is necessary to take measures against structural failure by assuming the existence of a crack at such a stress concentration site. It should be recognized that the strength design for a notch root radius ρ close to 0 is essentially against the common sense of the strength of materials [1, 2]. However, if a crack exists from the beginning, the nature of a stress singularity must be considered for element meshing. In order to pick up the behavior of a singularity in stress distribution for cracks and sharp notches, very fine element meshing has to be made until the singularity can be deduced from the calculated results. After confirming the singularity behavior in a log-log plot of stress σ and distance r from the crack or notch tip, the stress intensity factor K or the *singularity factor* γ (singular eigen value) of $r^{-\lambda}$ is determined.

As already explained in Chapter 2 of Part II, the stress distribution near the crack tip has a singularity of $r^{-0.5}$. If the notch root radius ρ at the notch tip is $\rho = 0$, as in Figure 7.4, in general the stress distribution has a singularity of the following form:

$$\sigma \propto r^{-\lambda} \tag{7.2}$$

The value of the singularity factor λ are different depending on the open angle ϕ of notches and the values are summarized in Table 7.1 and Figure 7.6 [3–5]. Since $\lambda = 0.456$ even for $\phi = 90°$, if notch root radius ρ is $\rho = 0$, these notches may be regarded actually equivalent to a crack for wide range of ϕ.

Table 7.1 Relationship between notch open angle (Figure 7.6) and eigen value λ (free boundary condition [3–5])

ϕ(Deg.)	λ	ϕ(Deg.)	λ	ϕ(Deg.)	λ
0	0.500000				
5	0.499994	65	0.484288	125	0.366960
10	0.499948	70	0.480145	130	0.347730
15	0.499822	75	0.475276	135	0.326417
20	0.499574	80	0.469604	140	0.302835
25	0.499164	85	0.463046	145	0.276782
30	0.498548	90	0.455516	150	0.248025
35	0.497678	95	0.446921	155	0.216302
40	0.496509	100	0.437161	160	0.181304
45	0.494990	105	0.426131	165	0.142669
50	0.493068	110	0.413722	170	0.099956
55	0.490684	115	0.399810	175	0.052629
60	0.487779	120	0.384269	180	0.000000

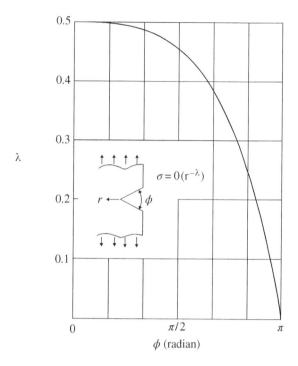

Figure 7.6   Variation of eigen value with open angle for sharp notch at free boundary of 2D plane stress problems [3–5]

Problems of Chapter 7

1. When a plate containing a circular hole as in Figure 7.7 is analyzed by FEM, explain the reason why the stress distribution (and stress concentration factor K_t) is independent of Young's modulus E and Poisson's ratio ν.

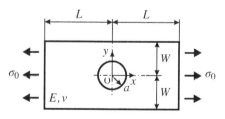

Figure 7.7

2. When a plate containing a circular hole as in Figure 7.8 is analyzed by FEM, explain the reason why the stress distribution (and stress concentration factor K_t) is dependent on Young's modulus E and Poisson's ratio ν.

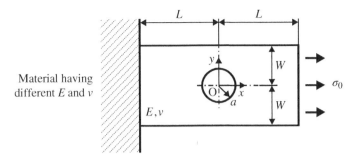

Figure 7.8

3. Figure 7.9a shows a central crack contained in a rectangular plate under tensile stress σ_0 at the plate ends. Figure 7.9b shows the element meshing near the crack. The size s of the minimum element is 1/10 of half the crack length c.

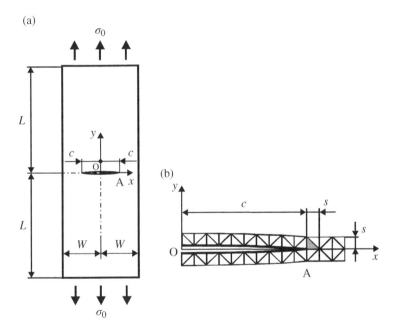

Figure 7.9   (a) $a/W = 0.5$, $L/W = 2.75$. (b) Element meshing near crack tip; $s/c = 0.1$, triangular constant strain element

The stress intensity factor for this crack was obtained by M. Isida [6] as follows:

$$K_I = 1.187\sigma_0\sqrt{\pi c}$$

The stress $\sigma_{y\,tip}$ at the crack tip element (gray-colored element) is determined by FEM as follows.

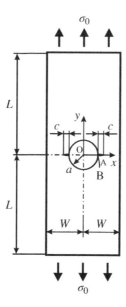

Figure 7.10 Crack emanating from circular hole. $c/a = 0.375$, $a/W = 0.4$, $(a+c)/W = 0.55$, $L/W = 2.75$

$$\sigma_{y\,tip} = 4.04\sigma_0$$

Next, the problem of Figure 7.10 was solved by FEM. The element meshing near the crack tip was made completely the same as Figure 7.9b. In this analysis, the stress $\sigma_{y\,tip}$ at the crack tip element in the y direction was calculated as:

$$\sigma_{y\,tip} = 9.15\sigma_0$$

In the next step, the stress distribution for the plate of Figure 7.10 which contains only the circular hole without crack was calculated with the same element meshing. In this analysis, the stress $\sigma_y{}^*$ in the y direction at the same element as the element at the crack tip of the previous problem was calculated as

$$\sigma_y{}^* = 1.96\sigma_0$$

Based on the above data, determine the stress intensity factor in terms of the correction factor F_I in the following form for the crack of Figure 7.10.

$$K_I = F_I \sigma_0 \sqrt{\pi(a+c)}$$

References

[1] Japan Atomic Energy Agency (2015) Home page. Available on: http://www.jaea.go.jp/; http://www.jnc.go.jp/zmonju/mj_home.html. Accessed on 12 July 2015.

[2] Yomiuri-shinbun Science Division (1996) *Accident Monju*, Mioshin Publishing, pp. 149–221.

[3] M. L. Williams (1951) Surface stress singularity resulting from various boundary conditions in angular corners of plates under bending, *Proc. 1st U.S. Nat. Congr. Appl. Mech.*, **1951**:325–329.

[4] M. L. Williams (1952) Stress singularity resulting from various boundary conditions in angular corners of plates in extension, *Trans. ASME, Ser. E J. Appl. Mech.*, **19**:526–528.

[5] Y. Murakami, ed. (1987) *Stress Intensity Factors Handbook*, vol. **1,2**, Pergamon, Oxford; (1992) *Stress Intensity Factors Handbook*, vol. **3**, Pergamon, Oxford and Soc. Materials Sci., Japan, Kyoto; (2001) *Stress Intensity Factors Handbook*, vol. **4,5**, Society of Materials Science, Japan, Kyoto and Elsevier, Oxford.

[6] M. Isida (1971) Effect of width and length on stress intensity factors of internally cracked plates under various boundary conditions, *Int. J. Fract.*, **7**:301–316.

8

Hollow Cylinder Subjected to Internal or External Pressure

8.1 Basic Solution

Although the problems of *circular hollow cylinder* under internal and external pressure (Figure 8.1) are not directly related to the problem of stress concentration, the solution is useful for the analysis of practical problems of stress concentration.

The solution for this problem is given by the following form [1] (see Chapter 6 of Part I).

$$\left. \begin{aligned} \sigma_r &= C_1 + \frac{C_2}{r^2} \\ \sigma_\theta &= C_1 - \frac{C_2}{r^2} \end{aligned} \right\} \tag{8.1}$$

C_1 and C_2 are constants to be determined by the boundary conditions at $r = a$ and $r = b$.

For internal pressure p_i and external pressure p_o, C_1 and C_2 are determined from the following boundary condition.

$$\sigma_r = -p_i \text{ at } r = a$$
$$\sigma_\theta = -p_o \text{ at } r = b$$

Thus, C_1 and C_2 are determined as follows:

$$\left. \begin{aligned} C_1 &= \frac{p_i a^2 - p_o b^2}{b^2 - a^2} \\ C_2 &= \frac{a^2 b^2 (p_o - p_i)}{b^2 - a^2} \end{aligned} \right\} \tag{8.2}$$

Theory of Elasticity and Stress Concentration, First Edition. Yukitaka Murakami.
© 2017 John Wiley & Sons, Ltd. Published 2017 by John Wiley & Sons, Ltd.

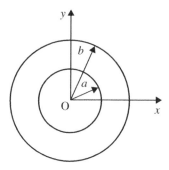

Figure 8.1 Hollow circular cylinder

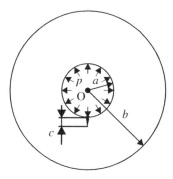

Figure 8.2 Small crack at inner wall of hollow cylinder with internal pressure $p (c \ll a)$

If only internal pressure acts and external pressure is 0, we simply put $p_o = 0$. If internal pressure is 0 and only external pressure acts, we put $p_i = 0$.

8.2 Crack at Inner Wall of Hollow Cylinder Under Internal Pressure

When a small crack exists at the inner wall of a hollow cylinder under internal pressure p, as in Figure 8.2, the method of determination of the stress intensity factor K_{I} will be explained.

When the small crack is absent, the tensile stress acting in the circumferential direction at the prospective location of the crack is approximately expressed by Equation 8.3 (put $r = a$, $p_i = p$, $p_o = 0$ in Equations 8.1 and 8.2; see also Equation 8.10):

$$\sigma_\theta = \frac{b^2 + a^2}{b^2 - a^2} p \tag{8.3}$$

This situation can be imagined as the illustration in Figure 8.3. If we regard Figure 8.2 is equivalent to Figure 2.9, K_{I} may be estimated by the following equation.

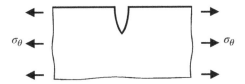

Figure 8.3 Tensile stress acting at inner wall of hollow cylinder

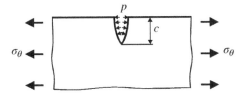

Figure 8.4 Tensile stress at inner wall of hollow cylinder

$$K_I \cong 1.12 \cdot \frac{b^2 + a^2}{b^2 - a^2} p \sqrt{\pi c} \tag{8.4}$$

However, it must be noted that there is an important difference between the problems of Figures 8.2 and 2.9. In the problem of Figure 8.2, the tensile stress σ_θ acts at the inner wall and at the same time the inner pressure p acts on the crack surface. Therefore, the actual situation is not like Figure 8.3 but can be modeled as Figure 8.4, in which the effect of the inner pressure on K_I is considered.

Consequently, the stress intensity factor K_I for the crack of Figure 8.2 can be estimated as follows.

$$K_I \cong 1.12 \cdot (\sigma_\theta + p) \sqrt{\pi c}$$

$$= 1.12 \cdot \left(\frac{b^2 + a^2}{b^2 - a^2} + 1 \right) p \sqrt{\pi c} \tag{8.5}$$

For the structural integrity of high pressure machine components which may contain small scratches and crack-like defects, the effect of inner pressure p on the stress intensity factor must be taken into consideration in addition to the circumferential stress σ_θ at the inner wall.

If defects at the inner wall are not like cracks but have a notch-like shape, the effect of the inner pressure p acting on a notch is a little different from a crack. This problem will be treated in Section 8.4.

8.3 Shrink Fit Problems

The solution for the problem of Figure 8.1 (Equation 8.1) can be applied also to *shrink fit* problems.

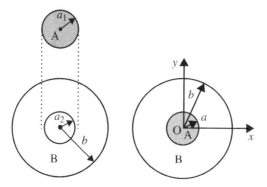

Figure 8.5   Shrink fitting of a solid disk into a hollow cylinder. (a) Before shrink fit. (b) After shrink fit

As shown in Figure 8.5a, let us determine the stress which occurs by shrink-fitting a disk with radius a_1 into a hollow disk having inner radius a_2 and outer radius b.

It is assumed that two disks have the identical Young's modulus E and Poisson's ratio ν. There is a slight difference between the outer diameter of the disk and the inner diameter of the hollow circular disk as $a_1 > a_2$. The nominal sizes of a_1 and a_2 are $a_1 \cong a$, $a_2 \cong a$; and the difference of the diameters $2a_1 - 2a_2 = \delta$ is termed the *shrink fit clearance*. The standard value of δ is about 3/1000 of diameter $2a$. As a_1 is larger than a_2, shrink fit is performed by heating and expanding the hollow circular disk, followed by fitting disk A into disk B when the inner diameter of B becomes larger than the diameter $2a_1$ of A. The stress after cooling to the ambient temperature can be determined by using Equation (8.1) as follows.

Solution 1   After shrink fit, the outer diameter of A is slightly decreased and the inner diameter of B is slightly increased. The shrink fit clearance δ is the sum of the decrease δ_1 in the diameter of A and the increase δ_2 in the inner diameter of B. Namely,

$$\delta = \delta_1 + \delta_2 \tag{8.6}$$

The disk A is subjected to an external pressure. *In this state, the stresses σ_r, σ_θ and other stresses inside disk A are equal everywhere in all directions* (see Appendix A.1.4 of Part II or Example problem 1.1 of Chapter 1 in Part I). Thus, denoting the outer pressure by p, the stresses inside A become everywhere as follows.

$$\left.\begin{array}{c} \sigma_r = -p \\ \sigma_\theta = -p \end{array}\right\} \tag{8.7}$$

Therefore, δ_1 is calculated with the strain ε_θ at the periphery of disk A (strains are equal everywhere inside A) by the following equation (*note that not ε_r but ε_θ is used*):

$$\delta_1 = 2a|\varepsilon_\theta|_{r=a} = 2a\left|\frac{1}{E}(\sigma_\theta - \nu\sigma_r)\right| = \frac{p(1-\nu)}{E}2a \tag{8.8}$$

Likewise, δ_2 is calculated by the stress state at the inner wall of B as follows:

$$\delta_2 = 2a\varepsilon_\theta\Big|_{r=a} = 2a\cdot\frac{1}{E}(\sigma_\theta - \nu\sigma_r)\Big|_{r=a} \tag{8.9}$$

σ_r and σ_θ at $r=a$ are given by Equations (8.1) and (8.2) as follows:

$$\left.\begin{aligned}\sigma_r &= -p \\ \sigma_\theta &= \frac{b^2+a^2}{b^2-a^2}p\end{aligned}\right\} \tag{8.10}$$

Hence,

$$\begin{aligned}\delta_2 &= 2a\cdot\frac{1}{E}\left[\frac{b^2+a^2}{b^2-a^2}p + \nu p\right] \\ &= 2a\cdot\frac{p}{E}\left[\frac{b^2+a^2}{b^2-a^2} + \nu\right]\end{aligned} \tag{8.11}$$

Here, considering $\delta = \delta_1 + \delta_2$,

$$\frac{p}{E}(1-\nu)2a + \frac{p}{E}2a\left[\frac{b^2+a^2}{b^2-a^2} + \nu\right] = \delta \tag{8.12}$$

$$\frac{p}{E}2a\left[(1-\nu) + \frac{b^2+a^2}{b^2-a^2} + \nu\right] = \delta$$

$$\frac{p}{E}2a\frac{2b^2}{b^2-a^2} = \delta$$

$$p = \frac{b^2-a^2}{2b^2}\cdot\frac{\delta}{2a}E \tag{8.13}$$

Thus, when Young's modulus E and Poisson's ration ν of two disks are identical, the pressure generated by shrink fitting is given by Equation 8.13 and the factors influencing p are a, b, δ/a and Young's modulus E. It is interesting to see that Poisson's ratio ν has no relationship with the shrink fit problem of Figure 8.5.

In the above calculations, it must be noted that the absolute value of ε_θ must be used. The reason is that $\varepsilon_\theta < 0$ inside the disk A; and if we calculate $\delta = \delta_1 + \delta_2$ by simply using $\delta_1 = a\varepsilon_\theta$, then p is evaluated smaller than the correct value, because δ_1 is erroneously calculated as negative and δ_2 is positive. This is a typical mistake.

Solution 2   Although Solution 1 is the standard method for the shrink fit problem of Figure 8.5 which considers the deformations of two disks A and B, it is possible to solve the problem only paying attention to the deformation of disk B without treating this problem as a two bodies problem. The method is to start from imaging the final state of the shrink fitted disks.

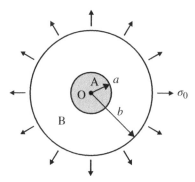

Figure 8.6   Applying tensile stress σ_0 to the periphery of disk B

If we imagine the final state of shrink fit, disk A is fitted into disk B as if they are one disk. In this state, let us apply tensile stress σ_0 to the periphery of disk B (Figure 8.6). Increasing the value of σ_0, the inner diameter of B is increased and finally disk A drops off.

At the final moment when A drops off, the pressure between A and B becomes 0. When the shrink fit pressure p (compressive stress $\sigma_r = -p$) between A and B is reduced to 0 by canceling with the tensile stress applied at the periphery of B, the inner radius of B is increased by δ from the initial size. Therefore, it is necessary only to determine the stress σ_0 to be applied in the radial direction at the outer periphery of B as a one body problem so that the inner diameter of B is increased by δ. At this moment, the condition $\sigma_r = -p + \sigma_0 = 0$ at $r = a$ is satisfied. The following is the calculation:

$$\left.\begin{aligned}
\sigma_r &= \frac{b^2}{b^2 - a^2}\sigma_0 - \frac{a^2 b^2}{(b^2 - a^2)\,r^2}\sigma_0 \\[2mm]
\sigma_\theta &= \frac{b^2}{b^2 - a^2}\sigma_0 + \frac{a^2 b^2}{(b^2 - a^2)\,r^2}\sigma_0
\end{aligned}\right\} \tag{8.14}$$

$$\delta = 2a\varepsilon_\theta = 2a\cdot\frac{1}{E}(\sigma_\theta - \nu\sigma_r) = \frac{\sigma_\theta}{E}2a = \frac{\sigma_0\cdot 4ab^2}{E(b^2 - a^2)} \quad \text{at } r = a \tag{8.15}$$

Hence,

$$\sigma_0 = \frac{b^2 - a^2}{2b^2}\frac{\delta}{2a}E \tag{8.16}$$

This solution agrees with Equation 8.13. This solution is simpler and easier than Solution 1 in which the calculation was done for both disks A and B. It should be noted to avoid misunderstanding that in the shrink fitted state the pressure between A and B is of course compressive and $\sigma_r = -\sigma_0$.

8.3.1 Shrink Fit of Shaft and Hollow Cylinder

In actual shrink fit problems, usually A is a shaft and B is a hollow cylinder and the axial length of A is longer than B. In such a situation, extremely high stress concentration occurs at the contact ends of shrink fit. Figure 8.7 shows examples of FEM analyses for such problems.

8.4 Other Related Problems

This section deals with several practical problems mutually related to previous chapters.

Example 1 Difference between a circular hole under internal pressure (Figure 8.8a) and a circular hole under remote uniaxial tension (Figure 8.8b)

The circumferential stress σ_θ at the internal wall of Figure 8.8a can be obtained by putting $b \to \infty$ in Equation 8.10.

$$\sigma_\theta = p \qquad (8.17)$$

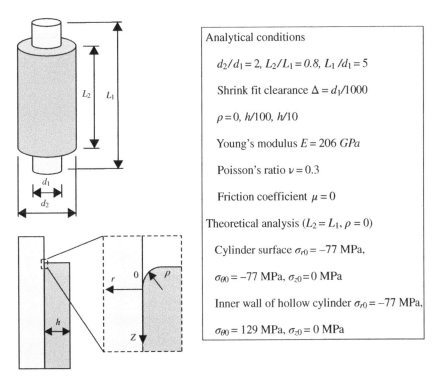

Analytical conditions

$d_2/d_1 = 2$, $L_2/L_1 = 0.8$, $L_1/d_1 = 5$

Shrink fit clearance $\Delta = d_1/1000$

$p = 0$, $h/100$, $h/10$

Young's modulus $E = 206$ GPa

Poisson's ratio $\nu = 0.3$

Friction coefficient $\mu = 0$

Theoretical analysis ($L_2 = L_1$, $\rho = 0$)

Cylinder surface $\sigma_{r0} = -77$ MPa,

$\sigma_{\theta0} = -77$ MPa, $\sigma_{z0} = 0$ MPa

Inner wall of hollow cylinder $\sigma_{r0} = -77$ MPa,

$\sigma_{\theta0} = 129$ MPa, $\sigma_{z0} = 0$ MPa

Figure 8.7 Stress concentration for different lengths of cylinder and hollow cylinder. (a) Stress distribution at surface of cylinder. (b) Stress distribution at inner wall of hollow cylinder. (c) Stress singularity for the case of $\rho = 0$

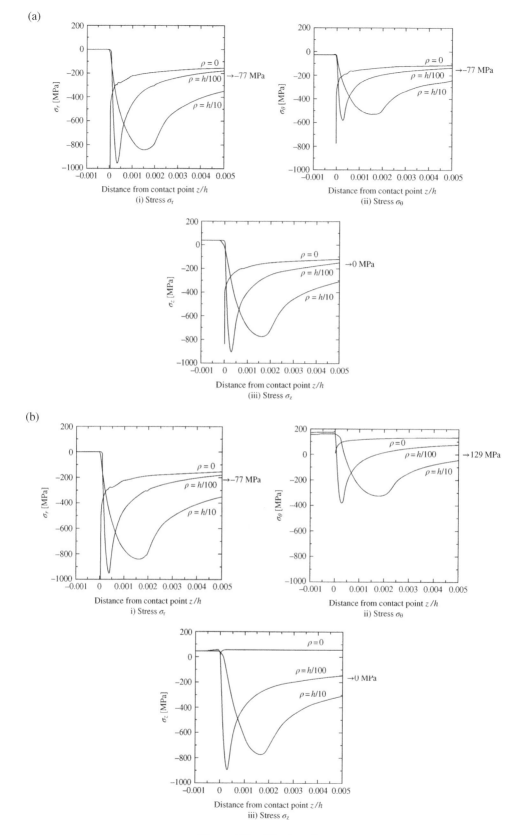

Figure 8.7 (*Continued*)

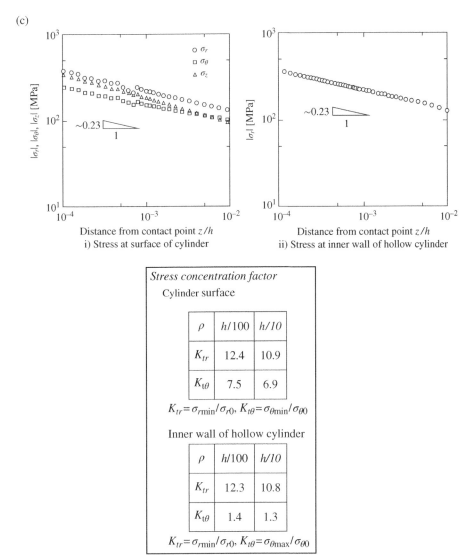

Figure 8.7   (*Continued*)

Alternatively, the circumferential stress at point B of Figure 8.8b is given by

$$\sigma_\theta = 3\sigma_0 \tag{8.18}$$

The distribution of σ_x along the y axis in Figure 8.8a is expressed by

$$\sigma_x = \frac{a^2}{y^2}\sigma_0 \tag{8.19}$$

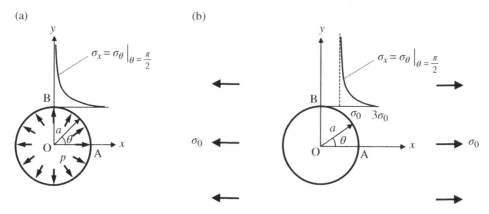

Figure 8.8   Comparison between circular hole under internal pressure p and circular hole under remote tensile stress σ_0 $(p=\sigma_0)$. (a) Circular hole in infinite plate under internal pressure p. (b) Circular hole under remote tensile stress σ_0

Figure 8.9   Stress state near point B of Figure 8.8a

Integrating σ_x from a to infinity, the resultant force acting for $y > a$ is calculated as follows:

$$\int_a^{\infty} \sigma_x dy = \sigma_0 a \tag{8.20}$$

However, in Figure 8.8b, integrating the quantity $(\sigma_x - \sigma_0)$ for $a > 0$ becomes $\sigma_0 a$. Namely, in Figure 8.8b,

$$\int_a^{\infty} (\sigma_x - \sigma_0) dy = \sigma_0 a \tag{8.21}$$

Thus, it can be understood easily that there should be at least a difference of σ_0 between the stress concentrations for these two cases. Where does the remaining σ_0 come from?

In Figure 8.8b, the tensile stress σ_0 in the x direction is released within $y = -a \sim a$ by the presence of the circular hole (the resultant force in the x direction is $\sigma_0 2a$). In Figure 8.8a too, the resultant force $\sigma_0 2a$ in the x direction is released within $y = -a \sim a$. However, in Figure 8.8a, the resultant force $\sigma_0 2a$ acts in the y direction within $x = -a \sim a$ as well as in the x direction. The magnification near point B is illustrated as Figure 8.9.

Recalling the chapters on contact stress (Chapter 13 of Part I and Chapter 13 of Part II), the distributed pressure produces the compressive stress $\sigma_x = -p$ in the direction perpendicular to the distribution of pressure.

In this way, the reason why we have $\sigma_\theta = p = \sigma_0$ at $r = a$ in the problem of Figure 8.8a can be understand as follows:

$\sigma_\theta \cong (\sigma_x - \sigma_0$ at point B of Figure 8.8b) – the compressive stress in Figure 8.9

$$= 3\sigma_0 - \sigma_0 - \sigma_0 = \sigma_0 \tag{8.22}$$

Example 2 Stress concentration of a semi-circular notch subjected to internal pressure (Figure 8.10)

The problem of Figure 8.10 is not completely identical with Figure 8.8a as explained above on the difference between a circular hole and a semi-circular notch. However, we can note some common nature from a mechanics viewpoint in both the problems. Therefore, taking the analysis for the problem of Figure 8.10 into consideration, the approximate value of the maximum tensile stress at the root of a semi-circular notch can be estimated as follows:

$$\sigma_{max} \cong p \tag{8.23}$$

If we analyze this problem by FEM, the maximum stress at the notch root is calculated as

$$\sigma_{max} = 0.73p \tag{8.24}$$

An approximate estimation like Equation 8.23 is useful to avoid simple mistakes by FEM.

Example 3 Stress concentration of V-shaped notch subjected to internal pressure (Figure 8.11)

By the same reason for the difference between parts a and b of Figure 8.8, the stress concentration of Figure 8.11 is similar to Figure 8.12 but not completely the same.

The crucial difference is that the remote stress in Figure 8.11 is 0 (Figure 8.13a) but it is σ_0 in the case of Figure 8.12 (Figure 8.13b). In Figure 8.13a, the resultant force calculated by integrating σ_x from $y = t$ to ∞ is pt for plate thickness 1, as in Equation 8.25:

$$\int_t^\infty \sigma_x dy = pt \tag{8.25}$$

In contrast, the resultant force in the problem of Figure 8.13b is

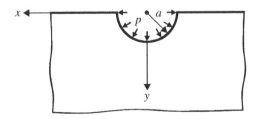

Figure 8.10

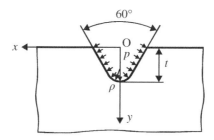

Figure 8.11   V-notch at the edge of semi-infinite plate under internal pressure p

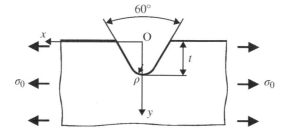

Figure 8.12   V-notch at the edge of semi-infinite plate under remote tensile stress σ_0

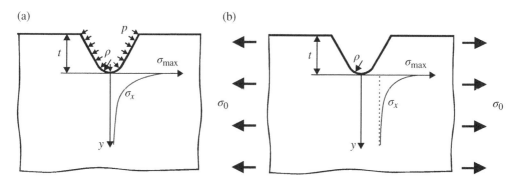

Figure 8.13   Difference of stress concentration between a V-notch subjected to pressure p and a V-notch subjected to remote tensile stress σ_0 $(p = \sigma_0)$. (a) Stress distribution ahead of notch subjected to pressure. (b) Stress distribution ahead of notch subjected to remote tensile stress σ_0

$$\int_{t}^{\infty} \sigma_x dy = \infty .$$

However, the integral for the quantity only above σ_0 is finite as the following equation:

$$\int_{t}^{\infty} (\sigma_x - \sigma_0) dy = \sigma_0 \cdot t \qquad (8.26)$$

The meaning of the above integral is interpreted as follows. In Figure 8.13a, if we imagine the right half section of V-shaped notch, it is illustrated like Figure 8.14 and the resultant force of the distribution of p is in equilibrium with the resultant force of σ_x acting along $y = t \sim \infty$.

On the other hand, the stress distribution along $y = t \sim \infty$ in Figure 8.13b is the superposition of the stress distribution $(\sigma_x = \sigma_0)$ from the edge of the semi-infinite plate without notch $(y = 0)$ to $y = \infty$ with the stress distribution produced by releasing by the introduction of notch the resultant force $\sigma_0 \cdot t$ which is supported by the part of the notch before introduction of notch. The released resultant force is redistributed at $y = t \sim \infty$ over the constant stress σ_0.

Therefore, referring the estimation methods of Example 1 and Example 2, the approximate stress concentration factor K_t for Figure 8.13a is estimated as follows:

$$K_t \cong \left(1 + 2\sqrt{\frac{t}{\rho}} \right) - 1 - 1$$

$$\cong 2\sqrt{\frac{t}{\rho}} - 1 \qquad (8.27)$$

For example, K_t is estimated as $K_t \cong 1.83$ for $\rho = 0.5$, $t = 1$. The value calculated by FEM is $K_t = 1.45$.

Figure 8.15 shows some other examples of comparisons between the simple estimation explained in the above and FEM. The accuracy of the approximation by Equation 8.27 increases with increasing K_t (i.e. increasing the value of t/ρ).

The estimation method explained above is fairly rough. The error of the estimation is caused by the ambiguity of the effect of the force component perpendicular to the direction of the

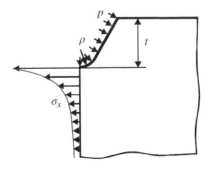

Figure 8.14

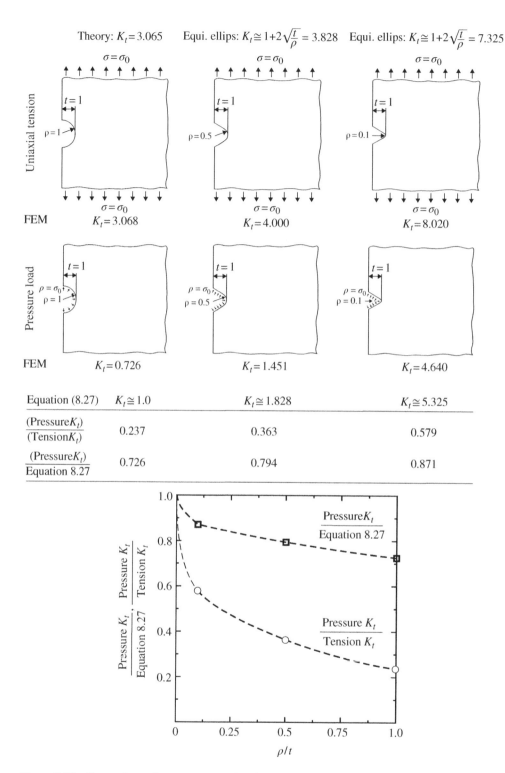

Figure 8.15 Comparison of stress concentration factor between a V-notch subjected to pressure and a V-notch subjected to uniaxial tensile stress: Simple estimation and FEM analysis

applied stress and also by the difference of the effect of a semi-infinite plate (free edge) due to the different shape of notches. This issue will be revealed in the following crack problems.

Example 4   Stress concentration of a crack subjected to pressure (Figure 8.16).

The stress intensity factor K_I for the crack in Figure 8.16 is $K_I = 1.12p\sqrt{\pi a}$. If the value of p is equal to the remote stress σ_0 of Figure 8.17, the stress intensity factors for these two problems are completely equal. This is a point different from Examples 2 and 3. However, it must be noted that the stress distribution σ_x at $y > a$ is different except for the very vicinity of the crack tip. The essence of the difference is the same as the discussion on Figure 8.13.

How does the difference of the stress distribution influence the strength evaluation? The stress σ_x along the y axis in Figure 8.16 decreases with increasing r as:

$$\sigma_x \cong \frac{K_I}{\sqrt{2\pi r}} \tag{8.28}$$

In the problem of Figure 8.17, although the major part of stress near crack tip can be approximated by Equation 8.28, σ_x converges to σ_0 with increasing r. Thus, as the plastic zone size at the crack tip increases, the difference between the two cases becomes noticeable. In other words, as shown in Figure 8.18, the state of the problem of Figure 8.17 is produced by superimposing the problem of a plate without crack under tensile stress (Figure 8.18b) on Figure 8.16.

Examples 1–4 will be useful for strength evaluation of pipes and hollow circular cylinders containing pits and cracks at the inner wall, as shown in Figures 8.19 and 8.20.

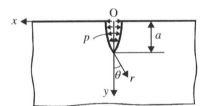

Figure 8.16   Edge crack subjected to pressure

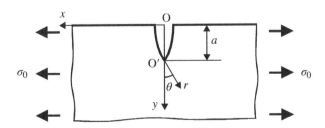

Figure 8.17   Edge crack in plate under remote tensile stress

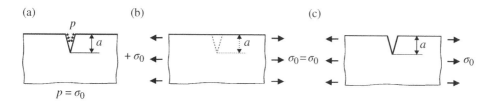

Figure 8.18 Difference of pressure load and uniform tension in crack problem. Equivalent loading condition: (a) + (b) = (c)

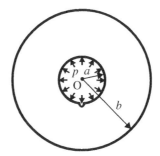

Figure 8.19 Small notch at inner wall of circular hollow cylinder under pressure p

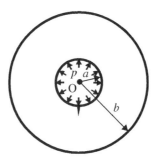

Figure 8.20 Small crack at inner wall of circular hollow cylinder under pressure p

Example 5

In order to determine the stress concentration for the notch in Figure 8.19, it is useful to imagine the stress state at the notch by magnifying the detail of the notch, as in Figure 8.21, so that the problem of Figure 8.19 can be approximated by the superposition of Figures 8.11 and 8.12.

In this approximation, the value of $\sigma_q{}^*$ in Figure 8.21 is calculated by Equation 8.1 with $r = a$ as follows:

$$\sigma_\theta^* = \frac{b^2 + a^2}{b^2 - a^2} p \tag{8.29}$$

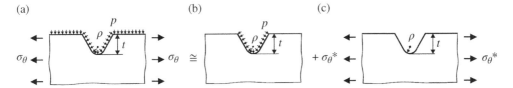

Figure 8.21 The method to estimate the stress concentration factor for the problem of Figure 8.19

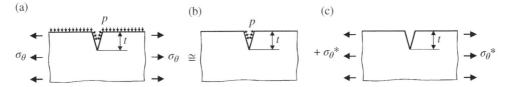

Figure 8.22 Estimation method of stress intensity factor for the problem of Figure 8.20

Thus, the maximum stress at the notch root of Figure 8.19 is estimated approximately by the following equation.

$$\sigma_{max} \cong \left(2\sqrt{\frac{t}{\rho}}-1\right)p+\left(1+2\sqrt{\frac{t}{\rho}}\right)\sigma_\theta^* \qquad (8.30)$$

Example 6
The crack problem of Figure 8.20 can also be solved in the same way (see Figure 8.22). The stress intensity factor is approximately estimated by the following equation:

$$K_I \cong 1.12\left(p+\sigma_\theta^*\right)\sqrt{\pi t} \qquad (8.31)$$

The essence of the above analysis is not to ignore the term of p in Equations 8.30 and 8.31. The error caused by overlooking the term p for pipe lines and pressure vessels with low pressure is not so serious. However, if the pressure is high, the mistake by overlooking p will bring about a fatal accident.

Problem of Chapter 8

1. Explain the reason why the stress intensity factor K_I for the crack in parts a–c of Figure 8.23 is 0.

(a)   (b)   (c)

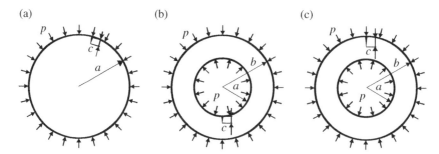

Figure 8.23   (a) Crack at periphery of circular disk subjected to external pressure p. (b) Crack at inner wall of hollow cylinder subjected to internal pressure p and external pressure p. (c) Crack at outer periphery of hollow cylinder subjected to internal pressure p and external pressure p

Reference

[1] S. P. Timoshenko and J. N. Goodier (1970) *Theory of Elasticity*, 3rd edn, McGraw-Hill, New York.

9

Circular Disk Subjected to Concentrated Point Forces

9.1 Basic Solution

The solution of a circular disk subjected to equal concentrated forces P per unit thickness at the two ends of a diameter, as shown in Figure 9.1, can be applied to various practical problems.

The stress σ_x along the y axis and σ_y along the x axis are expressed by the following equations (see Chapter 6 of Part I) [1].

σ_x along y axis:

$$\sigma_x = \frac{2P}{\pi D} \tag{9.1}$$

σ_y along x axis:

$$\sigma_y = \frac{2P}{\pi D}\left[1 - \frac{4D^4}{(D^2 + 4x^2)^2}\right] \tag{9.2}$$

Although the disk is subjected to a compressive load on the top and bottom of the disk, the unique nature of this problem is that σ_x along the diameter from top to bottom (y axis) is tensile and constant along the y axis. This nature is used for the strength testing of brittle materials such as concrete cylinders and rocks.

Theory of Elasticity and Stress Concentration, First Edition. Yukitaka Murakami.
© 2017 John Wiley & Sons, Ltd. Published 2017 by John Wiley & Sons, Ltd.

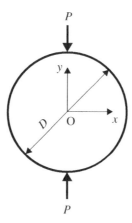

Figure 9.1

9.2 Circular Hole in Circular Disk

The stress concentration for a circular hole at the center of a circular hollow disk subjected to concentrated compressive loads, as shown in Figure 9.2, is dealt with as follows.

First of all, let us assume that the diameter of the hole d is sufficiently small compared to the diameter of the disk D ($d/D \ll 1$). In such a case, the stress state near the circular hole may be imagined as Figure 9.3. The stress $\sigma_x{}^*$ near the hole is approximated by Equation 9.1 and $\sigma_y{}^*$ by putting $x = 0$ in Equation 9.2.

Thus,

$$\sigma_x^* = \frac{2P}{\pi D} \tag{9.3}$$

$$\sigma_y^* = -\frac{6P}{\pi D} \tag{9.4}$$

Hence, this problem can be reduced to a 2D problem (see Chapter 6 of Part I). The maximum stress σ_{max} occurs at point B and the value is calculated as

$$\sigma_{max} = 3\sigma_x^* - \sigma_y^*$$
$$= \frac{12P}{\pi D} \tag{9.5}$$

It must be noted that σ_{max} is six times larger than $\sigma_x{}^*$.

On the other hand, the minimum stress (the maximum compressive stress) occurs at point A and the values is calculated as

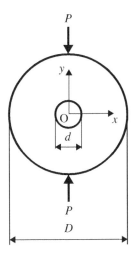

Figure 9.2 Circular hollow disk subjected to concentrated compressive loads

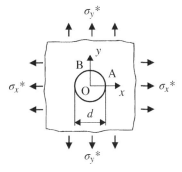

Figure 9.3 Stress field near circular hole for the case of $d/D \ll 1$

$$\sigma_{min} = 3\sigma_y^* - \sigma_x^*$$

$$= -\frac{20P}{\pi D} \tag{9.6}$$

This is compressive and the absolute value is 10 times larger than σ_x^*.

Thus, if a hole is present at the center of disk in the problem of Figure 9.2, the stress concentration is fairly large compared to the nominal stress along the section of the disk.

With increasing value of d/D, the values of σ_{max} and σ_{min} deviate from the values estimated above. Table 9.1 shows the result of a numerical analysis for $d/D = 0.05 \sim 0.8$ (See Part I). The stress concentration factor K_t is defined based on $2P/\pi D$.

Table 9.1 Stress concentration factor at points A and B in terms of d/D (stress without a hole: $2P/\pi D$). See Part I. Source: Murakami 2016 [2]. Reproduced with permission of John Wiley & Sons

d/D	K_{tA}	K_{tB}
0.05	6.095	−10.03
0.1	6.385	−10.11
0.2	7.598	−10.53
0.3	9.856	−11.56
0.4	13.68	−13.70
0.5	20.31	−17.80
0.6	32.84	−25.72
0.7	60.45	−42.99
0.8	140.7	−92.07

9.3 Various Concentrated Forces

As explained in the previous section, the stress concentration for a hole with size $d/D \ll 1$ at the center of a disk can be calculated by knowing the stress state at the center of the disk. Thus, the next step is to find the stress state at the center of a disk under various concentrated forces applied along the periphery of the disk.

The first example is to determine the stresses $\sigma_x{}^*$, $\sigma_y{}^*$ and $\tau_{xy}{}^*$ at the center of a disk subjected to three concentrated forces, $3P$, which are loaded at equal angles of $120°$ around the periphery of the disk, as shown in Figure 9.4. The solution is expressed by the following equations[1].

$$\sigma_x^* = -\frac{3P}{\pi D} \tag{9.7}$$

$$\sigma_y^* = -\frac{3P}{\pi D} \tag{9.8}$$

$$\tau_{xy}^* = 0 \tag{9.9}$$

The next example is the problem of four concentrated forces, $4P$, which are loaded at equal angles of $90°$ around the periphery of the disk. The solution of this problem is easier than the case of $3P$. The solution is only to superimpose the problem of Figure 9.1 onto the problem in which two concentrated forces, $2P$, are loaded at the two ends of the disk in the direction of the x

---

[1] See the solution in the Appendix of Part II.

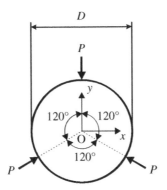

Figure 9.4

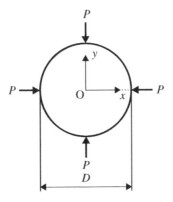

Figure 9.5

axis, as shown in Figure 9.5. The stresses $\sigma_x{}^*$ and $\sigma_y{}^*$ at the center are given as ($\tau_{xy}{}^*$ is naturally 0):

$$\sigma_x^* = -\frac{4P}{\pi D} \tag{9.10}$$

$$\sigma_y^* = -\frac{4P}{\pi D} \tag{9.11}$$

In general, when n concentrated compressive forces P are loaded at equal angles around the periphery of a disk, as in Figure 9.6, the stresses $\sigma_x{}^*$, $\sigma_y{}^*$ and $\tau_{xy}{}^*$ at the center are given by the following equations[2].

---

[2] See the solution in the Appendix of Part II.

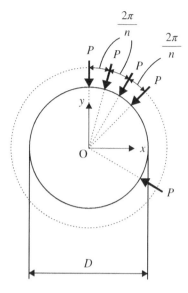

Figure 9.6

$$\sigma_x^* = -\frac{nP}{\pi D}$$

$$\sigma_y^* = -\frac{nP}{\pi D} \qquad (9.12)$$

$$\tau_{xy}^* = 0$$

It is interesting that the stress state at the center of a disk is always in hydrostatic compression regardless of n as long as $n > 2$ when the same magnitude of concentrated compressive forces P are loaded at equal distances around the outer periphery of the disk. Therefore, if a small hole with a sufficiently small diameter compared to disk diameter D is present at the center of a disk, the stress σ_θ^* at the edge of the hole is given by

$$\sigma_\theta^* = -\frac{2nP}{\pi D} \qquad (9.13)$$

By the way, when uniform pressure p is loaded along the outer periphery of a disk, as shown in Figure 9.7, the normal stress is $-p$ and the shear stress is 0 everywhere regardless of direction inside the disk (see Example problem 1.1 in Chapter 1 of Part I). The solution of this problem for the stress state at the center of the disk can be solved as an extension of the problem of Figure 9.6.

The stresses at the center of Figure 9.6 are given by Equation 9.12 as follows.

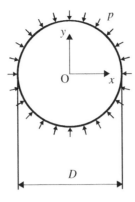

Figure 9.7

$$\sigma_x^* = \sigma_y^* = -\frac{n}{\pi D}\frac{\pi D}{n}p = -p$$

For finite n the stress state is hydrostatic compression $\sigma_x^* = \sigma_y^* = -p$ only near the center of the disk. Reducing $n \to \infty$, the state of Figure 9.6 approaches Figure 9.7 and the zone of the stress state $\sigma_x^* = \sigma_y^* = -p$ is spread to the whole disk.

Problems of Chapter 9

1. Determine the stress intensity factor K_{I} for the short crack contained in the disk of Figure 9.8.

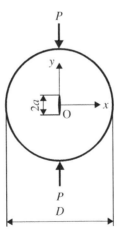

Figure 9.8   Crack contained in a disk subjected to concentrated forces. $a \ll D$, plate thickness: 1

2. Explain the reason that the stress state in the disk of Figure 9.7 is normal stress $\sigma = -p$ and shear stress $\tau = 0$ irrespective of direction.

3. Explain the reason why disk-shaped concretes and rocks compressed as shown in Figure 9.9 fracture by cracking like the dotted line.

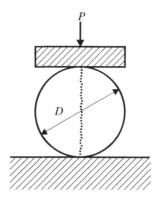

Figure 9.9

Reference

[1] S. P. Timoshenko and J. N. Goodier (1970) *Theory of Elasticity*, 3rd edn, McGraw-Hill, New York.
[2] Y. Murakami (2016) *Theory of Elasticity and Stress Concentration*, Part I , Wiley.

10

Stress Concentration by Point Force

Concentrated force is defined as a force applied at a point. In this chapter concentrated force is termed as *point force*. There is no actual situation in which force acts at a point. However, in the theory of elasticity it is possible to obtain the solution for a force acting at a point. Although the solution is completely theoretical and it may be regarded as unrealistic, the solution has an important nature and is applicable to various practical problems. In this chapter the basic solution and some applications will be introduced.

10.1 Point Force Acting at the Edge of Semi-Infinite Plate

Figures 10.1 and 10.2 shows examples of point force acting on the edge of a semi-infinite plate. The stress field for the problem of Figure 10.1 (see Chapter 6 of Part I) is given by

$$\sigma_r = -\frac{2P}{\pi} \cdot \frac{\cos \theta}{r} \tag{10.1}$$

$$\sigma_\theta = 0 \tag{10.2}$$

$$\tau_{r\theta} = 0 \tag{10.3}$$

The stress field for the problem of Figure 10.2 (see Chapter 6 of Part I) is given by

$$\sigma_r = -\frac{2Q}{\pi} \cdot \frac{\sin \theta}{r} \tag{10.4}$$

Theory of Elasticity and Stress Concentration, First Edition. Yukitaka Murakami.
© 2017 John Wiley & Sons, Ltd. Published 2017 by John Wiley & Sons, Ltd.

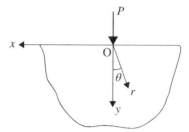

Figure 10.1

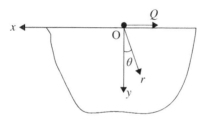

Figure 10.2

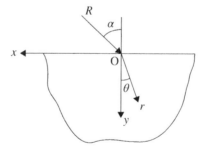

Figure 10.3

$$\sigma_\theta = 0 \tag{10.5}$$

$$\tau_{r\theta} = 0 \tag{10.6}$$

The stress field for an inclined point force, as shown in Figure 10.3, can be obtained by super-imposing Figures 10.1 and 10.2.

The stress fields of Figures 10.1 to 10.3 have a unique characteristic in that, in their $r-\theta$ coordinates, only the stress σ_r exists and other components are $\sigma_\theta = 0$ and $\tau_{r\theta} = 0$. Because of this nature, this unique stress field is named the *simple radial distribution*.

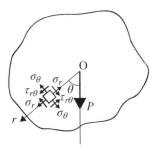

Figure 10.4 Point force acting in infinite plate

10.2 Point Force Acting in Infinite Plate

The stress field by a point force P acting in an infinite plate as shown in Figure 10.4 is given by the following equation.

$$\left.\begin{aligned}
\sigma_r &= -\frac{(3+\nu)}{4\pi}\frac{P\cos\theta}{r} \\
\sigma_\theta &= \frac{1-\nu}{4\pi}\frac{P\cos\theta}{r} \\
\tau_{r\theta} &= \frac{1-\nu}{4\pi}\frac{P\sin\theta}{r}
\end{aligned}\right\} \tag{10.7}$$

Unlike the problem of the previous section, the stress field of this problem is influenced by Poisson's ratio ν. This solution is reduced to an extreme case of solution of the problem of Section 10.3 by putting $a \to 0$.

10.3 Point Force Acting at the Inner Edge of a Circular Hole

The stresses produced by a point force P acting (per unit thickness) at the inner edge of a circular hole contained in an infinite plate are calculated by the following stress function [1].

$$\phi = -\frac{P}{\pi}\left\{\psi r\sin\theta - \frac{1}{4}(1-\nu)r\log r\cos\theta - \frac{1}{2}r\theta\sin\theta + \frac{a}{2}\log r - \frac{a^2}{8}(3-\nu)\frac{1}{r}\cos\theta\right\} \tag{10.8}$$

$$\sigma_r = -\frac{P}{\pi}\left\{\frac{\partial\psi}{\partial r}\sin\theta + \left(\frac{\partial^2\psi}{\partial\theta^2}\sin\theta + 2\frac{\partial\psi}{\partial\theta}\cos\theta\right)\frac{1}{r} - \frac{1}{4}(5-\nu)\frac{1}{r}\cos\theta + \frac{a}{2r^2} + \frac{a^2}{4}(3-\nu)\frac{1}{r^3}\cos\theta\right\}$$

$$\sigma_\theta = -\frac{P}{\pi}\left\{\frac{\partial^2\psi}{\partial r^2}r\sin\theta + 2\frac{\partial\psi}{\partial r}\sin\theta - \frac{1}{4}(1-\nu)\frac{1}{r}\cos\theta - \frac{a}{2r^2} - \frac{a^2}{4}(3-\nu)\frac{1}{r^3}\cos\theta\right\}$$

$$\tau_{r\theta} = \frac{P}{\pi}\left\{\frac{\partial}{\partial r}\left(\frac{\partial\psi}{\partial\theta}\right)\sin\theta + \frac{\partial\psi}{\partial r}\cos\theta + \frac{1}{4}(1-\nu)\frac{1}{r}\sin\theta - \frac{a^2}{4}(3-\nu)\frac{1}{r^3}\sin\theta\right\}$$

$$\tag{10.9}$$

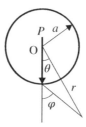

Figure 10.5 Point force P acting in radial direction at a point on the inner edge of a circular hole (plate thickness: 1)

where

$$\psi = \sin^{-1}\left(\frac{r\sin\theta}{\sqrt{r^2-2ra\cos\theta+a^2}}\right)$$

$$\frac{\partial\psi}{\partial r} = \frac{-a\sin\theta}{r^2-2ra\cos\theta+a^2}$$

$$\frac{\partial^2\psi}{\partial r^2} = \frac{2a\sin\theta(r-a\cos\theta)}{(r^2-2ra\cos\theta+a^2)^2}$$

$$\frac{\partial\psi}{\partial\theta} = \frac{r(r-a\cos\theta)}{r^2-2ra\cos\theta+a^2}$$ (10.10)

$$\frac{\partial^2\psi}{\partial\theta^2} = \frac{-ra\sin\theta(r+a)(r-a)}{(r^2-2ra\cos\theta+a^2)^2}$$

$$\frac{\partial}{\partial r}\left(\frac{\partial\psi}{\partial\theta}\right) = \frac{-a\{(r^2+a^2)\cos\theta-2ra\}}{(r^2-2ra\cos\theta+a^2)^2}$$

When the boundary conditions of 2D plane stress problems are given by forces or stresses, the stress distribution is in general independent of Young's modulus E and Poisson's ratio ν [2]. However, the problem of Figure 10.5 is an exceptional case in which Poisson's ratio ν influences the stress distribution. The stress field for the limiting case of $a \to 0$ is reduced to the one (Equation 10.7) produced by a point force P applied in infinite plate which is known being influenced by Poisson's ratio ν [1].

In this regard, there is no contradiction between two solutions.

10.4 Applications of Solutions

Although the solutions for the problems of Figures 10.1 to 10.5 look too simple at first glance and may be considered to be useless for practical applications, they can be applied to various practical problems, such as Figures 10.6 to 10.11. The basic procedure to obtain the stress distribution by the principle of superposition for the problems of distributed loading such as

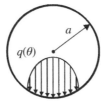

Figure 10.6   Circular hole subjected to distributed contact loading at inner periphery: applicable to stress concentration due to rivet hole

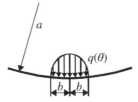

Figure 10.7   Stress distribution due to distributed loading acting on circular surface: Pressure distribution in $[-b, b]$ and $a \gg b$ given in Equation 10.9

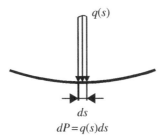

Figure 10.8   Application of the principle of superposition

Figures 10.6 and 10.7 is as follows. For this calculation, P in Equation 10.1 is replaced by distributed loading $q(s)ds$ acting on the infinitesimal division ds and the equation is integrated with respect to s.

In other applications, the solution can be applied to problems such as Figure 10.9a–c. It should be noted that the stress field in the problems of Figure 10.9a–c is not influenced by Poisson's ratio ν. When plural point forces are applied perpendicularly to the inner periphery of arbitrarily shaped holes (Figure 10.10) and the resultant forces of all applied forces in the x and y directions are in equilibrium, respectively, it is proved that the stress field is independent of Poisson's ratio ν [2].

If we put $P = 2a\sigma_0$ in the problem of Figure 10.9a, the maximum stress at the periphery of the hole occurs at points A and B of Figure 10.11 and the value is calculated as

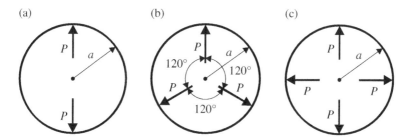

Figure 10.9 Point forces acting in radial direction at inner periphery of circular hole. The stress outside the circular hole is independent of Poisson's ratio ν

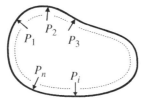

Figure 10.10 Point force acting perpendicularly on inner periphery of hole with arbitrary shape. If the resultant forces are in equilibrium in the x and y directions, respectively, the stress field is independent of Poisson's ν ratio

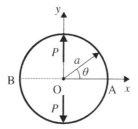

Figure 10.11   Stress at points A and B for $P = 2a\sigma_0$ are $\sigma_q\left(=\sigma_y\right) = \dfrac{4}{\pi}\sigma_0$

$$\sigma_{yA} = \frac{4}{\pi}\sigma_0$$

This value is a little larger than the stress $\sigma_\theta = \sigma_0$ for the case of uniform internal pressure p (in the case of $p = \sigma_0$) but smaller than $3\sigma_0$ due to stress concentration by remote tensile stress σ_0.

Although we may be able to obtain similar results on these problems by FEM, it must be noted that it is difficult to make systematically clear the crucial parameters which control the stress fields in terms of the size of the hole and loading configurations and also difficult to judge whether of not the stress field is influenced by Poisson's ratio ν.

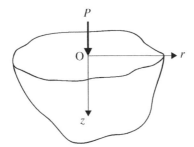

Figure 10.12

10.5   Point Force Acting on the Surface of Semi-Infinite Body

Although the problem of point force P acting on the surface of a semi-infinite body, as shown in Figure 10.12, looks similar to the 2D problem of Figure 10.1, there is an important difference.

In the 2D problem of Figure 10.1, the stress σ_x in the direction parallel to the edge of the semi-infinite plate $(y = 0)$ is $\sigma_x = 0$. However, in the 3D problem of Figure 10.12, the stress σ_r on the surface $(z = 0)$ is tensile, as shown by the following equation.

$$\sigma_r = \frac{P}{2\pi} \cdot \frac{1-2\nu}{r^2} \tag{10.11}$$

This nature is not well known among engineers. It must be noted that the stress on the surface of a semi-infinite body subjected to a compressive point force and locally distributed compressive forces is tensile, as shown above.

References

[1] S. P. Timoshenko and J. N. Goodier (1970) *Theory of Elasticity*, 3rd edn, McGraw-Hill, New York.
[2] Y. Kuranish (1970) *Theory of Elasticity*, reprinted version, Gendai Rikogakusha, Tokyo, p. 85.

11

Stress Concentration by Thermal Stress

Thermal stress occurs due to local temperature difference in a structure. Expansion and contraction are produced by temperature rise and drop, respectively. If the temperature is locally different, regions having different temperatures constrain free deformation to each other and the deviation from the strain corresponding to free deformation (zero stress state) causes thermal stress.

There are not many closed form solutions of stress concentration by thermal stress. As stress distribution under temperature distribution in actual situations is complex, the expression of stresses with closed form functions becomes difficult. Therefore, FEM is suitable for the analysis of thermal stress problems.

This chapter explains a few example of basic thermal stress concentration problems for inclusions. If inclusions contained in a matrix have a different Young's modulus E_I, Poisson's ratio ν_I and *coefficient of thermal expansion* α_I, residual stresses are generated around the inclusions when the material is cooled.

11.1 Thermal Residual Stress Around a 2D Circular Inclusion

This problem is similar to the problem of a circular cylinder in Chapter 8 of Part II. Therefore, the stress concentration factor can be calculated by solving the following basic equations:

$$\left. \begin{array}{l} \sigma_r = C_1 + \dfrac{C_2}{r^2} \\[3mm] \sigma_\theta = C_1 - \dfrac{C_2}{r^2} \end{array} \right\} \tag{11.1}$$

Theory of Elasticity and Stress Concentration, First Edition. Yukitaka Murakami.
© 2017 John Wiley & Sons, Ltd. Published 2017 by John Wiley & Sons, Ltd.

$$\left.\begin{array}{c} \varepsilon_r = \dfrac{1}{E}(\sigma_r - \nu\sigma_\theta) \\[3mm] \varepsilon_\theta = \dfrac{1}{E}(\sigma_\theta - \nu\sigma_r) \end{array}\right\}$$ (11.2)

$$\varepsilon_\theta = \frac{u}{r} \ (u : \text{Radial displacement, see Appendix A.1.5 of Part II})$$ (11.3)

When an inclusion, such as MnS, has a coefficient of thermal expansion α_I larger than that of the matrix, the inclusion contracts more than the matrix during cooling. Assuming perfect bonding at the interface between inclusion and matrix, stress at the interface and in the vicinity of the interface can be calculated as follows.

The free contraction displacement of the inclusion radius is

$$u_I = -\alpha_I \cdot \Delta T \cdot a$$ (11.4)

The free contraction displacement of the radius of the hole where the inclusion is embedded is

$$u_M = -\alpha_M \cdot \Delta T \cdot a$$ (11.5)

Since $|u_I| > |u_M|$, if the inclusion and the matrix are not bonded, the clearance $\delta = (|u_I| - |u_M|)$ is produced between the inclusion and the matrix. If the interface is perfectly bonded, the inclusion is subjected to tensile stress $\sigma_r = \sigma_0$ and the free contraction is restricted. On the other hand, the hole in the matrix is subjected to an identical tensile stress and the size of the hole becomes smaller than the size of the free contraction state. In the final state, the diameters of the inclusion and the hole becomes identical (Figure 11.1). The calculation for this process is as follows.

Stress, strain and displacement δ_I for the side of the inclusion:

$$\sigma_{rI} = \sigma_{\theta I} = \sigma_0 \ (\text{Stress in the } r \text{ direction at interface})$$

$$\left.\begin{array}{c} \varepsilon_\theta = \dfrac{1}{E_I}(\sigma_{\theta I} - \nu_I \sigma_{rI}) = \dfrac{1-\nu_I}{E_I}\sigma_0 \\[4mm] \varepsilon_\theta = \dfrac{\delta_I}{a} \\[4mm] \delta_I = \dfrac{1-\nu_I}{E_I}\cdot\sigma_0\cdot a \end{array}\right\}$$ (11.6)

(a)   (b)

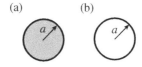

Figure 11.1 Deformation of inclusion and matrix due to temperature drop ΔT. (a) Circular inclusion: E_I, ν_I, α_I. (b) Matrix containing a circular hole: E_M, ν_M, α_M

Stress, strain and displacement δ_M for the side of the matrix:

$$\left.\begin{array}{ll} \sigma_{rM} = \sigma_0 = C_1 + \dfrac{C_2}{a^2} & \text{at } r = a \\[3mm] \sigma_{rM} = C_1 = 0 & \text{at } r = \infty \end{array}\right\} \qquad (11.7)$$

Hence,

$$\sigma_{rM} = \frac{C_2}{a^2}, \sigma_{\theta M} = -\frac{C_2}{a^2}, \qquad (11.8)$$

$$\varepsilon_\theta = \frac{\delta_M}{a}$$

$$\varepsilon_\theta = \frac{1}{E_M}\left(\sigma_{\theta M} - \nu_M \sigma_{rM}\right)$$

$$= -\frac{1}{E_M}\left(\frac{C_2}{a^2} + \nu_M \cdot \frac{C_2}{a^2}\right)$$

$$= -\frac{1 + \nu_M}{E_M} \frac{C_2}{a^2}$$

$$= -\frac{1 + \nu_M}{E_M}\sigma_0 \qquad (11.9)$$

$$\delta_M = -\frac{1 + \nu_M}{E_M} \cdot \sigma_0 \cdot a \qquad (11.10)$$

Since $\delta_I > 0$, $\delta_M < 0$, the radius of the inclusion becomes larger than the state of free contraction and the radius of the hole in the matrix becomes smaller than the state of free contraction. Considering $\delta_I + |\delta_M| = \delta$

$$\frac{1 - \nu_I}{E_I} \cdot \sigma_0 \cdot a + \frac{1 + \nu_M}{E_M} \cdot \sigma_0 \cdot a = \delta = \alpha_I \cdot \Delta T \cdot a - \alpha_M \cdot \Delta T \cdot a$$

It follows that the following tensile stress acts perpendicularly at the interface.

$$\sigma_0 = \frac{(\alpha_I - \alpha_M)\Delta T}{\dfrac{1 - \nu_I}{E_I} + \dfrac{1 + \nu_M}{E_M}} \qquad (11.11)$$

On the other hand, if the coefficient of thermal expansion of the inclusion α_I is smaller than that of the matrix α_M, compressive stress acts perpendicularly at the interface. In such a case, the circumferential stress along the interface of the matrix becomes tensile.

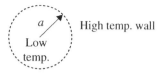

Figure 11.2   Local spot cooled by cold water jet on high temperature wall

11.2   Stress Concentration Due to Thermal Shock

Let's consider the thermal stress generated by quick cooling of a local spot with radius a which occurs by supplying cooling water in high speed to high temperature wall with temperature T_0 (Figure 11.2).

It is thought that due to heat transfer there is no discontinuous temperature difference between the interior of the local spot subjected to cooling water and the surrounding zone. However, in order to determine the upper bound of the thermal stress to be generated, we assume that the temperature becomes $T = 0$ at $r \leq a$ and $T = T_0$ at $r > a$ instantaneously. This situation is similar to the inclusion problem in the previous section. However, since the small spot with radius a shrinks instantaneously, it is assumed that the zone outside $r > 0$ cannot follow the deformation inside the cooled spot. As a result, the small spot $r \leq a$ is subjected to tension from the outside and tries to stay in its original size. Thus, denoting the tensile stress acting at the boundary between the low temperature zone and the high temperature zone by σ_0, we have

$$\varepsilon_\theta = \frac{1}{E}(\sigma_\theta - \nu\sigma_r) = \frac{1-\nu}{E}\sigma_0 \tag{11.12}$$

$$\delta = \alpha \cdot T_0 \cdot a \tag{11.13}$$

Considering $\varepsilon_\theta = \delta/a$ (see Chapter 2 of Part I and Appendix A.1.5 of Part II):

$$\sigma_0 = \frac{E}{1-\nu} \cdot \alpha \cdot T_0 \tag{11.14}$$

Considering the value of α for steels is the order of $\alpha \cong 1 \times 10^{-5}$ and Young's modulus is $E \cong 206\,\mathrm{GPa}$, it can be understood that quite a high tensile stress occurs in the small spot due to *thermal shock*, even for a wall temperature around $T_0 = 300\,°\mathrm{C}$.

In the above calculation, it can be understood from Appendix A.1.4 of Part II that $\sigma_r = \sigma_q = \sigma_0$ for $r \leq a$ (see Example problem 1.1 in Chapter 1 of Part I).

11.3   Thermal Residual Stress Around a Spherical Inclusion

There are various shapes of inclusions. MnS has a long cylindrical shape elongated by the rolling process. $Al_2O_3(CaO)_x$ maintains a spherical shape without deformation even after the rolling process. The tensile stress around a spherical inclusion due to external loading and thermal stress induced by the cooling process are summarized in Figures 11.3 and 11.4 [1].

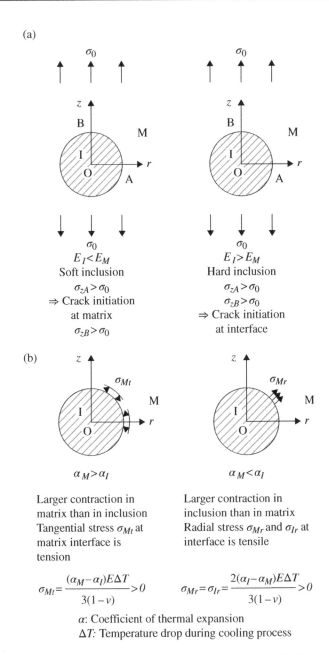

Figure showing stress around a spherical inclusion.

(a)

σ_0 σ_0

σ_0 σ_0

$E_I < E_M$ $E_I > E_M$

Soft inclusion   Hard inclusion

$\sigma_{zA} > \sigma_0$ $\sigma_{zA} > \sigma_0$

$\Rightarrow$ Crack initiation $\sigma_{zB} > \sigma_0$

at matrix $\Rightarrow$ Crack initiation

$\sigma_{zB} > \sigma_0$ at interface

(b)

$\alpha_M > \alpha_I$ $\alpha_M < \alpha_I$

Larger contraction in matrix than in inclusion Tangential stress σ_{Mt} at matrix interface is tension

Larger contraction in inclusion than in matrix Radial stress σ_{Mr} and σ_{Ir} at interface is tensile

$$\sigma_{Mt} = \frac{(\alpha_M - \alpha_I)E\Delta T}{3(1-v)} > 0$$

$$\sigma_{Mr} = \sigma_{Ir} = \frac{2(\alpha_I - \alpha_M)E\Delta T}{3(1-v)} > 0$$

α: Coefficient of thermal expansion

ΔT: Temperature drop during cooling process

Figure 11.3 Stress around a spherical inclusion (I: inclusion, M: matrix). (a) Stress around a spherical inclusion under remote tensile stress σ_0. (b) Stress around a spherical inclusion due to the difference in coefficients of thermal expansion (case for Young's modulus $E_I = E_M$, Poisson's ratio $v_I = v_M$)

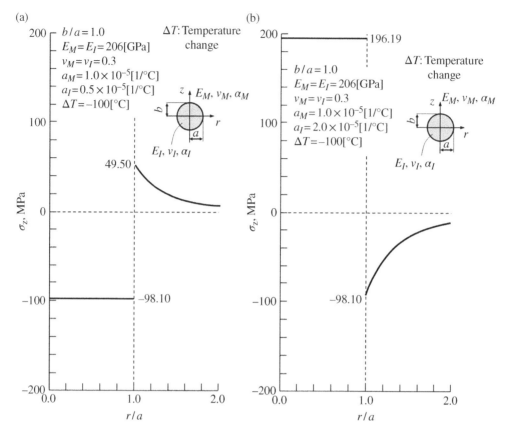

Figure 11.4 Distribution of stress σ_z on the plane $z = 0$ for a temperature change of $\Delta T = -100\,°C$ in an infinite body (matrix) containing a spherical inclusion. (a) $\alpha_I = 0.5 \times 10^{-5}$, $\alpha_M = 1.0 \times 10^{-5}$. (b) $\alpha_I = 2.0 \times 10^{-5}$, $\alpha_M = 1.0 \times 10^{-5}$ [1]. Source: Murakami 2002 [1]. Reproduced with permission of Elsevier

Reference

[1] Y. Murakami (2002) *Metal Fatigue: Effects of Small Defects and Nonmetallic Inclusions*, Elsevier, New York.

12

Stress Concentration Due to Dislocations

12.1 Basic Equations

Dislocation is a linear defect in a crystal. Therefore, the stress concentration due to dislocation can be interpreted as the stress concentration caused by a kind of geometrical discontinuity. The most typical dislocation is an *edge dislocation*, as shown in Figure 12.1.

If the line of an edge dislocation is along the z axis and the Burgers vector b is parallel to the x axis, the stress around the dislocation is given by Equation 12.1 [1].

$$
\left.
\begin{aligned}
\sigma_x &= -\frac{Gb}{2\pi(1-\nu)}\cdot\frac{y(3x^2+y^2)}{(x^2+y^2)^2} \\[2mm]
\sigma_y &= \frac{Gb}{2\pi(1-\nu)}\cdot\frac{y(x^2-y^2)}{(x^2+y^2)^2} \\[2mm]
\sigma_z &= -\frac{G\nu b}{\pi(1-\nu)}\cdot\frac{y}{x^2+y^2} \\[2mm]
\tau_{xy} &= \tau_{yx} = \frac{Gb}{2\pi(1-\nu)}\cdot\frac{x(x^2-y^2)}{(x^2+y^2)^2} \\[2mm]
\tau_{yz} &= \tau_{zx} = 0
\end{aligned}
\right\}
\qquad (12.1)
$$

The stress in the cylindrical coordinate is given by the following equation.

Theory of Elasticity and Stress Concentration, First Edition. Yukitaka Murakami.
© 2017 John Wiley & Sons, Ltd. Published 2017 by John Wiley & Sons, Ltd.

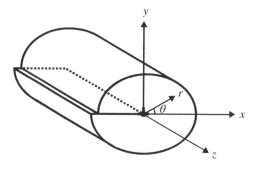

Figure 12.1   Displacement around edge dislocation

$$\left.\begin{array}{l} \sigma_r = \sigma_\theta = -\dfrac{Gb}{2\pi(1-\nu)}\dfrac{\sin\theta}{r} \\[2mm] \tau_{r\theta} = \dfrac{Gb}{2\pi(1-\nu)}\dfrac{\cos\theta}{r} \\[2mm] \sigma_z = \dfrac{-G\nu b}{\pi(1-\nu)}\dfrac{\sin\theta}{r} \end{array}\right\} \tag{12.2}$$

As seen in the above equations, the stress at the core of a dislocation becomes unbounded within elastic theory. Although it may not be realistic to discuss the exact stress at the core of a dislocation which exists in a domain within the atomic order, it can be presumed that an extremely high stress concentration occurs at least around the core of dislocation. It has been described in several chapters of this book that it is not a rational way of thinking to discuss fracture and yielding based on the stress at a point or a very small region. If we discuss fracture using only the maximum stress, we reach the unrealistic conclusion that the extremely high stress induced at the core of a dislocation by a small applied stress can immediately lead a material to fracture.

12.2   Stress Concentration Due to Slip Bands

Let us consider a stress concentration due to pile-up dislocations which are blocked by obstacles at $x = \pm a$, as shown in Figure 12.2. This problem is a model of a slip band generated in one grain among polycrystals. There is resistance against the movement to dislocations on a slip band plane. The resistance is termed the *friction force of dislocation*. Denoting the friction force by τ_f, the stress concentration produced at neighboring grains by the dislocation arrays on the *slip line* $(x = -a \sim a)$ under remote shear stress τ_0 is expressed by the following equation [1]:

$$\tau_{xy} = \frac{(\tau_0 - \tau_f)x}{\sqrt{x^2 - a^2}} + \tau_f \tag{12.3}$$

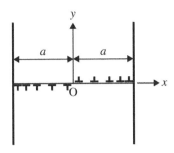

Figure 12.2   Pile-up dislocations on a slip line blocked by obstacles

For $\tau_f = 0$, Equation 12.3 is reduced to Equation 2.9. This means that, under the absence of friction force or dislocation motion, the slip band of Figure 12.2 is mechanically equivalent to a crack. Thus, as *slip bands* initiate with an increase in the applied stress τ_0, the stress distribution in the vicinity of the tip of the slip bands becomes similar to that of a crack.

Near the tip of the slip bands (or eventually a crack) at $x = a + r$, $r \ll a$, Equation 12.3 is expressed as

$$\tau_{xy} = \frac{(\tau_0 - \tau_f)(a + r)}{\sqrt{(a+r)^2 - a^2}} \cong \frac{(\tau_0 - \tau_f)\sqrt{a}}{\sqrt{2r}} = \frac{(\tau_0 - \tau_f)\sqrt{\pi a}}{\sqrt{2\pi r}} \tag{12.4}$$

The numerator of Equation 12.4 can be regarded as the stress intensity factor of Mode II, K_{II}. It follows that the stress intensity factor K_{II} for slip bands is defined as

$$K_{II} = (\tau_0 - \tau_f)\sqrt{\pi a} \tag{12.5}$$

From the above analysis, it is noted that the effect of *slip bands* to neighboring grains appears through stress $(\tau_0 - \tau_f)$ in Equation 12.5 and *grain size* $(2a)$.

As schematically shown in Figure 12.3, even if slip bands initiate first in one grain, the neighboring grains do not necessarily have slip planes in a crystallographically convenient direction for slip under an applied remote shear stress τ_0. In order to activate slip in the neighboring grains, K_{II} in Equation 12.5 needs to exceed a certain critical value K_{IIC}. It follows that the smaller the grain size, the higher the stress τ_0 necessary for activating slip or yielding of neighboring grains. In this case, grain size influences through the square root of grain size, that is $\sqrt{a}$. The rationality of the *Hall–Petch equation* [2, 3] (Equation 12.6) can be understood from the above viewpoint, because it relates the yield stress τ_Y (or σ_Y) of poly-crystalline materials with grain size (d) as the following equation:

$$\tau_Y = C_1 + \frac{C_2}{\sqrt{d}} \tag{12.6}$$

Equation 12.6 can be interpreted as $(\tau_Y - C_1)\sqrt{d} = \text{Constant}$.

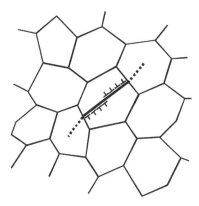

Figure 12.3 Stress concentration due to slip bands in poly-crystals

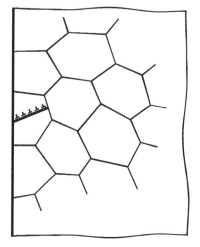

Figure 12.4 Slip bands in a surface grain

Based on the above concept that slip bands are mechanically equivalent to a crack, we can understand that grains at a material surface are likely to glide more easily compared to subsurface grains with the same size. Comparing Figures 12.3 and 12.4, a surface grain is mechanically equivalent to a subsurface grain of twice its size.

References

[1] H. Suzuki (1967) *Introduction to Dislocation Theory*, Agune, Tokyo.
[2] E. O. Hall (1951) The deformation and aging of mild steel : III Discussion of results, *Proc. Phys. Soc. Sec. B*, **64**:747–753.
[3] N. J. Petch (1953) The cleavage strength of polycrystals, *J. Iron and Steel Inst.*, **174**:25–28.

13

Stress Concentration Due to Contact Stress

13.1 Basic Nature of Contact Stress Field

Stress concentration due to *contact stress* is difficult to treat compared to other problems, because a consideration of stress concentration only due to contact stress cannot solve strength problems, especially fatigue strength problems. Especially in *fretting fatigue*, since cyclic tensile stress acts in addition to contact stress, the role of contact stress must be separately or mutually considered with the influence of other stresses.

When a uniform contact stress q acts on the edge of a wide plate, as shown in Figure 13.1, the stress σ_y within $x = -b \sim b$ at the edge of the plate $(y = 0)$ is obviously given as

$$\sigma_y = -q \tag{13.1}$$

However, the stress σ_x at the same edge of the plate is not well recognized. The value of σ_x is equal to σ_y (see Chapter 13 of Part I). Namely,

$$\sigma_x = -q \tag{13.2}$$

The nature of this stress field is the basis for the analysis of stress concentration at a small defect present at the edge of a plate under contact stress, as shown Figure 13.2.

When the contact stress acts through a lubricant between two bodies, the stress distribution near the defect will be like Figure 13.3. Namely, a pressure identical to the contact stress acts on the surface of the defect. Thus, the stress concentration of this problem is similar to that of a defect present at the inner wall of a hollow circular cylinder under internal pressure explained in Chapter 8 of Part II. But in the case of contact stress, it must be noted that the major stress σ_x is compressive. Therefore, the stress intensity factor K_I for a problem like Figure 13.3b is $K_I = 0$ or $K_I < 0$.

Theory of Elasticity and Stress Concentration, First Edition. Yukitaka Murakami.
© 2017 John Wiley & Sons, Ltd. Published 2017 by John Wiley & Sons, Ltd.

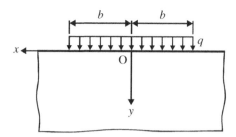

Figure 13.1   Uniform contact stress

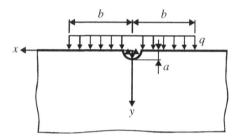

Figure 13.2

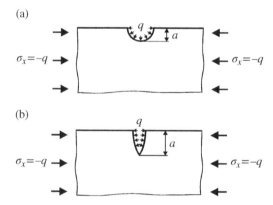

Figure 13.3   Stress field near defect. (a) Semi-circular notch. (b) Crack

In general, as the contact width is limited, as in Figure 13.4, the stress intensity factor K_I changes from a negative to a positive value due to the mutual influence of increase in crack size and primary stress σ_x. If no lubricant exists between two contacting bodies, the pressure q considered in the above analysis may be excluded from the influence factor to defect.

The way of thinking described above must be considered for FEM analysis as the basic mechanics factor.

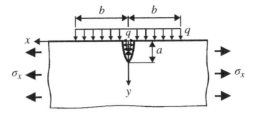

Figure 13.4

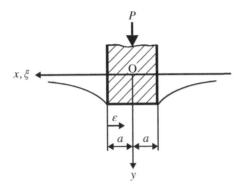

Figure 13.5   Compression by 2D rigid punch. P is the load per unit plate thickness

13.2   Stress Concentration Due to 2D Rigid Punch

Applying a compressive load P on a 2D *rigid punch* on the edge of an elastic plate, as shown in Figure 13.5, a constant displacement is made on the contact plane. The pressure distribution on the contact plane is given by the following equation (see Chapter 13 of Part I) [1].

$$q(\xi) = \frac{P}{\pi\sqrt{a^2-\xi^2}} \quad (|x| < a) \tag{13.3}$$

Putting $\xi = a - \varepsilon$, the above equation is reduced to

$$q(\varepsilon) = \frac{P}{\pi\sqrt{a^2-(a-\varepsilon)^2}} = \frac{P}{\pi\sqrt{2a\varepsilon-\varepsilon^2}}$$

$$\cong \frac{P}{\pi\sqrt{2a\varepsilon}} = \frac{P}{\sqrt{\pi a}}\frac{1}{\sqrt{2\pi\varepsilon}} = \frac{K_{\mathrm{I}}}{\sqrt{2\pi\varepsilon}} \tag{13.4}$$

where

$$K_{\mathrm{I}} = \frac{2}{\pi}q_0\sqrt{\pi a}, q_0 : \text{Average contact stress } P/2a \tag{13.5}$$

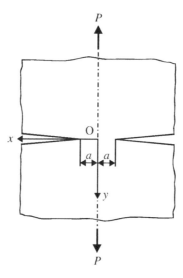

Figure 13.6

Thus, the pressure distribution under the rigid punch is the same as the stress singularity for a crack having the stress intensity factor $K_I = \dfrac{2}{\pi}q_0\sqrt{\pi a}$, though the stress is compressive.

Although the stress under the rigid punch of Figure 13.5 is compressive, if we imagine a problem like Figure 13.6 in which an infinite plate containing infinitely deep cracks are connected with the central part within width $2a$ in between and is subjected to a remote tensile load P, the stress distribution becomes tensile in contrast to Figure 13.5. The stress intensity factor K_I for this case has the same absolute value as $K_I = \dfrac{2}{\pi}\sigma_0\sqrt{\pi a}$, $\sigma_0 = \dfrac{P}{2a}$, where the stress ahead of the crack is positive.

13.3   Stress Concentration Due to 3D Circular Rigid Punch

As a 3D problem similar to the 2D problem (Figure 13.5), a circular rigid punch like Figure 13.7 is considered. The pressure distribution under the punch is given by (see Chapter 13 of Part I)

$$q(r) = \frac{P}{2\pi a\sqrt{a^2 - r^2}} \tag{13.6}$$

Expressing the stress singularity at infinitesimal distance ε from the outer boundary of the punch, the singularity similar to 2D problem occurs as

$$q(\varepsilon) \simeq \frac{q_0\sqrt{\pi a}}{2\sqrt{2\pi\varepsilon}} = \frac{K_I}{\sqrt{2\pi\varepsilon}} \tag{13.7}$$

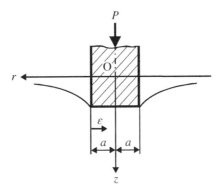

Figure 13.7 Compression by circular rigid punch

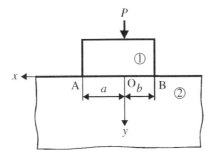

Figure 13.8

where

$$K_{\mathrm{I}} = \frac{1}{2} q_0 \sqrt{\pi a}, \quad q_0 : \text{Average contact stress } P/\pi a^2 \tag{13.8}$$

13.4 Stress Concentration Due to Contact of Two Elastic Bodies

Under the contact of two elastic bodies, as shown in Figure 13.8, a high stress concentration occurs at the contact ends A and B. The intensity of the stress concentration is influenced by rounding and coning of the corner at the end points A and B of the elastic body ① (Figure 13.9). The shrink fit problem shown in Section 8.3 is one such example.

The stress concentration at the contact area is influenced also by the friction coefficient. Since in general the stress concentration in practical problems is influenced by complicated boundary conditions of the contact area, the application of FEM is suitable for the analysis. However, a basic knowledge on the basic nature of a contact stress field and the stress singularity at the contact area is essentially necessary.

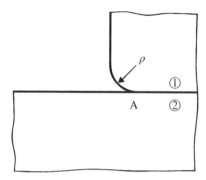

Figure 13.9   Magnification of part of A in Figure 13.8

Reference

[1] M. Sadowsky (1928) Zweidimensionale Probleme der Elastizitätstheorie, *Z. Angew. Math. Mech.*, **8**:107–121.

14

Strain Concentration

When stress and strain are in an elastic state, strains can be expressed with stresses based on Hooke's law. Thus, problems of strain concentration in an elastic state are equivalent to those of elastic stress concentration. However, once the stress states satisfy the yield condition of the material, the *strain concentration* factor deviates from the value of the stress concentration (Figure 14.1).

Elastic–plastic analysis is much more difficult compared to elastic analysis. There is no perfect and systematic strain concentration solutions for a notch under elastic–plastic conditions. Historically, some simple estimation methods of strain concentration have been proposed. One typical method is the *Neuber rule*. Once the notch tip zone yields, strain concentration increases and stress concentration decreases. Considering this nature of stress–strain behavior, Neuber proposed the following simple formula [1].

$$\sqrt{K_\sigma \cdot K_\varepsilon} = K_t \qquad (14.1)$$

where K_t is the elastic stress concentration factor, K_σ is the plastic stress concentration factor and K_ε is the plastic strain concentration factor. K_σ and K_ε are defined by the following equation:

$$K_\sigma = \sigma_{\max}/\sigma_n, \, K_\varepsilon = \varepsilon_{\max}/\varepsilon_n \qquad (14.2)$$

where σ_n is the nominal stress and ε_n is the nominal strain corresponding to σ_n on the $\sigma - \varepsilon$ curve. In Equation 14.1, within elastic plane stress conditions K_ε agrees with K_σ.

Neuber derived Equation 14.1 based on anti-plane shear (transverse shear) theory. Therefore, it must be noted that there is no theoretical basis for 2D plane stress problems, plane strain

Theory of Elasticity and Stress Concentration, First Edition. Yukitaka Murakami.
© 2017 John Wiley & Sons, Ltd. Published 2017 by John Wiley & Sons, Ltd.

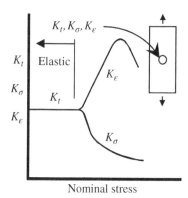

Figure 14.1 Change in stress concentration factor and strain concentration factor in terms of nominal stress increase

problems and 3D problems. There are many applications and extensions of the Neuber rule. However, nowadays we can obtain realistic and practically useful solutions by FEM.

Strain concentration is strongly influenced by *plastic constraint*, namely by the extension morphology of the *plastic zone*. Figure 14.2 shows FEM analyses of the growth of a plastic zone for the following four cases [2]:

(a) Tension of thin plate containing a notch (plane stress).
(b) Tension of thick plate containing a notch (plane strain).
(c) Tension of cylindrical specimen containing a notch.
(d) Torsion of cylindrical specimen.

In the cases of (b) and (c), an *elastic core* (Figure 14.2b, c) close to a tri-axial tension state is formed at the center of the specimen even after *general yielding*. These elastic cores are produced across a minimum section of the specimen as the result of *plastic constraint* and prevent growth of the plastic zone into the center of the specimen. As a result, the strain concentration at the notch tip under equal nominal stress is lower in the case of higher strain constraint.

Figure 14.3 shows the experimental observation of the internal growth of a plastic zone. The plastic zones were revealed by a special etching method.

Figure 14.4 shows the strains measured at different locations of notch root across the thickness. In the case of a thin plate, the strain is almost uniform across the thickness, and with increasing plate thickness, the values of strain become different at different locations on the notch root across the thickness. The maximum strain occurs at the midpoint of the notch root, as shown in Figure 14.4c. Figure 14.5 indicates that a difference of strains at the surface (the notch root edge) and the midpoint cannot be ignored. Compared to the case of a thin plate (Figure 14.5a), the strain at the notch root for a thick plate (Figure 14.5c) is lower under the same nominal stress. The plastic constraint is the reason for this difference.

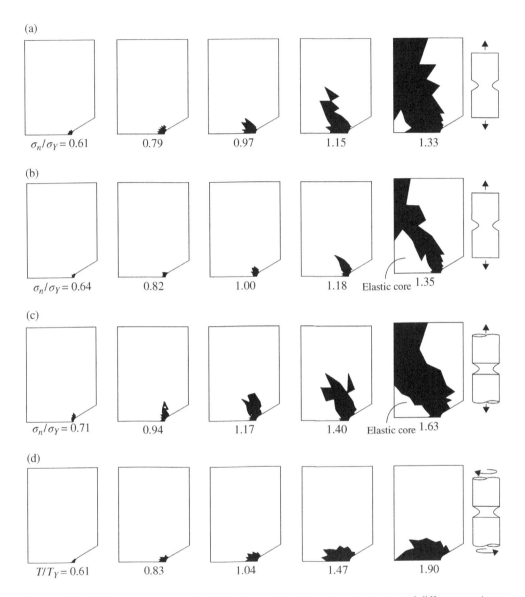

Figure 14.2 FEM analyses of growth of plastic zone for different loading types and different specimen shapes [2]. Only 1/4 of the specimens and plastic zones are shown from the problem symmetry. (a) Tension of notched plate (plane stress). (b) Tension of notched plate (plane strain). (c) Tension of notched cylindrical bar. (d) Torsion of notched cylindrical bar, $T_Y = (\pi d^3/12)(\sigma_Y/\sqrt{3})$. σ_n: Nominal stress; σ_Y: yield stress; d: diameter of minimum section; T_Y: general yield twisting moment. Source: Murakami 1977 [2]. Reproduced with permission of The Japan Society of Mechanical Engineers

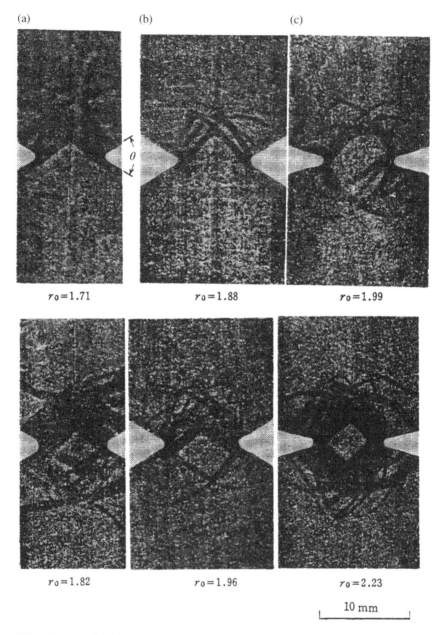

(a)   (b)   (c)

$r_0 = 1.71$ $r_0 = 1.88$ $r_0 = 1.99$

$r_0 = 1.82$ $r_0 = 1.96$ $r_0 = 2.23$

10 mm

Figure 14.3 Growth of yield zone of notched cylindrical bar (0.10% C annealed steel [2]). (a) $d = 8.0$, $D = 12.0$, $\rho = 0.4$, $\theta = 60°$. (b) $d = 8.0$, $D = 16.0$, $\rho = 0.4$, $\theta = 60°$. (c) $d = 8.0$, $D = 16.0$, $\rho = 0.4$, $\theta = 30°$. r_0: Nominal stress at minimum section/yield stress, σ_n/σ_Y; d: diameter of minimum section; D: outer diameter; ρ: notch root radius; θ: notch open angle. Source: Murakami 1977 [2]. Reproduced with permission of The Japan Society of Mechanical Engineers

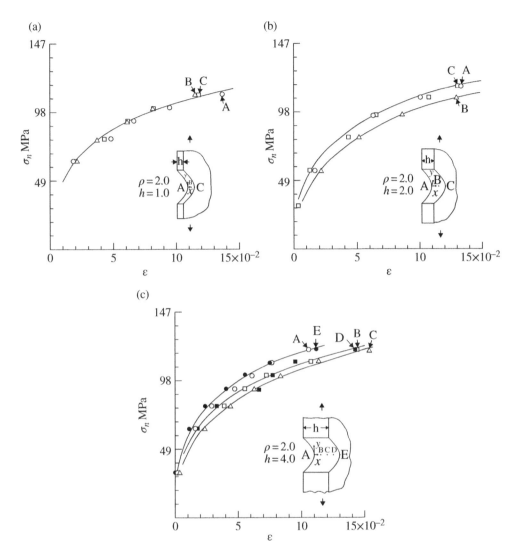

Figure 14.4   Strain in tensile direction at different locations of notch root [2]. In the case of a thick plate having a larger thickness h relative to the notch root radius ρ, the strain at the midpoint of the thickness is notably larger than that at the surface. (a) $h/\rho = 0.5$. (b) $h/\rho = 1.0$. (c) $h/\rho = 2.0$. h: Plate thickness; ρ: notch root radius; σ_n: nominal stress. Source: Murakami 1977 [2]. Reproduced with permission of The Japan Society of Mechanical Engineers

It must be noted that the point of the discussion of the thickness effect is *not the absolute size of the plate thickness*. The concept of *thick or thin* in the strength of materials must be judged by the relative value of thickness to notch root radius as shown in Figure 14.4.

Figure 14.6 compares the stress and strain distributions for the tension of a notched plate, the tension of a notched cylindrical bar and the torsion of a notched cylindrical bar. Looking at Figure 14.6 together with the results of Figures 14.3–14.5 and by imagining the plastic constraint produced by the growth pattern of the plastic zone, the reason for the difference between

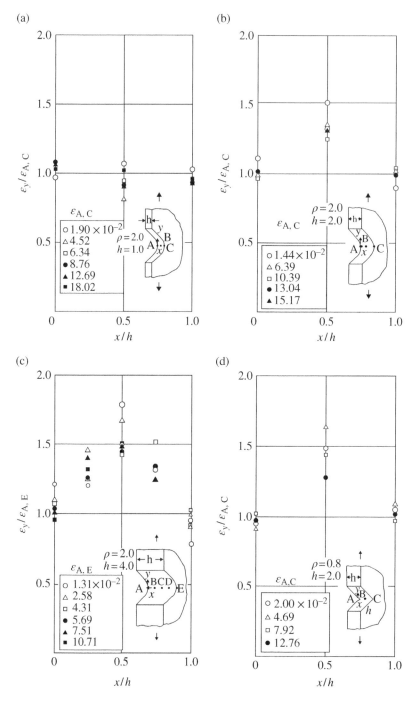

Figure 14.5   Strain in tensile direction at different locations of thickness at notch root [2]. (a) $h = 1.0$ mm, $\rho = 2.0$ mm. (b) $h = 2.0$ mm, $\rho = 2.0$ mm. (c) $h = 4.0$ mm, $\rho = 2.0$ mm. (d) $h = 2.0$ mm, $\rho = 0.8$ mm. Source: Murakami 1977 [2]. Reproduced with permission of The Japan Society of Mechanical Engineers

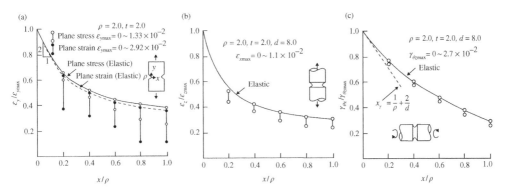

Figure 14.6   FEM analysis of strain distribution ahead of notch [2]. The vertical lines connecting the calculated values show the range of variation of calculated values. (a) Tension of notched plate. (b) Tension of notched bar. (c) Torsion of notched bar. Source: Murakami 1977 [2]. Reproduced with permission of The Japan Society of Mechanical Engineers

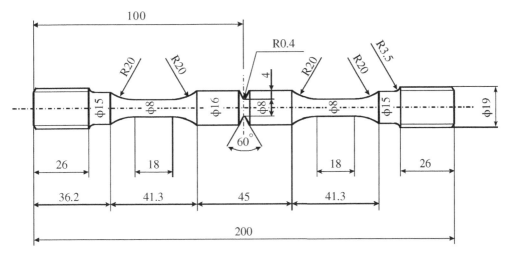

Figure 14.7   Specimen for comparison of tensile fracture strength at the notched part and at the smooth gage part with the same diameter (0.10% C annealed specimen). Notch root radius $\rho = 0.4$ mm, notch depth $t = 4$ mm, diameter of minimum section at notched part $d = 8$ mm, stress concentration factor at notched part $K_t = 7.32$

these three cases can be understood. Thus, the strain concentration and strain distribution at a notch are strongly influenced by plane stress state, plane strain state, loading modes and specimen shapes.

Example problem 14.1
Predict the location of a fracture by tension of the notched specimen of Figure 14.7 and explain the reason. Assume that the material is a ductile steel. The specimen has the same shape for both

the right and left half sides. The minimum diameter at the notched section and the diameter at the smooth gauge section are equal, $d = 8$ mm. The specimen has thread parts at both ends for a testing attachment.

Answer

As shown in Figure 14.8b, this specimen fractured at the smooth gage section after the necking (for detail, see Figure 14.9). During the tensile test, an elastic core with tri-axial tension, as shown in Figures 14.2 and 14.3, is developed at the central part of the notched section and then the maximum endurable load with plastic deformation became higher than that of the smooth gage section. Thus, for the strength design against static loading only the analysis of the notched part is not sufficient. On the other hand, it must be noted that *in the case of cyclic loading the notched part becomes the fracture origin*. Thus, analysis from the viewpoint of fatigue is necessary (Table 14.1).

(a)

(b)

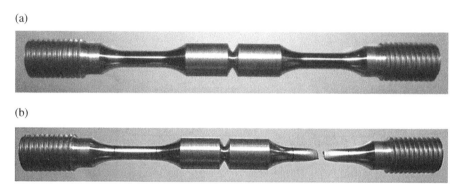

Figure 14.8 Pictures of specimen before and after tensile test. (a) Specimen before tensile test. (b) Specimen after tensile test

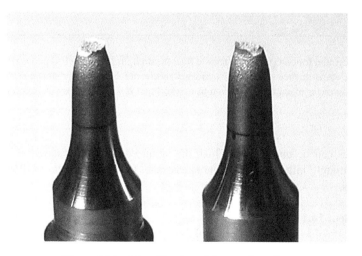

Figure 14.9 Magnification of fractured section

Table 14.1 Dimensions of specimen before and after tensile test, and mechanical properties

(a) Dimensions of specimen.

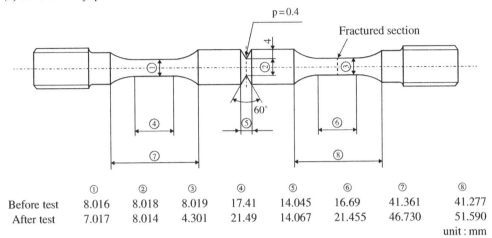

	①	②	③	④	⑤	⑥	⑦	⑧
Before test	8.016	8.018	8.019	17.41	14.045	16.69	41.361	41.277
After test	7.017	8.014	4.301	21.49	14.067	21.455	46.730	51.590

unit : mm

(b) Mechanical properties.

Upper yield stress (MPa)	Lower yield stress (MPa)	Ultimate tensile strength (MPa)	True tensile strength (MPa)	Fracture elongation (%)	Reduction of area (%)
256 (276)	208 (210)	390	910	53.1	71.2

Note: The upper and lower yield stress in parentheses show the values at the second yielding corresponding to either side of the smooth gage sections.

(c) Reference (standard mechanical properties of 0.10% C annealed steel).

Lower yield stress (MPa)	Ultimate tensile strength (MPa)	Reduction of area (%)
203	372	67.7

References

[1] H. Neuber (1961) Theory of stress concentration for shear-strained prismatical bodies with arbitrary nonlinear stress-strain law, *Trans. ASME, J, Appl. Mech.* **28**:544–550.
[2] H. Nisitani, Y. Murakami (1977) Effect of shapes, plate thickness and loading modes on the strain concentration at notches, *Jpn Soc. Mech. Eng.* **43**:426–436.

Answers and Hints for Part II Problems

Chapter 1

1. $\sigma_0 a$

   The force acting over the region of the circular hole ($y = 0 \sim a$) when the hole does not exist is released and this released force is supported by the region outside the hole, that is $y = a \sim \infty$. Since the solution is evident, an integral calculation is not necessary.

2. 0

   Although σ_y along the x axis is not zero, the stress varies from compression near $x = a$ to tension as x increases away from the edge of hole. Since $\sigma_y = 0$ at infinity, an integral calculation is not necessary. The integral should be 0 from the equilibrium condition.

3. (1) $\sigma_0 a$

   The same reason as Problem 1.

   (2) 0

   The same reason as Problem 2.

4. Assuming the stress concentration factor as K^*, the estimated value of the stress concentration is $K_t = \sim 3 \times K^*$.

5. As the stress and strain in the inclusion are uniform (see the note at the bottom of Table 1.2), the shape after deformation is an ellipse.

6. As the thin sheet deforms by following the deformation of the thick plate, the strains inside the thin sheet become the same as those of the thick plate. However, since the elastic moduli are different, the stresses inside the thin sheet are different from those of the thick plate. The stresses are determined by the following equations. The suffixes A and B denote the materials A and B.

Theory of Elasticity and Stress Concentration, First Edition. Yukitaka Murakami.
© 2017 John Wiley & Sons, Ltd. Published 2017 by John Wiley & Sons, Ltd.

$$\left. \begin{array}{l} \varepsilon_{xA} = \dfrac{1}{E_A}\left(\sigma_{xA} - \nu_A \sigma_{yA}\right) = \dfrac{\sigma_0}{E_A} \\[3mm] \varepsilon_{yA} = \dfrac{1}{E_A}\left(\sigma_{yA} - \nu_A \sigma_{xA}\right) = -\nu_A \cdot \dfrac{\sigma_0}{E_A} \end{array} \right\}$$

$$\left. \begin{array}{l} \varepsilon_{xB} = \varepsilon_{xA} \\[3mm] \varepsilon_{yB} = \varepsilon_{yA} \end{array} \right\}$$

$$\left. \begin{array}{l} \varepsilon_{xB} = \dfrac{1}{E_B}\left(\sigma_{xB} - \nu_B \sigma_{yB}\right) \\[3mm] \varepsilon_{yB} = \dfrac{1}{E_B}\left(\sigma_{yB} - \nu_B \sigma_{xB}\right) \end{array} \right\}$$

7. When a projection exists between A and B in Figure 1.44b, the deformation between A and B is constrained and remains smaller than the case with no projection. This situation is equivalent to the deformed state from the state of no projection by applying shear stress on the edge of the plate in the direction to recover the constrained length (to the direction of shortening the length AB). This shear stress produces tensile stress concentration near points A and B.

8. $\sigma_\theta^* = 2\sigma_0$. $\sigma_r^* = 2\sigma_0/(1+\nu)$

Chapter 2

1. The stress at crack tip is $\sigma_x = -\sigma_{x\infty}$ (see Section 1.2.1) even for the application of stress $\sigma_x = \sigma_{x\infty}$. Stress intensity factor is the quantity related to the singularity of $r^{-0.5}$ and a constant value of the stress at crack tip does not contributes to the value of stress intensity factor.

2. (1) $\sigma_0 a$

   The same reason as Problem 1 of Chapter 1 of Part II.

   (2) 0

   The same reason as Problem 2 of Chapter 1 of Part II.

3. For example, using the value for notch open angle 30°, the estimated values are

   $K_t = 5.339$ for $t/\rho = 4$

   $K_t = 7.264$ for $t/\rho = 8$

   Expressing K_t as $K_t \cong A + B\sqrt{\dfrac{t}{\rho}}$

   (Note: A and B are approximately constant, though A and B are not exactly constant values.)

   $$K_1 \cong \frac{1}{2}\sqrt{\pi}\lim_{\rho \to 0}\sqrt{\rho}\left(A + B\sqrt{\frac{t}{\rho}}\right)\sigma_0 = \frac{1}{2}\sqrt{\pi}B\sqrt{t}\sigma_0$$

Calculating B from K_t for $t/\rho = 4$ and $t/\rho = 8$,

$$A + B\sqrt{4} = 5.339$$

$$A + B\sqrt{8} = 7.264$$

And then,

$$B = \frac{7.264 - 5.339}{2\sqrt{2} - 2} = \frac{1.925}{0.8284} = 2.324$$

Thus,

$$K_I = \frac{1}{2} \times 2.324\sigma_0\sqrt{\pi t} = 1.162\sigma_0\sqrt{\pi t}$$

The exact solution is $K_I = 1.1215\sigma_0\sqrt{\pi t}$.

4. Calculate by the same method as Problem 3 of Chapter 2 of Part II.

$$K_t = A + B\sqrt{\frac{a}{\rho}} = 5.55 \text{ for } t/\rho = 16$$

$$K_t = A + B\sqrt{\frac{a}{\rho}} = 10.27 \text{ for } t/\rho = 64$$

$$A + B\sqrt{16} = 5.55$$

$$A + B\sqrt{64} = 10.27$$

$$4B = 10.27 - 5.55$$

$$B = 1.18$$

$$K_I = \frac{1}{2}\sqrt{\pi}\lim_{\rho \to 0}\sqrt{\rho}K_t\sigma_0 = \frac{1}{2}\sqrt{\pi}\lim_{\rho \to 0}\sqrt{\rho}\left(A + B\sqrt{\frac{a}{\rho}}\right)\cdot\frac{W}{W-a}\sigma$$

$$= \frac{1}{2}B\cdot\frac{W}{W-a}\sigma\sqrt{\pi a} = \frac{1}{2} \times 1.18 \times 2\sigma\sqrt{\pi a} = 1.18\sigma\sqrt{\pi a}$$

If we use the equation $K_I = F(\lambda)\sigma\sqrt{\pi a}$ in Figure 2.10 of Section 2.6, we can confirm the accuracy of the above calculation as follows.

$$F(\lambda) = \left(1 - 0.025\lambda^2 + 0.06\lambda^4\right)\sqrt{\sec(\pi\lambda/2)}$$

$$= \left(1 - 0.025 \times 0.5^2 + 0.06 \times 0.5^4\right)\sqrt{\sec(\pi/4)}$$

$$= (1 - 0.025 \times 0.25 + 0.06 \times 0.0625) \times 1.189$$

$$= (1 - 0.00625 + 0.00375) \times 1.189$$

$$= 1.186$$

Note the definition of σ_0 and σ.

5. For $a \ll W$, the equation in Figure 2.11 of Section 2.6 is used. x is the distance from the edge to the interior.

$$\sigma(x) = [1 - x/(W/2)] \cdot \sigma_{max} = [1 - x/(W/2)] \cdot M/(tW^2/6)$$

$$K_{\mathrm{I}} = \frac{6M}{tW^2}\sqrt{\pi a}[1.1215 - 0.6829 \times a/(W/2)]$$

$$= \frac{6M}{tW^2}\sqrt{\pi a}\left(1.1215 - 0.6829 \times \frac{2a}{W}\right)$$

6. If we magnify the crack tip of Figure 2.32 and imagine a small radius of curvature at the tip, the initial crack emanates in the direction of $\pm 46°$, as shown in Figure 2.31. However, as the crack grows by a small distance, the crack changes the propagation angle to the direction perpendicular to the maximum tensile stress. This direction is $70.5°$. With the decreasing radius of curvature at the extreme crack tip, the distance of propagation to $45°$ decreases and the actual propagation angle for the crack becomes $70.5°$. As the length of crack increases more, the propagation angle changes again under the influence of remote stress to the direction perpendicular to the remote tensile principal stress.

Chapter 3

1. $\pi a^2 p_0$
   This problem is similar to Problem 1 of Chapter 1.
2. (a) Midpoint of side AB and midpoint of side CD. (b) Midpoint of side BC. (c) Inner crack tip of the annular crack.
3. (a) Points A, B, C and D. (b) Point B
4. Case (1) $a = b$:

   From Equation 3.9, $K_{IB} = \dfrac{\sigma\sqrt{\pi b}}{E(k)}$.

   From Equation 3.7, $K_{Imax} \cong 0.5\sigma\sqrt{\pi\sqrt{area}}$.

   As $E(0) = \pi/2$ for $a = b$, $K_{IB} = \dfrac{2}{\pi}\sigma\sqrt{\pi b} = 0.6366\sigma\sqrt{\pi b}$.

   And $K_{Imax} \cong 0.5\sigma\sqrt{\pi\sqrt{area}} = 0.5\sigma\sqrt{\pi\sqrt{\pi b^2}} = 0.6657\sigma\sqrt{\pi b}$.
   Thus, the relative error $= 4.6\%$.
   Case (2) $a = 2b$:

   As $E(0.8660) = 1.211(0)$ for $a = 2b$, $K_{IB} = \dfrac{1}{1.211}\sigma\sqrt{\pi b} = 0.8258\sigma\sqrt{\pi b}$

   And $K_{Imax} \cong 0.5\sigma\sqrt{\pi\sqrt{area}} = 0.5\sigma\sqrt{\pi\sqrt{\pi(2b)b}} = 0.7916\sigma\sqrt{\pi b}$
   Thus, the relative error $= -4.1\%$.

Chapter 4

1. (a) $K_{IB} = \sigma_0\sqrt{\pi a}$, $K_{ID} = \sigma_0\sqrt{\pi c}$

 (b) $K_{IB} = \infty$, $K_{ID} = \sigma_0\sqrt{\pi(a+c)}$

2. (a) $K_{IB} = \sigma_0\sqrt{\pi a}$

 (b) $K_{IB} = \dfrac{1}{2}\sigma_0\sqrt{\pi a}$

 Since K_{ID} is also $K_{ID} = 1/2\sigma_0\sqrt{\pi a}$, the total stress intensity factor for twin cracks of $b/a \to 0$ is equal to $K_I = \sigma_0\sqrt{\pi a}$.

3. Machining additional the notch with $R3.5$, the corner between the grip end and the smooth section with 13 mm diameter has an effect equivalent to a larger radius. If machining only one corner radius R to reduce stress concentration, it is necessary to machine the specimen diameter smaller than 13 mm or to increase the diameter of the grip end; and this change in specimen shape and dimension creates other troubles, such as specimen failure outside the gauge section or the necessity of preparing a larger and heavier grip of the testing machine which prevents high speed operation of the testing machine.

4. (1) σ_x at point B_1 is larger than that at point B_0. The field of the stress σ_x around the central hole is lower than the fields of the stress σ_x around the two outside holes.

 (2) σ_y at point A_0 is larger than that at A_1. The field of the stress σ_y around the central hole is higher than the field of the stress σ_y around the two outside holes due to the superposition of the stress distribution by the outside two holes.

 (3) $2\sqrt{3}\sigma_0$. Use the concept of the equivalent ellipse.

Chapter 5

1. Since the notch root radius of Figure 5.9a and b are equal, the stress distributions for both the notches are equal when the maximum stresses at notch root are equal. Therefore,

$$K_t'\sigma_w' = K_t\sigma_w$$

$$\sigma_w' \cong \frac{3.065}{1+2\sqrt{\frac{t}{\rho}}}\sigma_w = \frac{3.065}{1+2\sqrt{4}}\sigma_w = \frac{3.065}{5}\sigma_w = 0.613\sigma_w$$

Chapter 6

1. Due to the concentrated forces at the four corners of the plate, the bending moment per unit length occurs along the dotted line in Figure P6.1a. The bending moment is expressed as

$$M = \frac{P \cdot \frac{s}{2}}{s} = \frac{1}{2}P$$

Therefore, Figure P6.1a is equivalent to Figure P6.1b. In this state, the maximum stress σ_{max} at the edge of the circular hole is calculated using Equations 6.1 and 6.3 as follows:

$$\sigma_{max} = K_t \cdot \frac{M}{\frac{h^2}{6}} = K_t \cdot \frac{6M}{h^2}$$

$$K_t = \frac{5+3v}{3+v} - \frac{1-v}{3+v} = \frac{4+4v}{3+v} = \frac{4(1+v)}{3+v}$$

(a)   (b)

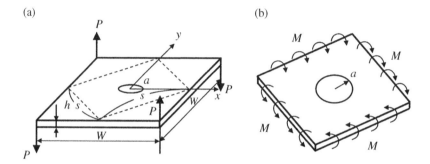

Figure P6.1

2. $K_t = \dfrac{5+3v}{3+v} + \dfrac{1-v}{3+v} = \dfrac{6+2v}{3+v} = 2$

$\sigma_{max} = K_t \cdot \dfrac{M}{\frac{h^2}{6}} = K_t \cdot \dfrac{6M}{h^2}$

Chapter 7

1. The use of Airy's stress function ϕ is known as one solution for elastic plane stress problems. ϕ is a bi-harmonic function which satisfies $\nabla^4\phi = 0$ and stresses are given as the second derivatives of ϕ in terms of the coordinates. Therefore, if boundary conditions are given by forces and stresses, the solution ϕ is independent from elastic moduli. In these cases too, strains and displacements are naturally influenced by Young's modulus E and Poisson's ratio v [1] (see Chapter 6 of Part I).

2. Stress is given by the second derivatives of ϕ. If boundary conditions are given by displacements or strains, the values of the second derivatives of ϕ are influenced by Young's modulus and Poisson's ratio. As a result, stresses are influenced by Young's modulus and Poisson's ratio.

3. Learning from the nature of stress concentration, stress concentration occurs due to a release of the force supported by the region of notches when the notches are absent or being moved to other region. The mechanism of stress concentration due to cracks is basically the same. The stress which exists ahead of a crack before crack initiation has no relationship with the stress singularity.

Based on this idea, in the problem of Figure 7.9, the following quantity of stress should have a relationship with stress singularity.

$$\left(\sigma_{y\text{tip}} - \sigma_0\right) = 4.04\sigma_0 - \sigma_0 = 3.04\sigma_0$$

Likewise, in the problem of Figure 7.10, the following quantity of stress should have a relationship with stress singularity.

$$\left(\sigma_{y\text{tip}} - \sigma_y^*\right) = 9.15\sigma_0 - 1.96\sigma_0 = 7.19\sigma_0$$

Since the stress intensity factor for the crack of Figure 7.9 is $K_I = 1.187\sigma_0\sqrt{\pi c}$, the stress intensity factor for the crack of Figure 7.10 can be estimated as follows:

$$K_I \cong \frac{7.19\sigma_0}{3.04\sigma_0} \times 1.187\sigma_0\sqrt{\pi c} = 2.81\sigma_0\sqrt{\pi c} = 2.81\sigma_0\sqrt{\pi(a+c)} \cdot \frac{\sqrt{c}}{\sqrt{a+c}}$$

$$K_I = 1.47\sigma_0\sqrt{\pi(a+c)}$$

Reference

[1] Y. Murakami (1976) *Trans. Japan. Soc. Mech. Engrs.*, Vol. 42–360, 2305–2315.

Chapter 8

1. See Appendix A.1.4 of Part II or Example problem 1.1 of Chapter 1 in Part I. The stresses everywhere in circular plates are as follows and there is no singularity.

$$\sigma_r = -p, \quad \sigma_\theta = -p, \quad \tau_{r\theta} = 0.$$

Chapter 9

1. $\dfrac{2P}{\pi D}\sqrt{\pi a}$
2. See Appendix A.1.4 of Part II.
3. Uniform tensile stress $2P/\pi D$ (Equation 9.1) acts along the diameter of the direction of the load P.

Appendices for Part II

A brief summary of useful principles and equations of theory of elasticity and stress concentration

A.1 Basis of the Theory of Elasticity

A.1.1 Rule of Shear Stress

Shear stresses act always at orthogonally intersecting planes as a set, as shown in Figure A1.1.

Shear stress action as shown in Figure A1.2 is impossible. This is the basic rule of shear stress irrespective of coordinates.

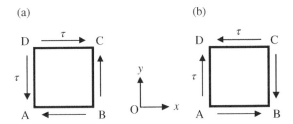

Figure A1.1 Possible set of shear stress actions

Theory of Elasticity and Stress Concentration, First Edition. Yukitaka Murakami.
© 2017 John Wiley & Sons, Ltd. Published 2017 by John Wiley & Sons, Ltd.

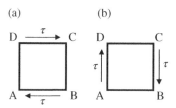

Figure A1.2   Impossible set of shear stress actions

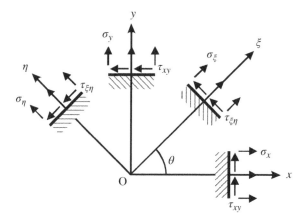

Figure A1.3

A.1.2   Stress Transformation Equations

A.1.2.1   2D Stress Transformation

$$\sigma_\xi = \sigma_x \cos^2\theta + \sigma_y \sin^2\theta + 2\tau_{xy}\cos\theta\sin\theta$$
$$\sigma_\eta = \sigma_x \sin^2\theta + \sigma_y \cos^2\theta - 2\tau_{xy}\cos\theta\sin\theta \qquad (A1.1)$$
$$\tau_{\xi\eta} = (\sigma_y - \sigma_x)\cos\theta\sin\theta + \tau_{xy}(\cos^2\theta - \sin^2\theta)$$

See Figure A1.3

A.1.2.2   3D Stress Transformation

$$\sigma_\xi = \sigma_x l_1^2 + \sigma_y m_1^2 + \sigma_z n_1^2 + 2(\tau_{xy}l_1 m_1 + \tau_{yz}m_1 m_1 + \tau_{zx}n_1 l_1)$$
$$\tau_{\xi\eta} = \sigma_x l_1 l_2 + \sigma_y m_1 m_2 + \sigma_z n_1 n_2 + \tau_{xy}(l_1 m_2 + l_2 m_1) + \tau_{yz}(m_1 n_2 + m_2 n_1) + \tau_{zx}(n_1 l_2 + n_2 l_1)$$
$$\tau_{\xi\zeta} = \sigma_x l_1 l_3 + \sigma_y m_1 m_3 + \sigma_z n_1 n_3 + \tau_{xy}(l_1 m_3 + l_3 m_1) + \tau_{yz}(m_1 n_3 + m_3 n_1) + \tau_{zx}(n_1 l_3 + n_3 l_1)$$

$$\vdots$$

$$(A1.2)$$

See Figure A1.4 and Table A1.1.

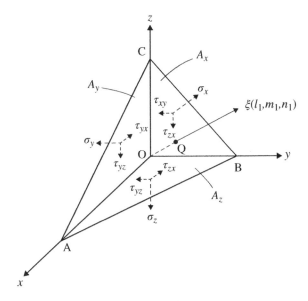

Figure A1.4

Table A1.1 Direction cosines. A direction cosine is the cosine for the angle between two axes

	x	y	z
ξ	l_1	m_1	n_1
η	l_2	m_2	n_2
ζ	l_3	m_3	n_3

A.1.3 Principal Stresses

A.1.3.1 Principal Stresses for 2D Problems

Principal stresses are defined as the normal stresses which act on planes with no shear stress (Figure A1.5b). The larger one is denoted by σ_1 and the smaller one is denoted by σ_2.

σ_1 and σ_2 are determined by

$$\begin{vmatrix} (\sigma_x - \sigma) & \tau_{xy} \\ \tau_{xy} & (\sigma_y - \sigma) \end{vmatrix} = 0 \tag{A1.3}$$

The roots of this determinant are given by

$$\sigma_1, \sigma_2 = \frac{(\sigma_x + \sigma_y) \pm \sqrt{(\sigma_x - \sigma_y)^2 + 4\tau_{xy}^2}}{2} \tag{A1.4}$$

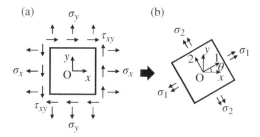

Figure A1.5

Once the principal stresses σ_1 and σ_2 are determined, the angles for the direction of the principal stresses, θ_1 and θ_2 (the principal axes), are determined by the following equations:

$$\theta_1 = \tan^{-1}\left(\frac{\sigma_1 - \sigma_x}{\tau_{xy}}\right)$$

$$\theta_2 = \tan^{-1}\left(\frac{\sigma_2 - \sigma_x}{\tau_{xy}}\right) \quad \text{or} \quad \theta_2 = \theta_1 + \frac{\pi}{2}$$

(A1.5)

The principal axes intersect orthogonally.

A.1.3.2 The Principal Stresses for 3D Problems

The principal stresses for 3D problems are also defined as the normal stresses which act on planes with no shear stress. Three principal stresses are the roots of the following determinant.

$$\begin{vmatrix} (\sigma_x - \sigma) & \tau_{xy} & \tau_{zx} \\ \tau_{xy} & (\sigma_y - \sigma) & \tau_{yz} \\ \tau_{zx} & \tau_{yz} & (\sigma_z - \sigma) \end{vmatrix} = 0$$

(A1.6)

A.1.4 *Stress in a Plate of Arbitrary Shape Subjected to External Pressure p along the Outer Periphery*

As shown in Figure A1.6, when a plate with no internal hole is subjected to a constant pressure p along the outer periphery, the normal stress inside the plate is $\sigma = -p$ and the shear stress is $\tau = 0$ everywhere irrespective of direction [1] (see Chapter 1 of Part I).

Consider a rectangular plate subjected to pressure p along the outer periphery, as shown in Figure A1.7. The stresses inside the rectangular plate in the x–y coordinate system are $\sigma_x = -p$, $\sigma_y = -p$, $\tau_{xy} = 0$. Draw a closed loop Γ with the same shape as Figure A1.6 inside the rectangular plate, as shown in Figure A1.7. Draw the normal line to the periphery at an arbitrary point B of Γ and define the line as the ξ axis. Take an arbitrary point O′ on the ξ axis and define the

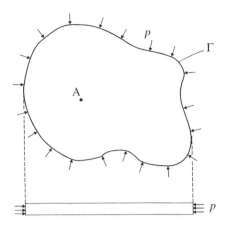

Figure A1.6   Plate of arbitrary shape subjected to external pressure p

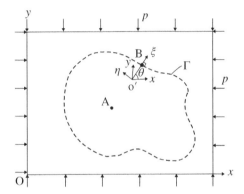

Figure A1.7   Rectangular plate subjected to external pressure

$\xi-\eta$ coordinate system with the origin O'. Denoting the angle between the x and ξ axes by θ, the stresses at point B are calculated by Equation A1.1 as follows.

$$\sigma_\xi = \sigma_x \cos^2\theta + \sigma_y \sin^2\theta + 2\tau_{xy}\cos\theta\cdot\sin\theta = -p\cos^2\theta - p\sin^2\theta + 0 = -p \tag{A1.7}$$

$$\tau_{\xi\eta} = \left(\sigma_y - \sigma_x\right)\cos\theta\cdot\sin\theta + \tau_{xy}\left(\cos^2\theta - \sin^2\theta\right) = 0 \tag{A1.8}$$

Since point B is not a special point on Γ, normal stress is $-p$ everywhere on Γ and shear stress is 0. This means that the boundary condition for a plate encircled by Γ drawn in the rectangular plate is identical to the boundary condition of Figure A1.6. Therefore, in order to determine the stress state at point A inside the plate of Figure A1.6, it is only necessary to determine the stress state at the same point inside Γ of Figure A1.7. As explained above, since point B is not a special point in the rectangular plate, the stress at point A can be expressed in the same manner as point B. Consequently, the stresses inside Γ are $\sigma = -p$ and $\tau = 0$ everywhere and in every direction.

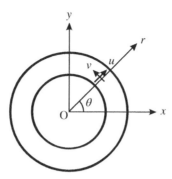

Figure A1.8

In this problem, the stress state inside the plate can be obtained only with the boundary condition and stress transformation equation. It should be noted that, in this problem, nothing about the material properties except for homogeneity is assumed. *The conclusion obtained from this problem is notably important to be memorized not only for elasticity problems but also for various plasticity problems.*

Applications of this problem appear in many problems throughout Parts I and II of this book.

A.1.5   Relationship between Strains and Displacements in Axially Symmetric Problems

Assume an axially symmetric hollow circular cylinder as shown Figure A1.8 which is subjected to axially symmetric loading such as internal pressure and external pressure. (Also see Chapter 2 of Part I.)

The displacements and strains at radius r are denoted as follows: u: displacement in r direction; v: displacement in θ direction;

ε_r: normal strain in r direction; ε_θ: normal strain in θ direction.

The relationship between strains and displacements is expressed as follows:

$$\left.\begin{aligned} \varepsilon_r &= \frac{du}{dr} \\ \varepsilon_\theta &= \frac{u}{r} \end{aligned}\right\} \tag{A1.9}$$

It must be noted in the above equation that ε_r is expressed with a derivative of u and ε_θ is directly related to u. Therefore, *in order to determine the displacement in radial direction u, it is important to **pay attention not to radial strain ε_r but to normal strain ε_θ in the θ direction***. This important nature of the displacement–strain relationship is not well known among engineers.

A.1.6   Hooke's Law of Isotropic and Homogeneous Elastic Material

E: Young's modulus; ν: Poisson's ratio; G: shear modulus.

$$G = \frac{E}{2(1+\nu)} \tag{A1.10}$$

A.1.6.1   2D Plane Stress ($\sigma_z = 0$)

$$\left.\begin{array}{l} \varepsilon_x = \dfrac{1}{E}\left(\sigma_x - \nu\sigma_y\right) \\[2mm] \varepsilon_y = \dfrac{1}{E}\left(\sigma_y - \nu\sigma_x\right) \\[2mm] \varepsilon_z = -\dfrac{\nu}{E}\left(\sigma_x + \sigma_y\right) \\[2mm] \gamma_{xy} = \dfrac{\tau_{xy}}{G} \end{array}\right\} \tag{A1.11}$$

A.1.6.2   2D Plane Plane Strain ($\varepsilon_z = 0$)

$$\left.\begin{array}{l} \varepsilon_x = \dfrac{1-\nu^2}{E}\left[\sigma_x - \dfrac{\nu}{1-\nu}\sigma_y\right] \\[2mm] \varepsilon_y = \dfrac{1-\nu^2}{E}\left[\sigma_y - \dfrac{\nu}{1-\nu}\sigma_x\right] \\[2mm] \gamma_{xy} = \dfrac{\tau_{xy}}{G}, \sigma_z = \nu\left(\sigma_x + \sigma_y\right) \end{array}\right\} \tag{A1.12}$$

A.1.6.3   2D Polar Coordinate System (r, θ)

$$\left.\begin{array}{ll} \varepsilon_r = \dfrac{1}{E}\left(\sigma_r - \nu\sigma_\theta\right) & \varepsilon_r = \dfrac{\partial u}{\partial r} \\[3mm] \varepsilon_\theta = \dfrac{1}{E}\left(\sigma_\theta - \nu\sigma_r\right) & \varepsilon_\theta = \dfrac{u}{r} + \dfrac{1}{r}\dfrac{\partial v}{\partial \theta} \\[3mm] \gamma_{r\theta} = \dfrac{\tau_{r\theta}}{G} & \gamma_{r\theta} = \dfrac{1}{r}\dfrac{\partial u}{\partial \theta} + \dfrac{\partial v}{\partial r} - \dfrac{v}{r} \end{array}\right\} \tag{A1.13}$$

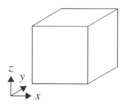

Figure A1.9

As shown in Section A1.5, the relationship between strain and displacement for axially symmetric problem is expressed as

$$\left.\begin{array}{c} \varepsilon_r = \dfrac{du}{dr} \\[2mm] \varepsilon_\theta = \dfrac{u}{r} \end{array}\right\} \tag{A1.14}$$

A.1.6.4   3D Orthogonal Coordinate System (x, y, z)

See Figure A1.9.

$$\left.\begin{array}{l} \varepsilon_x = \dfrac{1}{E}\left[\sigma_x - \nu\left(\sigma_y + \sigma_z\right)\right] \\[2mm] \varepsilon_y = \dfrac{1}{E}\left[\sigma_y - \nu\left(\sigma_z + \sigma_x\right)\right] \\[2mm] \varepsilon_z = \dfrac{1}{E}\left[\sigma_z - \nu\left(\sigma_x + \sigma_y\right)\right] \\[2mm] \gamma_{xy} = \dfrac{\tau_{xy}}{G}, \quad \gamma_{yz} = \dfrac{\tau_{yz}}{G}, \quad \gamma_{zx} = \dfrac{\tau_{zx}}{G} \end{array}\right\} \tag{A1.15}$$

A.1.6.5   3D Cylindrical Coordinate System (x, θ, z)

See Figure A1.10.

$$\left.\begin{array}{l} \varepsilon_r = \dfrac{1}{E}[\sigma_r - \nu(\sigma_\theta + \sigma_z)] \\[2mm] \varepsilon_\theta = \dfrac{1}{E}[\sigma_\theta - \nu(\sigma_z + \sigma_r)] \\[2mm] \varepsilon_z = \dfrac{1}{E}[\sigma_z - \nu(\sigma_r + \sigma_\theta)] \\[2mm] \gamma_{r\theta} = \dfrac{\tau_{r\theta}}{G}, \gamma_{\theta z} = \dfrac{\tau_{\theta z}}{G}, \gamma_{zr} = \dfrac{\tau_{zr}}{G} \end{array}\right\} \tag{A1.16}$$

Important point: Normal strain $\varepsilon = 0$ does not necessarily mean $\sigma = 0$ in the same direction.

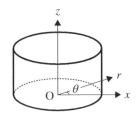

Figure A1.10

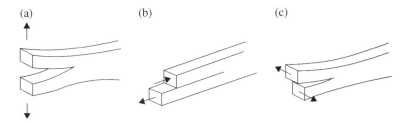

Figure A2.1 (a) Mode I (opening mode). (b) Mode II (in plane shear mode). (c) Mode III (out of plane shear mode)

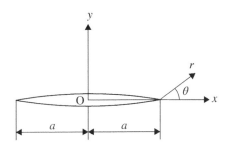

Figure A2.2 Crack in infinite body. The z axis is perpendicular to the page plane

A.2 Stress Field at Crack Tip

The equations of the stress fields in the vicinity of a crack tip expressed in terms of stress intensity factor in plane stress and out of plane shear problems are given as follows (Figure A2.1).

A.2.1 *Expression of Stresses and Displacements (u, v) in x, y, z Orthogonal Coordinate System*

A.2.1.1 Mode I (Opening Mode)

See Figure A2.2.

$$\sigma_x = \frac{K_I}{\sqrt{2\pi r}}\cos\frac{\theta}{2}\left(1-\sin\frac{\theta}{2}\cdot\sin\frac{3\theta}{2}\right)$$

$$\sigma_y = \frac{K_I}{\sqrt{2\pi r}}\cos\frac{\theta}{2}\left(1+\sin\frac{\theta}{2}\cdot\sin\frac{3\theta}{2}\right)$$

$$\tau_{xy} = \frac{K_I}{\sqrt{2\pi r}}\cos\frac{\theta}{2}\cdot\sin\frac{\theta}{2}\cdot\cos\frac{3\theta}{2} \qquad\qquad (A2.1)$$

$$u = \frac{K_I}{2G}\sqrt{\frac{r}{2\pi}}\cos\frac{\theta}{2}\left(\kappa-1+2\sin^2\frac{\theta}{2}\right)$$

$$\upsilon = \frac{K_I}{2G}\sqrt{\frac{r}{2\pi}}\sin\frac{\theta}{2}\left(\kappa+1-2\cos^2\frac{\theta}{2}\right)$$

υ: Displacement in y direction

A.2.1.2   Mode II (in-Plane Shear Mode)

$$\sigma_x = -\frac{K_{II}}{\sqrt{2\pi r}}\sin\frac{\theta}{2}\left(2+\cos\frac{\theta}{2}\cdot\cos\frac{3\theta}{2}\right)$$

$$\sigma_y = \frac{K_{II}}{\sqrt{2\pi r}}\sin\frac{\theta}{2}\cdot\cos\frac{\theta}{2}\cdot\cos\frac{3\theta}{2}$$

$$\tau_{xy} = \frac{K_I}{\sqrt{2\pi r}}\cos\frac{\theta}{2}\left(1-\sin\frac{\theta}{2}\cdot\sin\frac{3\theta}{2}\right) \qquad\qquad (A2.2)$$

$$u = \frac{K_{II}}{2G}\sqrt{\frac{r}{2\pi}}\sin\frac{\theta}{2}\left(\kappa+1+2\cos^2\frac{\theta}{2}\right)$$

$$\upsilon = -\frac{K_{II}}{2G}\sqrt{\frac{r}{2\pi}}\cos\frac{\theta}{2}\left(\kappa-1-2\sin^2\frac{\theta}{2}\right)$$

υ: Displacement in y direction

A.2.1.3   Mode III (out of Plane Shear Mode)

$$\tau_{xz} = -\frac{K_{III}}{\sqrt{2\pi r}}\sin\frac{\theta}{2}$$

$$\tau_{yz} = \frac{K_{III}}{\sqrt{2\pi r}}\cos\frac{\theta}{2} \qquad\qquad (A2.3)$$

$$w = \frac{2K_{III}}{G}\sqrt{\frac{r}{2\pi}}\sin\frac{\theta}{2}$$

where G: shear modulus, ν: Poisson's ratio and

$$\kappa = \begin{cases} (3-\nu)/(1+\nu) : \text{Plane stress} \\ 3-4\nu \quad\quad : \text{Plane strain} \end{cases} \tag{A2.4}$$

A.2.1.4   Stress Components in Polar Coordinate System

$$\left. \begin{aligned}
\sigma_r &= \frac{K_{\mathrm{I}}}{\sqrt{2\pi r}}\left(\frac{5}{4}\cos\frac{\theta}{2}-\frac{1}{4}\cos\frac{3\theta}{2}\right)+\frac{K_{\mathrm{II}}}{\sqrt{2\pi r}}\left(-\frac{5}{4}\sin\frac{\theta}{2}+\frac{3}{4}\sin\frac{3\theta}{2}\right) \\
\sigma_\theta &= \frac{K_{\mathrm{I}}}{\sqrt{2\pi r}}\left(\frac{3}{4}\cos\frac{\theta}{2}+\frac{1}{4}\cos\frac{3\theta}{2}\right)+\frac{K_{\mathrm{II}}}{\sqrt{2\pi r}}\left(-\frac{3}{4}\sin\frac{\theta}{2}-\frac{3}{4}\sin\frac{3\theta}{2}\right) \\
\tau_{r\theta} &= \frac{K_{\mathrm{I}}}{\sqrt{2\pi r}}\left(\frac{1}{4}\sin\frac{\theta}{2}+\frac{1}{4}\sin\frac{3\theta}{2}\right)+\frac{K_{\mathrm{II}}}{\sqrt{2\pi r}}\left(\frac{1}{4}\cos\frac{\theta}{2}+\frac{3}{4}\cos\frac{3\theta}{2}\right)
\end{aligned} \right\} \tag{A2.5}$$

A.3   Stress Concentration for a Circular or an Elliptical Hole with Diameter Approaching to Width of Strip ($a/W \to 1.0$)

The stress concentration at the limiting configuration of this problem was solved by W. T. Koiter [2]. Here, the stress concentration for the limiting case of a hole, which has the notch tip curvature of an m degree curve ($m > 1$), is determined (Figure A3.1).

Consider the half part of the configuration of a hole approaching the strip edge, as shown in Figure A3.2.

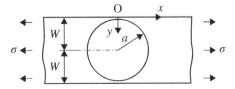

Figure A3.1   Circular or elliptical hole approaching the strip edge

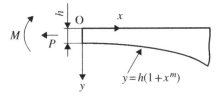

Figure A3.2   Limiting configuration of hole approaching the strip edge

P is the resultant force and M is the bending moment acting at the minimum remaining section of the strip. The width of the minimum section is denoted by h and the thickness of the plate is assumed 1. h is assumed sufficiently small compared to other sizes of the plate and the curvature of the hole is expressed as $y = h(1 + x^m)$. If h is sufficiently small, the stress distribution at the section near the minimum section may be assumed linear as the combined result of the resultant force P and the bending moment M. Thus, assuming the part near the minimum section as a beam, the basic equation of the strength of materials is applied to calculate the inclination of beam y' at the origin, as follows.

$$y' = \int_0^{x_0} \frac{M - P\{h(1 + x^m)/2 - h/2\}}{EI} dx \tag{A3.1}$$

where $x_0 \gg 1$. E is Young's modulus and $I = (1/12)h^3(1 + x^m)^3$. The following integral formula is used to calculate Equation A3.1.

$$I[p, q, m] = \int x^p (ax^m + b)^q dx$$

$$= -\frac{x^{p+1}(ax^m + b)^{q+1}}{(q+1)mb} + \frac{p + m + qm + 1}{(q+1)mb} I[p, q+1, m] \tag{A3.2}$$

Putting $I_2 = I[0, -3, m]$ and $I_1 = I[0, -2, m]$, and considering $x_0 \gg 1$, the relationship between I_2 and I_1 is derived as follows:

$$I_2 = -\frac{(1 - 2m)}{2m} I_1 \tag{A3.3}$$

Thus, using Equation A3.3, Equation A3.1 is expressed as

$$y' = \left\{ \left(M + \frac{h}{2}P \right) \left(\frac{2m-1}{2m} \right) - \frac{h}{2}P \right\} \frac{12I_1}{Eh^3} \tag{A3.4}$$

Since $y' = 0$ at the origin, M and P must satisfy the following relationship:

$$M = \frac{h}{2(2m-1)} P \tag{A3.5}$$

Consequently the stress concentration K_t at the edge of hole is given by

$$K_t = 1 + \frac{6M}{hP} = 1 + \frac{3}{2m-1} \tag{A3.6}$$

From Equation A3.6, $K_t = 2$ for circle and ellipse ($m = 2$). In the case of $m = \infty$, naturally $K_t = 1$, uniform tensile stress state.

A.4   Stress Concentration and Stress Distribution Due to Elliptic Hole in Infinite Plate Under Remote Stress $(\sigma_x, \sigma_y, \tau_{xy}) = (\sigma_{x\infty}, \sigma_{y\infty}, \tau_{xy\infty})$

When an infinite plate containing an elliptic hole with major axis a and minor axis b, as shown in Figure A4.1, is subjected to a remote uniform stress $(\sigma_{y\infty}, \sigma_{x\infty}, \tau_{xy\infty})$, the maximum or minimum normal stress σ_η along the edge of the hole is given by the following equation [3].

$$\sigma_\eta = \frac{(R^4 + 2R^2 - 1 - 2R^4\cos2\theta)\sigma_{x\infty} + (R^4 - 2R^2 - 1 + 2R^4\cos2\theta)\sigma_{y\infty} - 4R^4\tau_{xy\infty}\sin2\theta}{R^4 - 2R^2\cos2\theta + 1}$$

where θ is the value which satisfies the following equation.

$$\sigma_{x\infty}(R^6 - R^4 - R^2 + 1)\sin2\theta - \sigma_{y\infty}(R^6 + R^4 - R^2 - 1)\sin2\theta + 2\tau_{xy\infty}R^2\{2R^2 - (R^4 + 1)\cos2\theta\} = 0$$

and

$$R = \sqrt{\frac{a+b}{a-b}}$$

When the remote stress is only $\tau_{xy\infty}$, the maximum or minimum normal stress σ_η along the edge of the hole is given by the following equation [3].

$$\frac{\sigma_\xi + \sigma_\eta}{\tau_{xy\infty}} = \frac{\sigma_\eta}{\tau_{xy\infty}} = \frac{4R^4\sin2\theta}{2R^2\cos2\theta - R^4 - 1}$$

where θ is the value which satisfies the following equation.

$$\cos2\theta = \frac{2R^2}{R^4 + 1} = \frac{\lambda^2 - 1}{\lambda^2 + 1}$$

$$\lambda = a/b$$

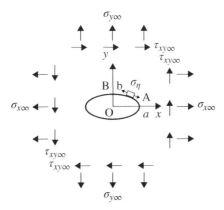

Figure A4.1

A.5   Verification of Stress σ_x^*, σ_y^*, τ_{xy}^* at a Disk Center Subjected to Periodical Point Forces Around the Outer Periphery

σ_x^*, σ_y^*, τ_{xy}^* are given in Equation A5.1.

$$\left.\begin{array}{l} \sigma_x^* = -\dfrac{nP}{\pi D} \\[3mm] \sigma_y^* = -\dfrac{nP}{\pi D} \\[3mm] \tau_{xy}^* = 0 \end{array}\right\} \tag{A5.1}$$

where $n \ge 3$ (Figure A5.1).

First of all, let us start from the easy problem of $n = 4$.

The stress at the center for Figure A5.2 can be determined by the superposition of a pair of point forces ① and another pair of point forces ②, using Equations 9.1 and 9.2 (or Equations 6.92 and 6.93 of Part I).

The stresses at the center due to the pair of point forces ① are

$$\left.\begin{array}{l} \sigma_x^* = \dfrac{2P}{\pi D} \\[3mm] \sigma_y^* = -\dfrac{6P}{\pi D} \end{array}\right\} \tag{A5.2}$$

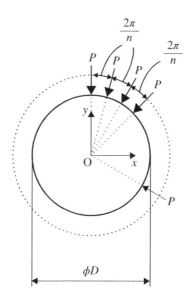

Figure A5.1   This is the same as Figure 10.6

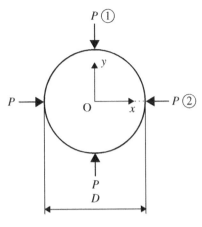

Figure A5.2

The stresses at the center due to the pair of point forces ② are

$$
\left.
\begin{aligned}
\sigma_x^* &= -\frac{6P}{\pi D} \\[2mm]
\sigma_y^* &= \frac{2P}{\pi D}
\end{aligned}
\right\}
\tag{A5.3}
$$

Superposing Equations A5.2 and A5.3, the stress at the center becomes

$$
\left.
\begin{aligned}
\sigma_x^* &= -\frac{4P}{\pi D} \\[2mm]
\sigma_y^* &= -\frac{4P}{\pi D}
\end{aligned}
\right\}
\tag{A5.4}
$$

Next is the case of $n = 3$. Regarding the problem of Figure A5.3, the shear stress at the center $(0, 0)$ is $\tau_{xy}^* = 0$ from the symmetry of the problem (from $\gamma_{xy}^* = 0$, $\tau_{xy}^* = G\gamma_{xy}^* = 0$).

Although the normal stresses at the center (σ_x^*, σ_y^*) are unknown, considering the nature of $\tau_{xy}^* = 0$, the normal stress σ_ξ^* in the radial direction of $\theta = -\frac{\pi}{6}$ is calculated by stress transformation as

$$
\sigma_\xi^* = \sigma_x^* \cos^2\left(-\frac{\pi}{6}\right) + \sigma_y^* \sin^2\left(-\frac{\pi}{6}\right)
\tag{A5.5}
$$

Since the y axis is the direction of the application of the same point force P, the stress σ_ξ^* should be equal to σ_y^*. It follows

$$
\sigma_y^* = \sigma_x^* \cos^2\frac{\pi}{6} + \sigma_y^* \sin^2\frac{\pi}{6}
\tag{A5.6}
$$

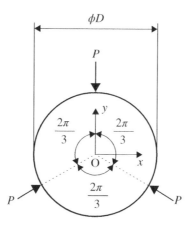

Figure A5.3

Thus,

$$\sigma_y{}^* \cos^2\frac{\pi}{6} = \sigma_x{}^* \cos^2\frac{\pi}{6} \tag{A5.7}$$

Consequently,

$$\sigma_y{}^* = \sigma_x{}^* \tag{A5.8}$$

This result describes that at the center of Figure A5.3 the stress state is hydrostatic compression as $\tau_{xy}{}^* = 0$ and $\sigma_x{}^* = \sigma_y{}^*$, namely the normal stress at the center has the same value not only in the x and y directions but also irrespective of direction.

Therefore, the stress at the center of Figure A5.3 is 1/2 of the stress at the center of Figure A5.4 for $n = 6$.

The stress $\sigma_y{}^*$ at the center of Figure A5.4 is determined by the superposition of the pairs of point forces ①, ②, ③. The influences of ② and ③ on $\sigma_y{}^*$ are equal. Therefore, superposing Equation A5.2 and the stresses in the y direction transformed from the stress due to pairs of forces ② and ③, $\sigma_y{}^*$ is given as follows.

$$\sigma_y{}^* = -\frac{6P}{\pi D} + 2 \times \left(-\frac{6P}{\pi D}\cos^2\frac{\pi}{3} + \frac{2P}{\pi D}\sin^2\frac{\pi}{3} \right)$$

$$= -\frac{6P}{\pi D} \tag{A5.9}$$

The nature of stress field $\sigma_x{}^* = \sigma_y{}^*$ is already described.
Consequently, the solution for the case of $n = 3$ is

$$\left. \begin{array}{l} \sigma_x^* = -\dfrac{3P}{\pi D} \\[2mm] \sigma_y^* = -\dfrac{3P}{\pi D} \end{array} \right\} \tag{A5.10}$$

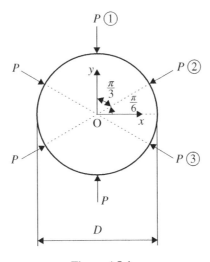

Figure A5.4

When n is an even number, all point forces make a pair in the diameter direction respectively and the stresses at the center can be determined using Equations A5.2 and A5.3, stress transformation and superposition. When n is an odd number, following the same method of the solution for $n = 3$, $2n$ forces are applied and the problem for the even number is first solved. The final solution is 1/2 of the result of this calculation.

After the solutions for $n = 3, 4$ are obtained, the solutions for the problems with multiple n are simply multiplied with the multiplication number for $n = 3, 4$. The specific nature of this problem is that the stress state at the center is hydrostatic, that is $\sigma_x{}^* = \sigma_y{}^*$ and $\tau_{xy}{}^* = 0$.

References

[1] Y. Murakami (1985) *Theory of Elasticity*, Yokendo, Tokyo.
[2] W. T. Koiter (1957) *Quart. Appl. Math.*, **15**:303.
[3] T. Kanezaki, K. Nagata and Y. Murakami (2007)*J. Solid Mech. Mat. Eng.*, *Japan Soc. Mech. Engrs.*, **1**:232–243.

Index

Printed and bound by CPI Group (UK) Ltd, Croydon, CR0 4YY

16/04/2025